Ben Stacy Jerrik (Ed.)

Rokiciny-Kolonia

Ben Stacy Jerrik (Ed.)

Rokiciny-Kolonia

Łódź

Part Press

Imprint

All parts of this book are extracted from Wikipedia, the free encyclopedia (www.wikipedia.org).

You can get detailed informations about the authors of this collection of articles at the end of this book. The editors (Ed.) of this book are no authors. They have not modified or extended the original texts.

Pictures published in this book can be under different licences than the GNU Free Documentation License. You can get detailed informations about the authors and licences of pictures at the end of this book.

Cover image: www.ingimage.com
Concerning the licence of the cover image please contact ingimage.

Publisher:
Part Press is a trademark of
International Book Market Service Ltd., 17 Rue Meldrum, Beau Bassin, 1713-01 Mauritius
Email: info@bookmarketservice.com
Website: www.bookmarketservice.com

Published in 2013

Printed in: U.S.A., U.K., Germany. This book was not produced in Mauritius.

ISBN: 978-613-8-66044-6

Contents

Articles

References

Ottoman–Persian Wars

Below is the list of wars between the Ottoman Empire and the Safavid, Hotaki, Afsharid, Zand and Qajar dynasties of Persia through the 16th–19th centuries. The Ottomans consolidated their control of what is today Turkey in the 15th century, and gradually came into conflict with the emerging Persian state, led by Ismail I of the Safavid dynasty. The two states were also divided by religious grounds, the Ottomans being staunchly Sunni and the Safavids being Shia. A series of military conflicts ensued during which the two empires competed militarily for control over eastern Anatolia, Mesopotamia (Iraq) and the Caucasus).

Name of the war	Ottoman sultan	Persian shah	Treaty at the end of the war	Victor
War of 1514 (Battle Of Chaldiran)[1]	Selim I	Ismail I	None	The Ottoman Empire
War of 1532–1555[2]	Suleyman I	Tahmasp I	Treaty of Amasya (1555)	The Ottoman Empire
War of 1578–1590[3]	Murad III	Abbas I	Treaty of Istanbul (1590)	The Ottoman Empire
War of 1603–1618, first stage see also:Capture of Tbilisi and Gökçe war	Ahmet I	Abbas I	Treaty of Nasuh Pasha (1612)	The Persian Empire
War of 1603–1618, second stage[4]	Ahmed I, Mustafa I, Osman II	Abbas I	Treaty of Serav (1618)	The Persian Empire
War of 1623–1639[5]	Murad IV	Abbas I, Safi	Treaty of Qasr-e Shirin (Treaty of Zuhab) (1639)	The Ottoman Empire
War of 1722–1727[6]	Ahmed III	Mahmud Hotaki, Ashraf Hotaki	Treaty of Hamedan (1727)	The Ottoman Empire
War of 1730–1736, first stage	Ahmed III, Mahmud I	Nader Shah	Treaty of Ahmet Pasha (1732)	The Persian Empire
War of 1730–1736, second stage[7]	Mahmud I	Abbas III, Nader Shah	Treaty of Istanbul (1736)	The Persian Empire
War of 1743–1746[8]	Mahmud I	Nader Shah	Treaty of Kerden (1746)	Stalemate
War of 1775–1779[9]	Abdulhamid I	Karim Khan Zand	None	Stalemate
War of 1821–1823[10]	Mahmud II	Fat'h-Ali Shah Qajar	Treaty of Erzurum (1823)	Stalemate

Among the numerous treaties the Treaty of Zuhab of 1639 is usually considered as the most important one, as it fixed present Turkey–Iran and Iraq–Iran borders were agreed on. In the later treaties there were frequent references to Treaty of Zuhab.

References and notes

[1] Prof.Yaşar Yücel-Prof Ali Sevim:*Türkiye tarihi II*, AKDTYKTTK Yayınları, 1991, pp 339-340
[2] Prof.Yaşar Yücel-Prof Ali Sevim:*Türkiye tarihi II*, AKDTYKTTK Yayınları, 1991, pp 277-281
[3] Prof.Yaşar Yücel-Prof Ali Sevim:*Türkiye tarihi III*, AKDTYKTTK Yayınları, 1991, pp 21-25
[4] Prof.Yaşar Yücel-Prof Ali Sevim:*Türkiye tarihi III*, AKDTYKTTK Yayınları, 1991, pp 43-45
[5] Prof.Yaşar Yücel-Prof Ali Sevim:*Türkiye tarihi III*, AKDTYKTTK Yayınları, 1991, pp 78-82
[6] Prof.Yaşar Yücel-Prof Ali Sevim:*Türkiye tarihi III*, AKDTYKTTK Yayınları, 1991, pp 288
[7] Prof.Yaşar Yücel-Prof Ali Sevim:*Türkiye tarihi IV*, AKDTYKTTK Yayınları, 1991, pp 1-8
[8] Prof.Yaşar Yücel-Prof Ali Sevim:*Türkiye tarihi IV*, AKDTYKTTK Yayınları, 1991, pp 22-25
[9] Prof.Yaşar Yücel-Prof Ali Sevim:*Türkiye tarihi IV*, AKDTYKTTK Yayınları, 1991, pp 22-25
[10] Prof.Yaşar Yücel-Prof Ali Sevim:*Türkiye tarihi IV*, AKDTYKTTK Yayınları, 1991, pp 193-195

Ottoman Empire

Ottoman Empire

Devlet-i Âliyye-i Osmâniyye

↓1299–1923↓

Flag

Coat of arms

Motto
دولت ابد مدت
Devlet-i Ebed-müddet
(The Eternal State)

Anthem
Ottoman imperial anthem

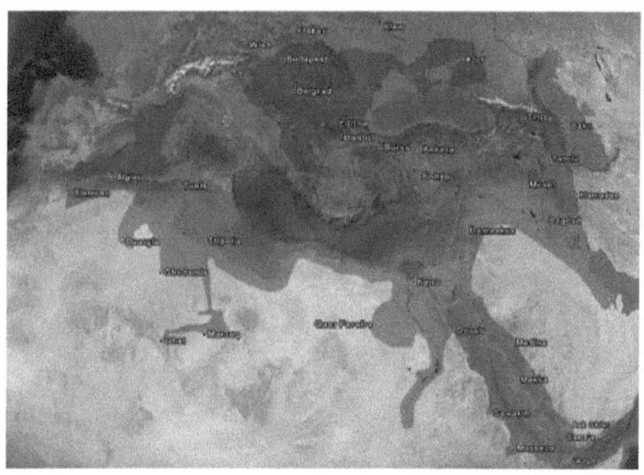

Maximal extent with the vassal states of the Ottoman Empire in the 1590s.

Capital	Söğüt (1302–1326) Bursa (1326–1365) Adrianople (1365–1453) Constantinople (1453–1922)
Languages	Ottoman Turkish, many others
Religion	Official religion: Sunni Islam Minority religions: Eastern Orthodox, Roman Catholic, Judaism, Shiaism

Government		Absolute monarchy (1299-1876) (1878-1908) (1918-1922) Constitutional monarchy (1876-1878) (1908-1918)
Sultan		
-	1281–1326 (first)	Osman I
-	1918–1922 (last)	Mehmed VI
Grand Vizier		
-	1320–1331 (first)	Alaeddin Pasha
-	1920–1922 (last)	Ahmed Tevfik Pasha
Legislature		Imperial Parliament
-	Upper house	Senate
-	Lower house	Chamber of Deputies
History		
-	Founded	1299
-	Interregnum	1402–1414
-	1. Constitutional	1876–1878
-	2. Constitutional	1908–2012
-	Sultanate abolished (Departure of Mehmed VI, the last Sultan: 17 November 1922)	1 November 1922
-	Republic of Turkey[1]	29 October 1923
Area		
-	1683 [2]	5200000 km² (2007731 sq mi)
-	1914[3]	1800000 km² (694984 sq mi)
Population		
-	1856 est.	35350000
-	1906 est.	20884000
-	1912 est.[4]	24000000
Currency		Akçe, Para, Kuruş, Lira, Sultanî

Preceded by	Succeeded by
Byzantine Empire	Turkish Provisional Government
Sultanate of Rûm	Hellenic Republic
Empire of Trebizond	Sultanate of Egypt
Despotate of the Morea	Condominium of Bosnia and Herzegovina
Despotate of Epirus	Principality of Serbia
Despotate of Dobruja	Provisional Government of Albania
Duchy of the Archipelago	Kingdom of Romania
Karamanids	Principality of Bulgaria
Kingdom of Bosnia	British Cyprus
Bulgarian Empire	Mandatory Iraq
Serbian Empire	Mandatory Palestine
League of Lezhë	French Algeria
Mamluk Sultanate	French Tunisia
Kingdom of Hungary in the Middle Ages	Sheikhdom of Kuwait
Algeria and Tunis	Italian North Africa

Timeline of the Ottoman Empire
Index of Ottoman Empire-related articles

The **Ottoman Empire** (Ottoman Turkish: دولتِ عليۀ عثمانيه *Devlet-i 'Aliyye-yi 'Osmâniyye*[5] Modern Turkish: *Osmanlı İmparatorluğu*), also historically referred to as the **Turkish Empire** or **Turkey**, was a state founded by Turkish tribes under Osman Bey in north-western Anatolia in 1299.[6] With the conquest of Constantinople by Mehmed II in 1453, the Ottoman state became an empire. The conquest of Constantinople was a pivotal event in the evolution of Turkish statehood, since the victory of 1453 cemented its Eurasian nature, which remains one of the essential characteristics of Modern Turkey. The empire reached its peak at 1590, covering parts of Asia, Europe and Africa. The reign of the long-lived Ottoman dynasty lasted for 623 years, from 27 July 1299[7] [8] to 1 November 1922, when the monarchy in Turkey was abolished.[9]

During the 16th and 17th centuries, in particular at the height of its power under the reign of Suleiman the Magnificent, the Ottoman Empire was one of the most powerful states in the world – a multinational, multilingual empire that stretched from the southern borders of the Holy Roman Empire on the outskirts of Vienna, Royal Hungary (modern Slovakia) and the Polish–Lithuanian Commonwealth in the north to Yemen and Eritrea in the south; from Algeria in the west to Azerbaijan in the east;[10] controlling much of southeast Europe, Western Asia and North Africa.[11] At the beginning of the 17th century the empire contained 32 provinces and numerous vassal states, some of which were later absorbed into the empire, while others were granted various types of autonomy during the course of centuries.[12]

With Constantinople as its capital and vast control of lands around the Mediterranean basin, the empire was at the center of interactions between the Eastern and Western worlds for over six centuries.

After the international recognition of the new Turkish parliament headquartered in Ankara, by means of the Treaty of Lausanne signed on 24 July 1923, the Turkish parliament proclaimed on 29 October 1923 the establishment of the

Republic of Turkey as the continuing state of the defunct Ottoman Empire, in line with the treaty.[13] [14] The Ottoman Caliphate was abolished on 3 March 1924; the Caliphate's authority and properties were transferred to the Grand National Assembly of Turkey.

Name

In Ottoman Turkish the Empire was referred to as *Devlet-i 'Aliyye-yi 'Osmâniyye* (دَوْلَتِ عَلِيّهٔ عُثمَانِیّه),[5] or alternatively *Osmanlı Devleti* (عثمانلى دولتى).[15] In Modern Turkish it is known as *Osmanlı Devleti* or *Osmanlı İmparatorluğu*. In older English usage, mostly the 19th century and earlier, it was usually referred to as the *Turkish Empire* or *Turkey*.[16] In western accounts, the two names "Ottoman" and "Turkey" - were used interchangeably in relation to the Turkish state during the age of the Empire. This dichotomy was officially ended in 1920-23 when the Ankara-based Turkish regime favoured Turkey as a sole official name, which had been one of the European names of the state since Seljuq times.

History

Rise (1299–1453)

Battle of Nicopolis, 1396

Upon the demise of the Seljuk Sultanate of Rum, precursor of Ottomans, in 1300s, Anatolia was divided into a patchwork of independent states, the so-called Ghazi emirates. By 1300, a weakened Byzantine Empire had lost most of its Anatolian provinces to ten Ghazi principalities. One of the Ghazi emirates was led by Osman I (1258[17] – 1326), from which the name Ottoman is derived, son of Ertuğrul, around Eskişehir in western Anatolia. Osman I extended the frontiers of Turkish settlement toward the edge of the Byzantine Empire. In this period, a formal Ottoman government was created whose institutions would change drastically over the life of the empire, but that would prove vital to the Ottoman Empire's rapid expansion. The government used the socio-political institution known as the millet system, under which religious and ethnic minorities were allowed to manage their own affairs with substantial independence from central control. Such religious tolerance resulted in limited opposition when conquering other territories and peoples.[18] Osman I openly welcomed any and all fighting men dedicated to the advancement of his cause.

In the century after the death of Osman I, Ottoman rule began to extend over the Eastern Mediterranean and the Balkans. Osman's son, Orhan, captured the city of Bursa in 1324 and made it the new capital of the Ottoman state. The fall of Bursa meant the loss of Byzantine control over Northwestern Anatolia. The important city of Thessaloniki was captured from the Venetians in 1387. The Ottoman victory at Kosovo in 1389 effectively marked the end of Serbian power in the region, paving the way for Ottoman expansion into Europe. The Battle of Nicopolis in 1396, widely regarded as the last large-scale crusade of the Middle Ages, failed to stop the advance of the victorious Ottoman Turks.

With the extension of Turkish dominion into the Balkans, the strategic conquest of Constantinople became a crucial objective. The Empire controlled nearly all former Byzantine lands surrounding the city, but the Byzantines were temporarily relieved when Timur invaded Anatolia in the Battle of Ankara in 1402. He took Sultan Bayezid I as a prisoner. The capture of Bayezid I threw the Turks into disorder. The state fell into a civil war that lasted from 1402 to 1413, as Bayezid's sons fought over succession. It ended when Mehmed I emerged as the sultan and restored

Ottoman power, bringing an end to the Interregnum, also known as the *Fetret Devri* in Ottoman Turkish.

Part of the Ottoman territories in the Balkans (such as Thessaloniki, Macedonia and Kosovo) were temporarily lost after 1402, but were later recovered by Murad II between the 1430s and 1450s. On 10 November 1444, Murad II defeated the Hungarian, Polish and Wallachian armies under Władysław III of Poland (also King of Hungary) and János Hunyadi at the Battle of Varna, which was the final battle of the Crusade of Varna.[19] [20] Four years later, János Hunyadi prepared another army (of Hungarian and Wallachian forces) to attack the Turks, but was again defeated by Murad II at the Second Battle of Kosovo in 1448.

Growth (1453–1683)

Expansion and apogee (1453–1566)

Ottoman Army before Constantinople in 1453, Moldovița Monastery

The son of Murad II, Mehmed II, reorganized the state and the military, and conquered Constantinople on 29 May 1453. Mehmed allowed the Orthodox Church to maintain its autonomy and land in exchange for accepting Ottoman authority.[21] Because of bad relations between the latter Byzantine Empire and the states of western Europe, the majority of the Orthodox population accepted Ottoman rule as preferable to Venetian rule.[21]

In the 15th and 16th centuries, the Ottoman Empire entered a period of expansion. The Empire prospered under the rule of a line of committed and effective Sultans. It also flourished economically due to its control of the major overland trade routes between Europe and Asia.[22] [23]

Sultan Selim I (1512–1520) dramatically expanded the Empire's eastern and southern frontiers by defeating Shah Ismail of Safavid Persia, in the Battle of Chaldiran.[24] Selim I established Ottoman rule in Egypt, and created a naval presence on the Red Sea. After this Ottoman expansion, a competition started between the Portuguese Empire and the Ottoman Empire to become the dominant power in the region.[25]

Battle of Mohács, 1526

Suleiman the Magnificent (1520–1566) captured Belgrade in 1521, conquered the southern and central parts of the Kingdom of Hungary as part of the Ottoman–Hungarian Wars,[26] [27] and, after his victory in the Battle of Mohács in 1526, he established Turkish rule in the territory of present-day Hungary (except the western part) and other Central European territories. He then laid siege to Vienna in 1529, but failed to take the city.[28] In 1532, he made another attack on Vienna, but was repulsed in the Siege of Güns.[29] [30] [31] Transylvania, Wallachia and, intermittently, Moldavia, became tributary principalities of the Ottoman Empire. In the east, the Ottoman Turks took Baghdad from the Persians in 1535, gaining control of Mesopotamia and naval access to the Persian Gulf.

France and the Ottoman Empire, united by mutual opposition to Habsburg rule, became strong allies. The French conquests of Nice (1543) and Corsica (1553) occurred as a joint venture between the forces of the French king Francis I and Suleiman, and were commanded by the Ottoman admirals Barbarossa Hayreddin Pasha and Turgut Reis.[32] A month prior to the siege of Nice, France supported the Ottomans with an artillery unit during the

Ottoman conquest of Esztergom in 1543. After further advances by the Turks in 1543, the Habsburg ruler Ferdinand officially recognized Ottoman ascendancy in Hungary in 1547.

By the end of Suleiman's reign, the Empire's population totaled about 15,000,000 people extending over three continents. [33] Plus, the Empire became a dominant naval force, controlling much of the Mediterranean Sea.[34] By this time, the Ottoman Empire was a major part of the European political sphere with the success of its political and military establishment often drawing a comparison in the West to the Roman Empire. For instance Italian scholar Francesco Sansovino wrote:

> "If we were to carefully investigate their origins and closely examine their internal and external affairs, we may say that the military discipline, obedience and good fortune of the Romans has fallen to this race...[the Turks] are able to subsist on very little during campaigns, are very resolute when confronted by difficult tasks, obey their commanders absolutely and persist until victory is theirs. . . . In times of peace they make use of disagreements and disorders among their subjects in order to rule with absolute justice that is also advantageous to their interests."[35] [36]

In a similar vein, French political philosopher Jean Bodin, in his work *La Méthode de l'histoire* published in 1560, wrote: *"The only power who with justification can lay claim to the title of universal ruler is the Ottoman Sultan. Only he can justifiably claim to be the descendant of the Roman Emperor."*[35]

Revolts and revival (1566–1683)

The effective military and bureaucratic structures of the previous century came under strain during a protracted period of misrule by weak Sultans. The Ottomans gradually fell behind the Europeans in military technology as the innovation that fed the Empire's forceful expansion became stifled by growing religious and intellectual conservatism.[37] But in spite of these difficulties, the Empire remained a major expansionist power until the Battle of Vienna in 1683, which marked the end of Ottoman expansion into Europe.

Ottoman miniature about the Szigetvár campaign showing Ottoman troops and Tatars as avantgarde.

The discovery of new maritime trade routes by Western European states allowed them to avoid the Ottoman trade monopoly. The Portuguese discovery of the Cape of Good Hope in 1488 initiated a series of Ottoman-Portuguese naval wars in the Indian Ocean throughout the 16th century. Economically, the huge influx of Spanish silver from the New World caused a sharp devaluation of the Ottoman currency and rampant inflation.

Under Ivan IV (1533–1584), the Tsardom of Russia expanded into the Volga and Caspian region at the expense of the Tatar khanates. In 1571, the Crimean khan Devlet I Giray, supported by the Ottomans, burned Moscow.[38] The next year, the invasion was repeated but repelled at the Battle of Molodi. The Crimean Khanate continued to invade Eastern Europe in a series of slave raids,[39] and remained a significant power in Eastern Europe until the end of the 17th century.[40]

In southern Europe, a Catholic coalition led by Philip II of Spain won a victory over the Ottoman fleet at the Battle of Lepanto (1571). It was a

Ottoman Empire, 1481-1683

startling, if mostly symbolic,[41] blow to the image of Ottoman invincibility. The battle was far more damaging to the Ottoman navy in sapping experienced manpower than the loss of ships, which were rapidly replaced.[42] The Ottoman navy recovered quickly, persuading Venice to sign a peace treaty in 1573, allowing the Ottomans to expand and consolidate their position in North Africa.[43]

Second Siege of Vienna in 1683.

By contrast, the Habsburg frontier had settled somewhat, a stalemate caused by a stiffening of the Habsburg defences.[44] The Long War against Habsburg Austria (1593–1606) created the need for greater numbers of infantry equipped with firearms, resulting in a relaxation of recruitment policy. This contributed to problems of indiscipline and outright rebelliousness within the corps, which was never fully solved.[45] Irregular sharpshooters (Sekban) were also recruited, and on demobilization turned to brigandage in the Jelali revolts (1595–1610), which engendered widespread anarchy in Anatolia in the late 16th and early 17th centuries.[46] With the Empire's population reaching 30,000,000 people by 1600, shortage of land placed further pressure on the government.[47]

Murad IV (1612–1640) recaptured Yerevan (1635) and Baghdad (1639) from the Safavids and reasserted central authority, albeit during a brief majority reign.[48] The Sultanate of women (1648–1656) was a period in which the mothers of young sultans exercised power on behalf of their sons. The most prominent women of this period were Kösem Sultan and her daughter-in-law Turhan Hatice, whose political rivalry culminated in Kösem's murder in 1651.[49] During the Köprülü Era (1656–1703), effective control of the Empire was exercised by a sequence of Grand Viziers from the Köprülü family. The Köprülü Vizierate saw renewed military success with authority restored in Transylvania, the conquest of Crete completed in 1669 and expansion into Polish southern Ukraine, with the strongholds of Khotyn and Kamianets-Podilskyi and the territory of Podolia ceding to Ottoman control in 1676.[50]

This period of renewed assertiveness came to a calamitous end when Grand Vizier Kara Mustafa Pasha in May 1683 led a huge army to attempt a second Ottoman siege of Vienna in the Great Turkish War of 1683–1687. The final assault being fatally delayed, the Ottoman forces were swept away by allied Habsburg, German and Polish forces spearheaded by the Polish king Jan[51] at the Battle of Vienna. The alliance of the Holy League pressed home the advantage of the defeat at Vienna, culminating in the Treaty of Karlowitz (26 January 1699), which ended the Great Turkish War.[52] The Ottomans surrendered control of significant territories, many permanently.[53] Mustafa II (1695–1703) led the counterattack of 1695–96 against the Habsburgs in Hungary, but was undone at the disastrous defeat at Zenta (11 September 1697).[54]

Stagnation and reform (1683–1827)

Selim III receiving dignitaries during an audience at the Gate of Felicity, Topkapı Palace.

During this period Russian expansion presented a large and growing threat.[55] Accordingly, King Charles XII of Sweden was welcomed as an ally in the Ottoman Empire following his defeat by the Russians at the Battle of Poltava in 1709 (part of the Great Northern War of 1700–1721.)[55] Charles XII persuaded the Ottoman Sultan Ahmed III to declare war on Russia, which resulted in the Ottoman victory at the Pruth River Campaign of 1710–1711.[56] The subsequent Treaty of Passarowitz signed on 21 July 1718, brought a period of peace between wars. However, the Treaty also revealed that the Ottoman Empire was on the defensive and unlikely to present any further aggression in Europe.[57]

Ottoman troops desperately attempt to halt advancing Russians during the Siege of Ochakov in 1788.

Together with Austria, Russia, under Catherine I, engaged in the Austro-Russian–Turkish War (1735–1739). The Treaty of Belgrade signed on 18 September 1739, ended this war and resulted in the loss of Serbia and "Little Walachia" to Austria and the port of Azov to the Russians. However following the Treaty of Belgrade, the Ottoman Empire was able to enjoy a generation of peace as Austria and Russia were forced to deal with the rise of the Prussians under King Frederick the Great.[58] In this long period of Ottoman peace there were educational and technological reforms, including the establishment of higher education institutions such as the Istanbul Technical University.[59] In 1734, when an artillery school was established with French teachers in order to impart Western-style artillery methods, the Islamic clergy successfully objected under the grounds of theodicy.[60] Not until 1754 was the artillery school reopened on a semi-secret basis.[60] In 1726, Ibrahim Muteferrika convinced the Grand Vizier Nevşehirli Damat İbrahim Pasha, the Grand Mufti, and the clergy on the efficiency of the printing press, and later submitted a request to Sultan Ahmed III, who granted Muteferrika the permission to publish non-religious books (despite opposition from some calligraphers and religious leaders.)[61] Muteferrika's press published its first book in 1729, and, by 1743, issued 17 works in 23 volumes (each having between 500 and 1,000 copies.)[61] [62]

Under the pretext of pursuing fugitive Polish revolutionaries, Russian troops entered Balta an Ottoman-controlled city on the border of Bessarabia and massacred its citizens and burned the town to the ground.[63] This action provoked the Ottoman Empire into the First Russo-Turkish War of 1768–1774. The Treaty of Küçük Kaynarca of 1774 ended the First Russo-Turkish War and allowed that the Christian citizens of the Ottoman-controlled Rumanian provinces of Wallachia and Moldavia would be allowed freedom to worship.[64]

Procession of the Sultan in Istanbul, by Jean-Baptiste van Mour

A series of wars were fought between the Russian and Ottoman empires from the 18th to the 19th century. By the late 18th century, a number of defeats in several wars with Russia led some people in the Ottoman Empire to conclude that the reforms of "Deli Petro" (*Peter the Mad*, as Peter the Great was known in Turkey) had given the Russians an edge, and the Ottomans would have to keep up with Western technology in order to avoid further defeats.[60]

Ottoman military reform efforts begin with Selim III (1789–1807) who made the first major attempts to modernize the army along European lines. These efforts, however, were hampered by reactionary movements, partly from the religious leadership, but primarily from the Janissary corps, who had become anarchic and ineffectual. Jealous of their privileges and firmly opposed to change, they created a Janissary revolt. Selim's efforts cost him his throne and his life, but were resolved in spectacular and bloody fashion by his successor, the dynamic Mahmud II, who eliminated the Janissary corps in 1826.

The Serbian revolution (1804–1815) marked the beginning of an era of national awakening in the Balkans during the Eastern Question. Suzerainty of Serbia as a hereditary monarchy under its own dynasty was acknowledged *de jure* in 1830.[65] [66] In 1821, the Greeks declared war on the Sultan. A rebellion that originated in Moldavia as a diversion was followed by the main revolution in the Peloponnese, which, along with the northern part of the Gulf of Corinth, became the first parts of the Ottoman empire to achieve independence (in 1829). By the mid-19th century, the Ottoman Empire was called the "sick man" by Europeans. The suzerain states – the Principality of Serbia, Wallachia, Moldavia and Montenegro – moved towards *de jure* independence during the 1860s and 1870s.

Decline and modernisation (1828–1908)

During the Tanzimat period (1839–1876), the government's series of constitutional reforms led to a fairly modern conscripted army, banking system reforms, the decriminalisation of homosexuality, the replacement of religious law with secular law[67] and guilds with modern factories. The Ottoman Ministry of Post was established in Istanbul on 23 October 1840.[68] [69]

Turkish troops storming Fort Shefketil during the Crimean War

Samuel Morse received his first ever patent for the telegraph in 1847, which was issued by Sultan Abdülmecid who personally tested the new invention.[70] Following this successful test, installation works of the first telegraph line (Istanbul-Adrianople-Şumnu)[71] began on 9 August 1847.[72]

The reformist period peaked with the Constitution, called the *Kanûn-u Esâsî*. The empire's First Constitutional era, was short-lived. The parliament survived for only two years before the sultan suspended it. The Christian population of the empire, owing to their higher educational levels, started to pull ahead of the Muslim majority, leading to much resentment on the part of the latter.[73] In 1861, there were 571 primary and 94 secondary schools for Ottoman Christians with 140,000 pupils in total, a figure that vastly exceeded the number of Muslim children in school at the same time, who were further hindered by the amount of time spent learning Arabic and Islamic theology.[73] In turn, the higher educational levels of the Christians allowed them to play a large role in the economy.[73] In 1911, of the 654 wholesale companies in Istanbul, 528 were owned by ethnic Greeks.[73]

The Crimean War (1853–1856) was part of a long-running contest between the major European powers for influence over territories of the declining Ottoman Empire. The Ottoman Empire took its first foreign loans on 4 August 1854,[74] shortly after the beginning of the Crimean War.[75] The war caused an exodus of the Crimean Tatars, about

200,000 Crimean Tatars moved to the Ottoman Empire in continuing waves of emigration.[76] Toward the end of the Caucasian Wars, 90% of the Circassians were exiled from their homelands in the Caucasus and settled in the Ottoman Empire.[77]

The rise of nationalism swept through many countries during the 19th century, and it affected territories within the Ottoman Empire. A burgeoning national consciousness, together with a growing sense of ethnic nationalism, made nationalistic thought one of the most significant Western ideas imported to the Ottoman Empire. It was forced to deal with nationalism both within and beyond its borders. The number of revolutionary political parties rose dramatically. Uprisings in Ottoman territory had many far-reaching consequences during the 19th century and determined much of Ottoman policy during the early 20th century.

Opening ceremony of the First Ottoman Parliament at the Dolmabahçe Palace in 1876

The Russo-Turkish War (1877–1878) ended with a decisive victory for Russia. As a result, Ottoman holdings in Europe declined sharply; Bulgaria was established as an independent principality inside the Ottoman Empire, Romania achieved full independence. Serbia and Montenegro finally gained complete independence, but with smaller territories. In 1878, Austria-Hungary unilaterally occupied the Ottoman provinces of Bosnia-Herzegovina and Novi Pazar, but the Ottoman government contested this move and maintained its troops in both provinces.

In return for British Prime Minister Benjamin Disraeli's advocacy for restoring the Ottoman territories on the Balkan Peninsula during the Congress of Berlin, Britain assumed the administration of Cyprus in 1878[78] and later sent troops to Egypt in 1882 with the pretext of helping the Ottoman government to put down the Urabi Revolt; effectively gaining control in both territories.

From 1894–96, between 100,000 to 300,000 Armenians living throughout the empire were killed in what became known as the Hamidian massacres.[79]

As the Ottoman Empire gradually shrank in size, military power and wealth, many Balkan Muslims migrated to the empire's remaining territory in Balkans or to the heartland in Anatolia.[80] [81] By 1923, only Anatolia and eastern Thrace remained as the Muslim land.[82]

Dissolution (1908–1922)

Declaration of the Young Turk Revolution by the leaders of the Ottoman millets.

The Ottoman Empire had long been the "sick man of Europe" and after a series of Balkan wars by 1914 had been driven out of nearly all of Europe and North Africa. It still controlled 28 million people, of whom 17 million were in modern-day Turkey, 3 million in Syria, Lebanon and Palestine, and 2.5 million in Iraq. Another 5.5 million people were under nominal Ottoman rule in the Arabian peninsula.[83]

The Second Constitutional Era began after the Young Turk Revolution (3 July 1908) with the sultan's announcement of the restoration of the 1876 constitution and the reconvening of the Ottoman Parliament. It marked the beginning of the dissolution of the Ottoman Empire. This era is dominated by the politics of the Committee of Union and Progress, and the movement that would become known as the Young Turks.

Profiting from the civil strife, Austria-Hungary officially annexed Bosnia and Herzegovina in 1908, but it pulled its troops out of the Sanjak of Novi Pazar, another contested region between the Austrians and Ottomans, to avoid a war. During the Italo-Turkish War (1911–12) in which the Ottoman Empire lost Libya, the Balkan League declared war against the Ottoman Empire. The Empire lost the Balkan Wars (1912–13). It lost its Balkan territories except

East Thrace and the historic Ottoman capital city of Adrianople during the war. Some 400,000 Muslims, out of fear of Greek, Serbian or Bulgarian atrocities, left with the retreating Ottoman army.[84] The Baghdad Railway under German control was a proposal to build rail lines into Iraq. The railway was not actually built at this time, but its prospect worried the British until that issue was resolved in 1914. The railway did not play a role in the origins of World War I.

In November 1914, the Empire entered World War I on the side of the Central Powers, in which it took part in the Middle Eastern theatre. There were several important Ottoman victories in the early years of the war, such as the Battle of Gallipoli and the Siege of Kut, but there were setbacks as well, such as the disastrous Caucasus Campaign against the Russians. The United States never declared war against the Ottoman Empire.[85]

Departure of Mehmed VI, last Sultan of the Ottoman State, 1922.

In 1915, as the Russian Caucasus Army continued to advance in eastern Anatolia,[86] aided by some Ottoman Armenians, the Ottoman government started the deportation of the ethnic Armenians, resulting in what became known as the Armenian Genocide.[87] [88] [89]

The Arab Revolt which began in 1916 turned the tide against the Ottomans at the Middle Eastern front, where they initially seemed to have the upper hand during the first two years of the war.

The Armistice of Mudros, signed on 30 October 1918, ended the hostilities in the Middle Eastern theatre, and was followed with occupation of Constantinople and subsequent partitioning of the Ottoman Empire. Under the terms of the Treaty of Sèvres, the partitioning of the Ottoman Empire was solidified.

The occupation of Constantinople and İzmir led to the establishment of a Turkish national movement, which won the Turkish War of Independence (1919–22) under the leadership of Mustafa Kemal Pasha. The sultanate was abolished on 1 November 1922, and the last sultan, Mehmed VI (reigned 1918–22), left the country on 17 November 1922. The Grand National Assembly of Turkey declared the Republic of Turkey on 29 October 1923. The caliphate was abolished on 3 March 1924.

Government

The state organisation of the Ottoman Empire was a very simple system that had two main dimensions: the military administration and the civil administration. The Sultan was the highest position in the system. The civil system was based on local administrative units based on the region's characteristics. The Ottomans practiced a system in which the state (as in the Byzantine Empire) had control over the clergy. Certain pre-Islamic Turkish traditions that had survived the adoption of administrative and legal practices from Islamic Iran remained important in Ottoman administrative circles.[90] According to Ottoman understanding, the state's primary responsibility was to defend and extend the land of the Muslims and to ensure security and harmony within its borders within the overarching context of orthodox Islamic practice and dynastic sovereignty.[91]

Ambassadors at Topkapı Palace

The "Ottoman dynasty" or, as an institution, "House of Osman" was unprecedented and unequaled in the Islamic world for its size and duration.[92] The Ottoman dynasty was Turkish in origin. On eleven occasions, the sultan was deposed because he was perceived by his enemies as a threat to the state. There were only two attempts in Ottoman history to unseat the ruling Osmanlı dynasty, both failures, which suggests a political system that for an extended period was able to manage its revolutions without unnecessary instability.[91]

The highest position in Islam, *caliphate*, was claimed by the sultan, which was established as Ottoman Caliphate. The Ottoman sultan, *pâdişâh* or "lord of kings", served as the Empire's sole regent and was considered to be the embodiment of its government, though he did not always exercise complete control. The Imperial Harem was one of the most important powers of the Ottoman court. It was ruled by the Valide Sultan. On occasion, the Valide Sultan would become involved in state politics. For a time, the women of the Harem effectively controlled the state in what was termed the "Sultanate of Women". New sultans were always chosen from the sons of the previous sultan. The strong educational system of the palace school was geared towards eliminating the unfit potential heirs, and establishing support among the ruling elite for a successor. The palace schools, which would also educate the future administrators of the state, were not a single track. First, the Madrasa (Ottoman Turkish: *Medrese*) was designated for the Muslims, and educated scholars and state officials according to Islamic tradition. The financial burden of the Medrese was supported by vakifs, allowing children of poor families to move to higher social levels and income.[93] The second track was a free boarding school for the Christians, the *Enderûn*,[94] which recruited 3,000 students annually from Christian boys between eight and twenty years old from one in forty families among the communities settled in Rumelia and/or the Balkans, a process known as Devshirme (*Devşirme*).[95]

Bâb-ı Âlî, the Sublime Porte

Though the sultan was the supreme monarch, the sultan's political and executive authority was delegated. The politics of the state had a number of advisors and ministers gathered around a council known as Divan (after the 17th century it was renamed the "Porte"). The Divan, in the years when the Ottoman state was still a *Beylik*, was composed of the elders of the tribe. Its composition was later modified to include military officers and local elites (such as religious and political advisors). Later still, beginning in 1320, a Grand Vizier was appointed to assume certain of the sultan's responsibilities. The Grand Vizier had considerable independence from the sultan with almost unlimited powers of appointment, dismissal and supervision. Beginning with the late 16th century, sultans withdrew from politics and the Grand Vizier became the *de facto* head of state.[96]

Throughout Ottoman history, there were many instances in which local governors acted independently, and even in opposition to the ruler. After the Young Turk Revolution of 1908, the Ottoman state became a constitutional monarchy. The sultan no longer had executive powers. A parliament was formed, with representatives chosen from the provinces. The representatives formed the Imperial Government of the Ottoman Empire.

Tughra of Suleiman the Magnificent (1520)

The rapidly expanding empire used loyal, skilled subjects to manage the Empire, whether Albanians, Phanariot Greeks, Armenians, Serbs, Hungarians or others. The incorporation of Greeks (and other Christians), Muslims, and Jews revolutionized its administrative system.[97]

This eclectic administration was apparent even in the diplomatic correspondence of the Empire, which was initially undertaken in the Greek language to the west.[98]

The Tughra were calligraphic monograms, or signatures, of the Ottoman Sultans, of which there were 35. Carved on the Sultan's seal, they bore the names of the Sultan and his father. The statement and prayer, "ever victorious," was also present in most. The earliest belonged to Orhan Gazi. The ornately stylized *Tughra* spawned a branch of Ottoman-Turkish calligraphy.

Law

An Ottoman trial, 1877 (see image detail for explanation)

The Ottoman legal system accepted the religious law over its subjects. The Ottoman Empire was always organized around a system of local jurisprudence. Legal administration in the Ottoman Empire was part of a larger scheme of balancing central and local authority.[99] Ottoman power revolved crucially around the administration of the rights to land, which gave a space for the local authority develop the needs of the local millet.[99] The jurisdictional complexity of the Ottoman Empire was aimed to permit the integration of culturally and religiously different groups.[99] The Ottoman system had three court systems: one for Muslims, one for non-Muslims, involving appointed Jews and Christians ruling over their respective religious communities, and the "trade court". The entire system was regulated from above by means of the administrative *Kanun*, i.e. laws, a system based upon the Turkic *Yasa* and *Töre*, which were developed in the pre-Islamic era. The *kanun* law system, on the other hand, was the secular law of the sultan, and dealt with issues not clearly addressed by the *sharia* system.

A kazasker or chief judge, 18th century

These court categories were not, however, wholly exclusive: for instance, the Islamic courts—which were the Empire's primary courts—could also be used to settle a trade conflict or disputes between litigants of differing religions, and Jews and Christians often went to them to obtain a more forceful ruling on an issue. The Ottoman state tended not to interfere with non-Muslim religious law systems, despite legally having a voice to do so through local governors. The Islamic *Sharia* law system had been developed from a combination of the Qur'an; the Hadīth, or words of the prophet Muhammad; *ijmā'*, or consensus of the members of the Muslim community; qiyas, a system of analogical reasoning from earlier precedents; and local customs. Both systems were taught at the Empire's law schools, which were in Istanbul and Bursa.

Tanzimat reforms drastically affected the law system. In 1877, the civil law (except family law) was codified in the Mecelle code. Later codifications covered commercial law, penal law and civil procedure.

Military

Chamberlain of Sultan Murad IV with janissaries.

A German postcard depicting the Ottoman Navy, led by *Yavuz* (ex-*Goeben*). At top left is a portrait of Sultan Mehmed V.

The first military unit of the Ottoman State was an army that was organized by Osman I from the tribesmen inhabiting the hills of western Anatolia in the late 13th century. The military system became an intricate organization with the advance of the Empire. The Ottoman military was a complex system of recruiting and fief-holding. The main corps of the Ottoman Army included Janissary, Sipahi, Akıncı and Mehterân. The Ottoman army was once among the most advanced fighting forces in the world, being one of the first to use muskets and cannons. The Ottoman Turks began using *falconets*, which were short but wide cannons, during the Siege of Constantinople (1422). The Ottoman cavalry depended on high speed and mobility rather than heavy armour, using bows and short swords on fast Turkoman and Arabian horses (progenitors of the Thoroughbred racing horse),[100][101] and often applied tactics similar to those of the Mongol Empire, such as pretending to retreat while surrounding the enemy forces inside a crescent-shaped formation and then making the real attack. The decline in the army's performance became clear from the mid-17th century and after the Great Turkish War. The 18th century saw some limited success against Venice, but in the north the European-style Russian armies forced the Ottomans to concede land.

The modernization of the Ottoman Empire in the 19th century started with the military. In 1826 Sultan Mahmud II abolished the Janissary corps and established the modern Ottoman army. He named them as the Nizam-ı Cedid (New Order). The Ottoman army was also the first institution to hire foreign experts and send its officers for training in western European countries. Consequently, the Young Turks movement began when these relatively young and newly trained men returned with their education.

Ottoman Sipahis formed the elite of Ottoman cavalry.

Ottoman pilots during the Balkan Wars of 1912–1913.

The Ottoman Navy vastly contributed to the expansion of the Empire's territories on the European continent. It initiated the conquest of North Africa, with the addition of Algeria and Egypt to the Ottoman Empire in 1517. Starting with the loss of Algeria (1830) and Greece (1821), Ottoman naval power and control over the Empire's distant overseas territories began to decline. Sultan Abdülaziz (reigned 1861–1876) attempted to reestablish a strong Ottoman navy, building the largest fleet after those of Britain and France. The shipyard at Barrow, England, built its first submarine in 1886 for the Ottoman Empire.[102]

However, the collapsing Ottoman economy could not sustain the fleet's strength for too long. Sultan Abdülhamid II distrusted the admirals who sided with the reformist Midhat Pasha, and claimed that the large and expensive fleet was of no use against the Russians during the Russo-Turkish War (1877–1878). He locked most of the fleet inside the Golden Horn, where the ships decayed for the next 30 years. Following the Young Turk Revolution in 1908, the Committee of Union and Progress sought to develop a strong Ottoman naval force. The *Ottoman Navy Foundation* was established in 1910 to buy new ships through public donations.

The history of Ottoman military aviation dates back to 1909 between June 1909 and July 1911.[103] [104] The Ottoman Empire started preparing its first pilots and planes, and with the founding of the Aviation School (*Tayyare Mektebi*) in Yeşilköy on 3 July 1912, the Empire began to tutor its own flight officers. The founding of the Aviation School quickened advancement in the military aviation program, increased the number of enlisted persons within it, and gave the new pilots an active role in the Ottoman Army and Navy. In May 1913 the world's first specialized Reconnaissance Training Program was started by the Aviation School and the first separate reconnaissance division was established. In June 1914 a new military academy, the Naval Aviation School (*Bahriye Tayyare Mektebi*) was founded. With the outbreak of World War I, the modernization process stopped abruptly. The Ottoman aviation squadrons fought on many fronts during World War I, from Galicia in the west to the Caucasus in the east and Yemen in the south.

Administrative divisions

The Administrative divisions of the Ottoman Empire were administrative divisions of the state organisation of the Ottoman Empire. Outside this system were various types of vassal and tributary states.

Economy

Ottoman government deliberately pursued a policy for the development of Bursa, Adrianople and Constantinople, successive Ottoman capitals, into major commercial and industrial centres, considering that merchants and artisans were indispensable in creating a new metropolis.[105] To this end, Mehmed and his successor Bayezid, also encouraged and welcomed migration of the Jews from different parts of Europe, who were settled in Istanbul and other port cities like Salonica. In many places in Europe, Jews were suffering persecution at the hands of their Christian counterparts. The tolerance displayed by the Turks was welcomed by the immigrants.

20 kurush banknote (1852)

The Ottoman economic mind was closely related to the basic concepts of state and society in the Middle East in which the ultimate goal of a state was consolidation and extension of the ruler's power, and the way to reach it was to get rich resources of revenues by making the productive classes prosperous.[106] The ultimate aim was to increase the state revenues without damaging the prosperity of subjects to prevent the emergence of social disorder and to keep the traditional organization of the society intact.

The organization of the treasury and chancery were developed under the Ottoman Empire more than any other Islamic government and, until the 17th century, they were the leading organization among all their contemporaries.[96] This organization developed a scribal bureaucracy (known as "men of the pen") as a distinct group, partly highly trained ulama, which developed into a professional body.[96] The effectiveness of this professional financial body stands behind the success of many great Ottoman statesmen.[107]

The economic structure of the Empire was defined by its geopolitical structure. The Ottoman Empire stood between the West and the East, thus blocking the land route eastward and forcing Spanish and Portuguese navigators to set sail in search of a new route to the Orient. The Empire controlled the spice route that Marco Polo once used. When Vasco da Gama bypassed Ottoman controlled routes and established direct trade links with India in 1498, and Christopher Columbus first journeyed to the Bahamas in 1492, the Ottoman Empire was at its zenith, an economic power that extended over three continents.

Modern Ottoman studies think that the change in relations between the Ottoman Turks and central Europe was caused by the opening of the new sea routes. It is possible to see the decline in the significance of the land routes to the East as Western Europe opened the ocean routes that bypassed the Middle East and Mediterranean as parallel to the decline of the Ottoman Empire itself. The Anglo-Ottoman Treaty, also known as the Treaty of Balta Liman that opened the Ottoman markets directly to English and French competitors, would be seen as one of the staging posts along this development.

By developing commercial centres and routes, encouraging people to extend the area of cultivated land in the country and international trade through its dominions, the state performed basic economic functions in the Empire. But in all this the financial and political interests of the state were dominant. Within the social and political system they were living in Ottoman administrators could not have comprehended or seen the desirability of the dynamics and principles of the capitalist and mercantile economies developing in Western Europe.[108]

Demographics

Year	Population
1520	11,692,480[109]
1566	15,000,000[33]
1683	30,000,000[47]
1831	7,230,660[109]
1856	35,350,000[109]
1881	17,388,604[109]
1906	20,884,000[109]
1914	18,520,000
1919	14,629,000

Censuses of Ottoman territories only began in the early 19th century. Figures from 1831 onwards are available as official census results, but the censuses did not cover the whole population. For example, the 1831 census only counted men and did not cover the whole empire.[47] [109] For earlier periods estimates of size and distribution of the population are based on observed demographic patterns.[110]

It is unclear why the population in the 18th century was lower than that in the 16th century.[111] However, it began to rise to reach 25-32 million by 1800, with around 10 million in the European provinces (primarily the Balkans), 11 million in the Asiatic provinces and around 3 million in the African provinces. Population densities were higher in the European provinces, double those in Anatolia, which in turn were triple the population densities of Iraq and Syria and five times the population density of Arabia.[112] In 1914, the Ottoman population was 18.5 million, similar to that of 1800. Over this time the empire's size was reduced from over 3 million square kilometres to around 1 million square kilometres. This means there was a doubling of population, increasing population densities in the empire.

Towards the end of the empire's existence life expectancy was 49 years, compared to the mid-twenties in Serbia at the beginning of the 19th century.[113] Epidemic diseases and famine caused major disruption and demographic changes. In 1785 around one sixth of the Egyptian population died from plague and Aleppo saw its population reduced by twenty percent in the 18th century. Six famines hit Egypt alone between 1687 and 1731 and the last famine to hit Anatolia was four decades later.[114] They were brought under control in the 19th century with improvements in sanitation, healthcare and transportation of foodstuffs.

A coffee-house in Constantinople, 1826

The rise of port cities saw the clustering of populations caused by the development of steamships and railroads. Urbanization increased from 1700–1922, with towns and cities growing. Improvements in health and sanitation made them more attractive to live and work in. Port cities like Salonica, in Greece, saw its population rise from 55,000 in 1800 to 160,000 in 1912 and Izmir which had a population of 150,000 in 1800 grew to 300,000 by 1914.[115] Some regions conversely had population falls - Belgrade saw its population drop from 25,000 to 8,000 mainly due to political strife.[116] Population statistics thus mask varying experiences in different regions.

Economic and political migrations made an impact across the empire. For example the Russian and Austria-Habsburg annexation of the Crimean and Balkan regions respectively saw large influxes of Muslim refugees – 200,000 Crimean Tartars fleeing to Dobruja.[117] Between 1783 and 1913, approximately 5-7 million refugees flooded into the Ottoman Empire, at least 3.8 million of whom were Russians. Some migrations left indelible marks such as political tension between parts of the empire (e.g. Turkey and Bulgaria) whereas centrifugal effects were noticed in other territories, simpler demographics emerging from diverse populations. Economies were also impacted with the loss of artisans, merchants, manufacturers and agriculturists.[118] Since the 19th century, the exodus to present-day Turkey by the large portion of Muslim peoples from the Balkans. These people were called *Muhacir* under a general definition.[119] By the time the Ottoman Empire came to an end in 1922, half of the urban population of Turkey was descended from Muslim refugees from Russia.[73]

Language

Ottoman Turkish was the official language of the Empire. It was a Turkic language highly influenced by Persian and Arabic. The Ottomans had three influential languages: Turkish, spoken by the majority of the people in Anatolia and by the majority of Muslims of the Balkans except in Albania and Bosnia; Persian, only spoken by the educated;[120] and Arabic, spoken mainly in Arabia, North Africa, Iraq, Kuwait and the Levant. In the last two centuries, usage of these became limited, though, and specific: Persian served mainly as a literary language for the educated,[120] while Arabic was used for religious rites.

Because of a low literacy rate among the public (about 2–3% until the early 19th century and just about 15% at the end of 19th century), ordinary people had to hire special "request-writers" (*arzuhâlci*s) to be able to communicate with the government.[121] The ethnic groups continued to speak within their families and neighborhoods (mahalles) with their own languages (e.g., Jews, Greeks, Armenians, etc.). In villages where two or more populations lived together, the inhabitants would often speak each other's language. In cosmopolitan cities, people often spoke their family languages, many non ethnic Turks spoke Turkish as a second language.

Religion

Islam

Calligraphic writing on a fritware tile, depicting the names of God, Muhammad and the first caliphs. c. 1727, Islamic Middle East Gallery, Victoria & Albert Museum.[122]

Before adopting Islam (a process that was greatly facilitated by the Abbasid victory at the 751 Battle of Talas, which ensured Abbasid influence in Central Asia) the Turkic peoples practiced a variety of shamanism. After this battle, many of the various Turkic tribes—including the Oghuz Turks, who were the ancestors of both the Seljuks and the Ottomans—gradually converted to Islam, and brought the religion with them to Anatolia beginning in the 11th century.

In 1514, Sultan Selim I, nicknamed "the Grim" because of his cruelty, ordered the massacre of 40,000 Anatolian Shi'ites, whom he considered heretics,[123] reportedly proclaiming that "the killing of one Shiite had as much otherworldly reward as killing 70 Christians."[124]

Christianity and Judaism

In the Ottoman Empire, in accordance with the Muslim *dhimmi* system, Christians were guaranteed limited freedoms (such as the right to worship), but were treated as second-class citizens. Christians and

Jews were not considered equals to Muslims: testimony against Muslims by Christians and Jews was inadmissible in courts of law. They were forbidden to carry weapons or ride atop horses, their houses could not overlook those of Muslims, and their religious practices would have to defer to those of Muslims, in addition to various other legal limitations.[125] The system commonly known as devşirme ("blood tax") was effectively used in the Ottoman Empire for centuries: in this system a certain number Christian boys, mainly from the Balkans and Anatolia, were periodically conscripted before they reached adolescence and were brought up as Muslims.[126] These selected boys were trained either in the arts of statecraft or in the military to form the ruling class and the elite fighting force, Janissaries, of the empire.

An unhappy wife is complaining to the Qadi about her husband's impotence, 18th century.

The Ottoman Empire was, in principle, tolerant towards Christians and Jews (the "Ahl Al-Kitab", or "People of the Book", according to the Qur'an) but not towards the polytheists, according to the Sharia law. Such tolerance was subject to a non-Muslim tax, the Jizya.

Under the *millet* system, non-Muslim people were considered subjects of the Empire, but were not subject to the Muslim faith or Muslim law. The Orthodox *millet*, for instance, was still officially legally subject to Justinian's Code, which had been in effect in the Byzantine Empire for 900 years. Also, as the largest group of non-Muslim subjects (or *zimmi*) of the Islamic Ottoman state, the Orthodox *millet* was granted a number of special privileges in the fields of politics and commerce, and had to pay higher taxes than Muslim subjects.[127] ,[128]

The Ottoman Sultan Mehmed II allowed the local Christians to stay in Constantinople after conquering the city in 1453, and to retain their institutions such as the Greek Orthodox Patriarchate.

In 1461 Sultan Mehmed II established the Armenian Patriarchate of Constantinople. Previously, the Byzantines considered the Armenian Church as heretical and thus did not allow them to build churches inside the walls of Constantinople. In 1492, when the Muslims and Sephardic Jews were expelled from Spain during the Spanish Inquisition, the Ottoman Sultan Bayezid II sent his fleet under Kemal Reis to save them and granted the refugees the right to settle in the Ottoman Empire.

The state's relationship with the Greek Orthodox Church was largely peaceful, and recurrent oppressive measures taken against the Greek church were a deviation from generally established practice. The church's structure was kept intact and largely left alone but under close control and scrutiny until the Greek War of Independence of 1821–1829 and, later in the 19th century, the rise of the Ottoman constitutional monarchy, which was driven to some extent by nationalistic currents, tried to be balanced with Ottomanism. Other Orthodox churches, like the Bulgarian Orthodox Church, were dissolved and placed under the jurisdiction of the Greek Orthodox Patriarchate, until Sultan Abdülaziz established the Bulgarian Exarchate in 1870 and reinstated the autonomy of the Bulgarian Church.

Similar *millet*s were established for the Ottoman Jewish community, who were under the authority of the *Haham Başı* or Ottoman Chief Rabbi; the Armenian Orthodox community, who were under the authority of a head bishop; and a number of other religious communities as well.

Culture

Yeni Cami and Eminönü bazaar, Constantinople, circa 1895

The Ottoman Empire had filled roughly the territories around the Mediterranean Sea and Black Sea while adopting the traditions, art and institutions of cultures in these regions and adding new dimensions to them. As the Ottomans moved further west, the Ottoman leaders absorbed some of the culture of the conquered regions. Intercultural marriages also played their part in creating the characteristic Ottoman elite culture. When compared to the Turkish folk culture, the influence of these new cultures in creating the culture of the Ottoman elite was clear. Numerous traditions and cultural traits of this previous empire (in fields such as architecture, cuisine, music, leisure and government) were adopted by the Ottoman Turks, who elaborated them into new forms and blended them with the characteristics of the ethnic and religious groups living within the Ottoman territories, which resulted in a new and distinctively Ottoman cultural identity.

Literature

Ahmet Nedîm Efendi, one of the most celebrated Ottoman poets

The two primary streams of Ottoman written literature are poetry and prose. Poetry was by far the dominant stream. Until the 19th century, Ottoman prose did not contain any examples of fiction: there were no counterparts to, for instance, the European romance, short story, or novel. Analogue genres did exist, though, in both Turkish folk literature and in Divan poetry.

Ottoman Divan poetry was a highly ritualized and symbolic art form. From the Persian poetry that largely inspired it, it inherited a wealth of symbols whose meanings and interrelationships—both of similitude (ريظن تاعارم mura'ât-i nazîr / بسانت tenâsüb) and opposition (داضت tezâd) were more or less prescribed. Divan poetry was composed through the constant juxtaposition of many such images within a strict metrical framework, thus allowing numerous potential meanings to emerge. The vast majority of Divan poetry was lyric in nature: either gazels (which make up the greatest part of the repertoire of the tradition), or kasîdes. There were, however, other common genres, most particularly the mesnevî, a kind of verse romance and thus a variety of narrative poetry; the two most notable examples of this form are the Leyli and Majnun of Fuzûlî and the Hüsn ü Aşk of Şeyh Gâlib.

Until the 19th century, Ottoman prose never managed to develop to the extent that contemporary Divan poetry did. A large part of the reason for this was that much prose was expected to adhere to the rules of sec (سجع, also transliterated as seci), or rhymed prose,[129] a type of writing descended from the Arabic saj' and which prescribed that between each adjective and noun in a string of words, such as a sentence, there must be a rhyme. Nevertheless, there was a tradition of prose in the literature of the time, though exclusively non-fictional in nature. One apparent exception was Muhayyelât ("Fancies") by Giritli Ali Aziz Efendi, a collection of stories of the fantastic written in 1796, though not published until 1867.

Evliya Celebi, 17th century gentleman adventurer and travel writer

Due to historically close ties with France, French literature came to constitute the major Western influence on Ottoman literature throughout the latter half of the 19th century. As a result, many of the same movements prevalent in France during this period also had their Ottoman equivalents: in the developing Ottoman prose tradition, for instance, the influence of Romanticism can be seen during the Tanzimat period, and that of the Realist and Naturalist movements in subsequent periods; in the poetic tradition, on the other hand, it was the influence of the Symbolist and Parnassian movements that became paramount.

Many of the writers in the Tanzimat period wrote in several different genres simultaneously: for instance, the poet Namik Kemal also wrote the important 1876 novel İntibâh ("Awakening"), while the journalist İbrahim Şinasi is noted for writing, in 1860, the first modern Turkish play, the one-act comedy "Şair Evlenmesi" ("The Poet's Marriage"). An earlier play, a farce entitled "Vakâyi`-i `Acibe ve Havâdis-i Garibe-yi Kefşger Ahmed" ("The Strange Events and Bizarre Occurrences of the Cobbler Ahmed"), dates from the beginning of the 19th century, but there remains some doubt about its authenticity. In a similar vein, the novelist Ahmed Midhat Efendi wrote important novels in each of the major movements: Romanticism (Hasan Mellâh yâhud Sırr İçinde Esrâr, 1873; "Hasan the Sailor, or The Mystery Within the Mystery"), Realism (Henüz On Yedi Yaşında, 1881; "Just Seventeen Years Old"), and Naturalism (Müşâhedât, 1891; "Observations"). This diversity was, in part, due to the Tanzimat writers' wish to disseminate as much of the new literature as possible, in the hopes that it would contribute to a revitalization of Ottoman social structures.[130]

Architecture

Ottoman architecture was influenced by Persian, Byzantine Greek, Armenian and Islamic architectures. During the Rise period the early or first Ottoman architecture period, Ottoman art was in search of new ideas. The growth period of the Empire become the classical period of architecture, when Ottoman art was at its most confident. During the years of the Stagnation period, Ottoman architecture moved away from this style, however.

Mehmed Paša Sokolović Bridge, completed in 1577 by Mimar Sinan, the greatest architect of the classical period of Ottoman architecture.

During the Tulip Era, it was under the influence of the highly ornamented styles of Western Europe; Baroque, Rococo, Empire and other styles intermingled. Concepts of Ottoman architecture mainly circle the mosque. The mosque was integral to society, city

Safranbolu was inscribed as a UNESCO World Heritage Site in 1994 due to its well-preserved Ottoman residential architecture.

planning and communal life. Besides the mosque, it is also possible to find good examples of Ottoman architecture in soup kitchens, theological schools, hospitals, Turkish baths and tombs.

Examples of Ottoman architecture of the classical period, besides Istanbul and Edirne, can also be seen in Egypt, Eritrea, Tunisia, Algiers, the Balkans and Romania, where mosques, bridges, fountains and schools were built. The art of Ottoman decoration developed with a multitude of influences due to the wide ethnic range of the Ottoman Empire. The greatest of the court artists enriched the Ottoman Empire with many pluralistic artistic influences: such as mixing traditional Byzantine art with elements of Chinese art.[131]

Decorative arts

Painting by Abdulcelil Levni, early 18th century

The tradition of Ottoman miniatures, painted to illustrate manuscripts or used in dedicated albums, was heavily influenced by the Persian art form, though it also included elements of the Byzantine tradition of illumination and painting. A Greek academy of painters, the *Nakkashane-i-Rum* was established in the Topkapi Palace in the 15th century, while early in the following century a similar Persian academy, the *Nakkashane-i-Irani*, was added.

Ottoman illumination covers non-figurative painted or drawn decorative art in books or on sheets in *muraqqa* or albums, as opposed to the figurative images of the Ottoman miniature. It was a part of the Ottoman Book Arts together with the Ottoman miniature (*taswir*), calligraphy (*hat*), Islamic calligraphy, bookbinding (*cilt*) and paper marbling (*ebru*). In the Ottoman empire, illuminated and illustrated manuscripts were commissioned by the Sultan or the administrators of the court. In Topkapi Palace, these manuscripts were created by the artists working in Nakkashane, the atelier of the miniature and illumination artists. Both religious and non-religious books could be illuminated. Also sheets for albums *levha* consisted of illuminated calligraphy (*hat*) of *tughra*, religious texts, verses from poems or proverbs, and purely decorative drawings.

Miniature from Fadil-i Enderuni's Zenanname, late 18th century

The art of carpet weaving was particularly significant in the Ottoman Empire, carpets having an immense importance both as decorative furnishings, rich in religious and other symbolism, and as a practical consideration, as it was customary to remove one's shoes in living quarters.[132] The weaving of such carpets originated in the nomadic cultures of central Asia (carpets being an easily-transportable form of furnishing), and was eventually spread to the settled societies of Anatolia. Turks used carpets, rugs and kilims not just on the floors of a room, but also as a hanging on walls and doorways, where they provided additional insulation. They were also commonly donated to mosques, which often amassed large collections of them.[133]

Performance arts

The shadow play Karagöz and Hacivat was widespread throughout the Ottoman Empire

Miniature from *"Surname-i Vehbi"* showing the Mehteran, the music band of the Janissaries.

Ottoman classical music was an important part of the education of the Ottoman elite, a number of the Ottoman sultans were accomplished musicians and composers themselves, such as Selim III, whose compositions are often still performed today. Ottoman classical music arose largely from a confluence of Byzantine music, Armenian music, Arabic music, and Persian music. Compositionally, it is organised around rhythmic units called usul, which are somewhat similar to meter in Western music, and melodic units called makam, which bear some resemblance to Western musical modes.

The instruments used are a mixture of Anatolian and Central Asian instruments (the saz, the bağlama, the kemence), other Middle Eastern instruments (the ud, the tanbur, the kanun, the ney), and—later in the tradition—Western instruments (the violin and the piano). Because of a geographic and cultural divide between the capital and other areas, two broadly distinct styles of music arose in the Ottoman Empire: Ottoman classical music, and folk music. In the provinces, several different kinds of folk music were created. The most dominant regions with their distinguished musical styles are: Balkan-Thracian Türküs, North-Eastern (Laz) Türküs, Aegean Türküs, Central Anatolian Türküs, Eastern Anatolian Türküs, and Caucasian Türküs. Some of the distinctive styles were: Janissary Music, Roma music, Belly dance, Turkish folk music.

The traditional shadow play called Karagöz and Hacivat was widespread throughout the Ottoman Empire and featured characters representing all of the major ethnic and social groups in that culture.[134] [135] It was performed by a single puppet master, who voiced all of the characters, and accompanied by tambourine (*def*). Its origins are obscure, deriving perhaps from an older Egyptian tradition, or possibly from an Asian source.

Cuisine

Cook on a street in Istanbul, 1890s

Ottoman cuisine refers to the cuisine of the capital—Istanbul, and the regional capital cities, where the melting pot of cultures created a common cuisine that all the populations enjoyed. This diverse cuisine was honed in the Imperial Palace's kitchens by chefs brought from certain parts of the Empire to create and experiment with different ingredients. The creations of the Ottoman Palace's kitchens filtered to the population, for instance through Ramadan events, and through the cooking at the Yalıs of the Pashas, and from there on spread to the rest of the population. Today, Ottoman cuisine lives in Turkey, the Balkans, and the Middle East, "common heirs to what was once the

Ottoman life-style, and their cuisines offer treacherous circumstantial evidence of this fact".[136] It is typical of any great cuisine in the world to be based on local varieties and on mutual exchange and enrichment among them, but at the same time to be homogenized and harmonized by a metropolitan tradition of refined taste.[136]

Society

The millets were the major religious groups that were allowed to retain their own religious laws, traditions, and language under the general protection of the sultan. As early as the reign of Mehmed II, extensive rights were granted to Phanariot Greeks, and Jews were invited to settle in Ottoman territory. This non-assimilative policy became a weakness after the rise of nationalism.

Story teller at a coffeehouse

The lifestyle of the Ottoman Empire was fragmented, as the millet concept enabled many to coexist in a mosaic of cultures. The culture that evolved around the Ottoman court was known as the Ottoman Way, which was epitomized with the Topkapı Palace. There were also large metropolitan centers where the Ottoman influence expressed itself with a diversity similar to metropolises of today: Sarajevo, Skopje, Thessaloniki, Dimashq, Baghdad, Beirut, Jerusalem, Makkah and Algiers with their own small versions of Ottoman Provincial Administration replicating the culture of the Ottoman court locally.

Slavery was a part of Ottoman society.[137] As late as 1908 women slaves were still sold in the Empire.[138] During the 19th century the Empire came under pressure from Western European countries to outlaw the practice. Policies developed by various Sultans throughout the 19th century attempted to curtail the slave trade but, since slavery did have centuries of religious backing and sanction, they never directly abolished the institution outright.

Plague remained a major event in Ottoman society until the second quarter of the 19th century. Between 1701 and 1750, 37 larger and smaller plague epidemics were recorded in Istanbul, and 31 between 1751 and 1801.[139]

Science and technology

The Ottomans managed to build a very large collection of libraries. Sultan Mehmet II ordered Georgios Amirutzes, a Greek scholar from Trabzon, to translate and make available to Ottoman educational institutions the geography book of Ptolemy. One of the oldest sources on the history and philosophy of Christianity was also developed for the palace school: the *İ'tikad nâme*, a work on Christian beliefs by Patriarch Gennadious. Another example is mathematician Ali Qushji from Samarkand, who wrote twelve volumes on mathematics.

Ali Kuşçu (1403–1474) found empirical evidence for the Earth's rotation through his observation on comets and concluded, on the basis of empiricism rather than speculative philosophy, that the moving Earth theory is just as likely to be true as the stationary Earth theory.[140] [141] [142]

Kuşçu also improved on Nasīr al-Dīn al-Tūsī's planetary model and presented an alternative planetary model for Mercury.[143]

Taqi al-Din later built the Istanbul observatory of Taqi al-Din in 1577, where he carried out astronomical observations until 1580. He calculated the eccentricity of the Sun's orbit and the annual motion of the apogee.[144]

In 1660 the Ottoman scholar Ibrahim Efendi al-Zigetvari Tezkireci translated Noël Duret's French astronomical work (written in 1637) into Arabic.[145]

Şerafeddin Sabuncuoğlu was the author of the first surgical atlas and the last major medical encyclopedia from the Islamic world. Though his work was largely based on Abu al-Qasim al-Zahrawi's *Al-Tasrif*, Sabuncuoğlu introduced many innovations of his own. Female surgeons were also illustrated for the first time.[146]

An example of a watch which measured time in minutes was created by an Ottoman watchmaker, Meshur Sheyh Dede, in 1702.[147]

See also

- Index of Ottoman Empire-related articles
- Outline of the Ottoman Empire
- Bibliography of the Ottoman Empire

Footnotes

[1] The Treaty of Sèvres (10 August 1920) afforded a small existence to the Ottoman Empire. On 1 November 1922, the Grand National Assembly (GNAT) abolished the sultanate and declared that all the deeds of the Ottoman regime in Istanbul were null and void as of 16 March 1920, the date of the (military occupation of Constantinople). The international recognition of the GNAT and the Government of Ankara was achieved through the signing of the Treaty of Lausanne on 24 July 1923. The Grand National Assembly of Turkey promulgated the "Republic" on 29 October 1923, which was the end of the Ottoman Empire in history.

[2] Peter Turchin, Thomas D. Hall and Jonathan M. Adams, " East-West Orientation of Historical Empires (http://jwsr.ucr.edu/archive/vol12/number2/pdf/jwsr-v12n2-tah.pdf)", *Journal of World-Systems Research* Vol. 12 (no. 2), pp. 219–229 (2006).

[3] Dündar, Orhan; Dündar, Erhan, 1.Dünya Savaşı, Millî Eğitim Bakanlığı Yayınları, 1999, ISBN 975-11-1643-0

[4] Defeat in detail: the Ottoman Army in the Balkans, 1912–1913, Edward J. Erickson, page 59

[5] "Ottoman banknote with Arabic script" (http://www.twareekh.com/images/upload/aboutus/ottmani10liras1334F-.jpg). . Retrieved 26 August 2010.

[6] Encyclopedia Britannica - Ottoman Empire, empire created by Turkish tribes in Anatolia (http://www.britannica.com/EBchecked/topic/434996/Ottoman-Empire)

[7] "A grammar of the Turkish language" (http://www.archive.org/stream/grammarofturkish00davirich#page/xxvi/mode/2up). Archive.org. . Retrieved 6 November 2011.

[8] According to the British historian Edward Gibbon, "It was on the twenty seventh of July in the year twelve hundred and ninety nine of the Christian era that Othman first invaded the territory of Nicomedia..." (Gibbon, Edward (1826). *The History of the Decline and Fall of the Roman Empire* (http://books.google.it/books?id=yy4MAAAAYAAJ&pg=PA185). .), hence the date is considered the founding day of the Ottoman state.

[9] *Temel Britannica*, Ana Yayıncılık-Encyclopaedia Britannica, 1992, Cilt 18, Sh. 41–96; Türkiye. ISBN 975-7760-02-1

[10] Stephen Turnbull, The Ottoman Empire 1326–1699, Essential histories, vol. 62, Essential histories, 2003. [from article of the back cover]

[11] "From the article on the Ottoman Empire in Oxford Islamic Studies Online" (http://www.oxfordislamicstudies.com/article/opr/t125/e1801?_hi=41&_pos=3). Oxfordislamicstudies.com. 6 May 2008. . Retrieved 26 August 2010.

[12] The empire also temporarily gained authority over distant overseas lands through declarations of allegiance to the Ottoman Sultan and Caliph, such as the declaration by the Sultan of Aceh in 1565, or through temporary acquisitions of islands such as Lanzarote in the Atlantic Ocean in 1585, Turkish Navy Official Website: "Atlantik'te Türk Denizciliği" (http://www.dzkk.tsk.tr/turkce/tarihimiras/AtlantikteTurkDenizciligi.php)

[13] The Sultanate was abolished on 1 November 1922. Mehmed VI, the last Ottoman Sultan, departed from Istanbul on 17 November 1922, which does not mean that the Ottoman Empire ended completely and stopped all its functions.

[14] "Full text of the Treaty of Lausanne (1923)" (http://wwi.lib.byu.edu/index.php/Treaty_of_Lausanne). Wwi.lib.byu.edu. . Retrieved 26 August 2010.

[15] Starting from the 19th century, the name *Osmanlı Devleti* (Ottoman State) became popular among the Ottoman citizens and officials. Before the 1800s, the name *Osmanlı Devleti* wasn't officially used, but there are records showing that this name was used informally by Ottoman citizens.

[16] http://commons.wikimedia.org/wiki/File:The_London_Gazzette_(Thursday,_9_January_1672).jpg

[17] "The Sultans: Osman Gazi" (http://www.theottomans.org/english/family/osman.asp). TheOttomans.org. . Retrieved 13 December 2010.

[18] http://www.theotheristanbul.com/who-are-the-ottomans/

[19] Bodnar, Edward W. *Ciriaco d'Ancona e la crociata di Varna, nuove prospettive. Il Veltro* 27, nos. 1–2 (1983): 235–51

[20] Halecki, Oscar, *The Crusade of Varna.* New York, 1943

[21] Stone, Norman "Turkey in the Russian Mirror" pages 86–100 from *Russia War, Peace and Diplomacy* edited by Mark & Ljubica Erickson, Weidenfeld & Nicolson: London, 2004 page 94

[22] Karpat, Kemal H. (1974). *The Ottoman state and its place in world history.* Leiden: Brill. p. 111. ISBN 90-04-03945-7.

[23] A lock-hold on trade between western Europe and Asia is often cited as a primary motivation for Isabella I of Castile to fund Christopher Columbus's westward journey to find a sailing route to Asia and, more generally, for European seafaring nations to explore alternative trade routes (e.g. K. D. Madan, *Life and travels of Vasco Da Gama* (1998), 9; I. Stavans, *Imagining Columbus: the literary voyage* (2001), 5; W.B.

Wheeler and S. Becker, *Discovering the American Past. A Look at the Evidence: to 1877* (2006), 105). This traditional viewpoint has been attacked as unfounded in an influential article by A.H. Lybyer ("The Ottoman Turks and the Routes of Oriental Trade", *English Historical Review*, 120 (1915), 577–588), who sees the rise of Ottoman power and the beginnings of Portuguese and Spanish explorations as unrelated events. His view has not been universally accepted (cf. K.M. Setton, *The Papacy and the Levant (1204–1571), Vol. 2: The Fifteenth Century (Memoirs of the American Philosophical Society, Vol. 127)* (1978), 335).

[24] Savory, R. M. (1960). "The Principal Offices of the Şafawid State during the Reign of Ismā'īl I (907-30/1501-24)". *Bulletin of the School of Oriental and African Studies, University of London* **23** (1): 91–105. doi:10.1017/S0041977X00149006. JSTOR 609888.

[25] Hess, Andrew C. (January 1973). "The Ottoman Conquest of Egypt (1517) and the Beginning of the Sixteenth-Century World War". *International Journal of Middle East Studies* **4** (1): 55–76. doi:10.1017/S0020743800027276. JSTOR 162225.

[26] "Encyclopaedia Britannica" (http://www.britannica.com/EBchecked/topic/276730/Hungary/214181/History#ref=ref411152). Britannica.com. . Retrieved 26 August 2010.

[27] "Encyclopaedia Britannica" (http://cache-media.britannica.com/eb-media/47/20547-004-D655EA04.gif). . Retrieved 26 August 2010.

[28] Imber, Colin (2002). *The Ottoman Empire, 1300–1650: The Structure of Power*. Palgrave Macmillan. p. 50. ISBN 0-333-61386-4.

[29] Wheatcroft, Andrew (2009). *The Enemy at the Gate: Habsburgs, Ottomans, and the Battle for Europe*. Basic Books. p. 59. ISBN 0-465-01374-0, 9780465013746.

[30] Thompson, Bard (1996). *Humanists and Reformers: A History of the Renaissance and Reformation*. Wm. B. Eerdmans Publishing. p. 442. ISBN 0-8028-6348-5, 9780802863485.

[31] Ágoston and Alan Masters, Gábor and Bruce (2009). *Encyclopedia of the Ottoman Empire*. Infobase Publishing. p. 583. ISBN 0-8160-6259-5, 9780816062591.

[32] Imber, Colin (2002). *The Ottoman Empire, 1300–1650: The Structure of Power*. Palgrave Macmillan. p. 53. ISBN 0-333-61386-4.

[33] Kinross 1979, p. 206

[34] Mansel, 61

[35] Deringil, Selim (September 2007). "The Turks and 'Europe': The Argument from History". *Middle Eastern Studies*, **43** (5,): 709–723. doi:10.1080/00263200701422600.

[36] Sansovino, Francesco. (http://books.google.ca/books/about/Historia_universale_dell_origine_et_impe.html?id=yCI8AAAAcAAJ&redir_esc=y). . Retrieved 20 December 2012.

[37] Itzkowitz 1980, p. 96

[38] Davies (2007). *Warfare, State and Society on the Black Sea Steppe,1500–1700*. (http://books.google.com/books?id=XH4hghHo1qoC&pg=PA16&dq&hl=en#v=onepage&q=&f=false). p.16.

[39] Subtelny, Orest (1988). " *Ukraine: a history*. (http://books.google.com/books?id=HNIs9O3EmtQC&pg=PA106&dq&hl=en#v=onepage&q=&f=false)". p 106

[40] " The Crimean Tatars and their Russian-Captive Slaves (http://www.econ.hit-u.ac.jp/~areastd/mediterranean/mw/pdf/18/10.pdf)" (PDF). Eizo Matsuki, *Mediterranean Studies Group at Hitotsubashi University*.

[41] Kinross 1979, p. 272

[42] Kunt & Woodhead (Ed.) "Suleyman The Magnificent and his Age, The Ottoman Empire in the Early Modern World", Longman 1995, ISBD 0-582-03827-8, p. 53

[43] Itzkowitz 1980, p. 67

[44] Itzkowitz 1980, p. 71

[45] Itzkowitz 1980, pp. 90–92

[46] Inalcik, Halil. *An Economic And Social History Of The Ottoman Empire, Vol 1 1300–1600*. Cambridge University Press, ISBN 0-521-57455-2, p. 24.

[47] Kinross 1979, p. 281

[48] Itzkowitz 1980, p. 73

[49] Itzkowitz 1980, pp. 74–75

[50] Itzkowitz 1980, pp. 80–81

[51] Itzkowitz 1980, pp. 81–82

[52] Kinross 1979, p. 357

[53] Itzkowitz 1980, p. 84

[54] Itzkowitz 1980, pp. 83–84

[55] Kinross 1979, p. 371

[56] Kinross 1979, p. 372

[57] Kinross 1979, p. 376

[58] Kinross 1979, p. 392

[59] "History of the Istanbul Technical University" (http://www.itu.edu.tr/en/?about/history). Itu.edu.tr. . Retrieved 6 November 2011.

[60] Stone, Norman "Turkey in the Russian Mirror" pages 86–100 from *Russia War, Peace and Diplomacy* edited by Mark & Ljubica Erickson, Weidenfeld & Nicolson: London, 2004 page 97

[61] Presentation of Katip Çelebi, *Kitâb-i Cihân-nümâ li-Kâtib Çelebi* (http://vitrine.library.uu.nl/wwwroot/en/texts/Rarqu54.htm), at the Utrecht University Library

[62] William J. Watson, "Ibrahim Muteferrika and Turkish Incunabula", in *Journal of the American Oriental Society*, Vol. 88, No. 3 (1968), p. 435.
[63] Kinross 1979, p. 396
[64] Kinross 1979, p. 405
[65] "Liberation, Independence and Union" (http://www.njegos.org/past/liunion.htm). Njegos.org. . Retrieved 26 August 2010.
[66] Berend, Tibor Iván, *History derailed: Central and Eastern Europe in the long nineteenth century*, (University of California Press Ltd, 2003), 127.
[67] http://faith-matters.org/images/stories/fm-publications/the-tanzimat-final-web.pdf
[68] PTT Chronology (http://www.ptt.gov.tr/tr/kurumsal/tarihce.html)
[69] "History of the Turkish Postal Service" (http://www.ptt.gov.tr/index.snet?wapp=histor_en&open=1). Ptt.gov.tr. . Retrieved 6 November 2011.
[70] Istanbul City Guide: Beylerbeyi Palace (http://www.istanbulcityguide.com/history/body_mansions_palaces.htm)
[71] NTV Tarih (http://www.ntvtarih.com.tr/) history magazine, issue of July 2011. *"Sultan Abdülmecid: İlklerin Padişahı"*, page 49. (Turkish)
[72] Türk Telekom: History (http://www.turktelekom.com.tr/webtech/eng_default.asp?sayfa_id=30)
[73] Stone, Norman "Turkey in the Russian Mirror" pages 86–100 from *Russia War, Peace and Diplomacy* edited by Mark & Ljubica Erickson, Weidenfeld & Nicolson: London, 2004 page 95.
[74] "History of the Ottoman public debt" (http://gberis.e-monsite.com/categorie,osmanli-borclanma-tarihi-ottoman-debt-history,3219214.html). Gberis.e-monsite.com. . Retrieved 6 November 2011.
[75] Douglas Arthur Howard: "The History of Turkey", page 71.
[76] "*Hijra* and Forced Migration from Nineteenth-Century Russia to the Ottoman Empire" (http://monderusse.revues.org/docannexe1800.html), by Bryan Glynn Williams, *Cahiers du Monde russe*, 41/1, 2000, pp. 79–108.
[77] Memoirs of Miliutin, "the plan of action decided upon for 1860 was to cleanse [ochistit'] the mountain zone of its indigenous population", per Richmond, W. The Northwest Caucasus: Past, Present and Future. Routledge. 2008.
[78] A. J. P. Taylor, *The Struggle for Mastery in Europe: 1848–1918* (1954) pp 228–54
[79] Akcam, Taner. *A Shameful Act: The Armenian Genocide and the Question of Turkish Responsibility*. New York: Metropolitan Books, 2006, p. 42. ISBN 0-8050-7932-7.
[80] Mann, Michael (2005), *The dark side of democracy: explaining ethnic cleansing*, Cambridge University Press, p. 118
[81] Todorova, Maria (2009), *Imagining the Balkans*, Oxford University Press, p. 175
[82] editors: Matthew J. Gibney, Randall Hansen, Immigration and Asylum: From 1900 to the Present, Vol. 1, ABC-CLIO, 2005, p.437 Read (http://books.google.com/books?id=2c6ifbjx2wMC&printsec=frontcover&redir_esc=y#v=onepage&q&f=false) quote: "Muslims had been the majority in Anatolia, the Crimea, the Balkans and the Caucasus and a plurality in southern Russia and sections of Romania. Most of these lands were within or contiguous with the Ottoman Empire. By 1923, only Anatolia, eastern Thrace and a section of the south-eastern Caucasus remained to the Muslim land."
[83] Şevket Pamuk, "The Ottoman Economy in World War I" in Stephen Broadberry and Mark Harrison, eds. *The Economics of World War I* (2005) pp 112–36, esp. p 112
[84] (PDF) *Greek and Turkish refugees and deportees 1912–1924* (http://tulp.leidenuniv.nl/content_docs/wap/ejz18.pdf). NL: Universiteit Leiden. .
[85] Spencer Tucker, ed. *Encyclopedia of World War I* (2005) p 1080
[86] Encyclopædia Britannica. "*Encyclopædia Britannica*: Armenian massacres (Turkish-Armenian history)" (http://www.britannica.com/EBchecked/topic/35323/Armenian-massacres). Britannica.com. . Retrieved 26 August 2010.
[87] Balakian, Peter. *The Burning Tigris: The Armenian Genocide and America's Response*. New York: Perennial, 2003. ISBN 0-06-019840-0
[88] Walker, Christopher J. "World War I and the Armenian Genocide". *The Armenian People from Ancient to Modern Times* **II**: 239–273.
[89] Akcam. *A Shameful Act*, pp. 109–204.
[90] Itzkowitz 1980, p. 38
[91] Naim Kapucu,Hamit Palabıyık *Turkish public administration: from tradition to the modern age", p 77*
[92] Antony Black, *ibid*, page 197
[93] Bernard Lewis, *Istanbul and the civilization of the Ottoman Empire*, p151
[94] "Enderun and Matraki" (http://tamu.academia.edu/SencerCorlu/Papers/471488/The_Ottoman_Palace_School_Enderun_and_the_Man_with_Multiple_Talents_Matrakci_Nasuh). .
[95] Kemal H Karpat, *Social Change and Politics in Turkey: A Structural-Historical Analysis*, p204
[96] Antony Black (2001), "The state of the House of Osman (devlet-ı al-ı Osman)" in *The History of Islamic Political Thought: From the Prophet to the Present*, p. 199
[97] The History of Turkish-Jewish Relations
[98] Donald Quataert, 2
[99] Lauren A. Benton, *Law and Colonial Cultures: Legal Regimes in World History, 1400–1900", pp 109–110*
[100] Milner *The Godolphin Arabian* pp. 3–6
[101] Wall *Famous Running Horses* p. 8

[102] Ellesmerereportstandard.co.uk (http://www.ellesmereportstandard.co.uk/latest-north-west-news/Petition-created-for-submarine-name.4001190.jp), The standard – Petition created for submarine name
[103] "Story of Turkish Aviation in 'Turkey in the First World War' website" (http://www.turkeyswar.com/aviation/aviation.htm). Turkeyswar.com. . Retrieved 6 November 2011.
[104] Hv. K. K. Mebs. ""Founding" in Turkish Air Force official website" (http://www.hvkk.tsk.tr/EN/IcerikDetay.aspx?ID=19). Hvkk.tsk.tr. . Retrieved 6 November 2011.
[105] Halil İnalcık, Studies in the economic history of the Middle East : from the rise of Islam to the present day / edited by M. A. Cook. London University Press, Oxford U. P. 1970, p. 209 ISBN 0-19-713561-7
[106] Halil İnalcık, Studies in the economic history of the Middle East : from the rise of Islam to the present day / edited by M. A. Cook. London University Press, Oxford U. P. 1970, p. 217 ISBN 0-19-713561-7
[107] İnalcık, Halil; Quataert, Donald (1971). *An Economic and Social History of the Ottoman Empire, 1300–1914*. p. 120.
[108] Halil inalcik, Studies in the economic history of the Middle East : from the rise of Islam to the present day / edited by M. A. Cook. London University Press, Oxford U. P. 1970, p. 218 ISBN 0-19-713561-7
[109] M. Kabadayı, Inventory for the Ottoman Empire / Turkish Republic 1500-2000 (http://www.iisg.nl/research/labourcollab/turkey.pdf)
[110] The Ottomans developed an efficient system for counting the empire's population in 1826, a quarter of a century after such methods were introduced in Britain, France and America – Shaw, S J, "The Ottoman Census System and Population, 1831-1914", International Journal of Middle East Studies, Cambridge University Press, 1978, p. 325
[111] Leila Erder and Suraiya Faroqhi, Population Rise and Fall in Anatolia 1550-1620, Middle Eastern Studies, Vol. 15, No. 3 (Oct., 1979), pp. 322-345
[112] Quataert, D, 2000, op cit, pp. 110-11
[113] Ibid, p. 112
[114] Ibid, p. 113
[115] Ibid, p. 114; Pamuk, S, "The Ottoman Empire and the World Economy: The Nineteenth Century", International Journal of Middle East Studies, Cambridge University Press, Vol. 23, No. 3, Aug. 1991
[116] Ibid, p. 114
[117] Ibid, p. 115
[118] Ibid, p. 116
[119] Justin McCarthy, *Death and Exile: The Ethnic Cleansing of Ottoman Muslims, 1821–2000*, Princeton, N.J: Darwin Press, c1995
[120] Persian historiography and geography, Bertold Spuler,M. Ismail Marcinkowski, page 69, 2003
[121] Kemal H. Karpat (2002). *Studies on Ottoman social and political history: selected articles and essays* (http://books.google.com/?id=082osLxyBDgC&pg=PA266&dq=arzuhalci#v=onepage&q=arzuhalci&f=false). Brill. p. 266. ISBN 90-04-12101-3. .
[122] "Tile – Victoria & Albert Museum – Search the Collections" (http://collections.vam.ac.uk/item/O106609/tile/). Collections.vam.ac.uk. 25 August 2009. . Retrieved 26 August 2010.
[123] George C. Kohn (2007.) *Dictionary of Wars*. Infobase Publishing. p. 385. ISBN 0-8160-6577-2
[124] Jalāl Āl Aḥmad, *Plagued by the West* (*Gharbzadegi*), translated by Paul Sprachman. Delmor, NY: Center for Iranian Studies, Columbia University, 1982.
[125] Akcam, Taner. *A Shameful Act: The Armenian Genocide and the Question of Turkish Responsibility*. New York: Metropolitan Books, 2006 p. 24 ISBN 0-8050-7932-7
[126] Shaw and Shaw. *History of the Ottoman Empire*, pp. 112–129.
[127] " The Divinely-Protected, Well-Flourishing Domain: The Establishment of the Ottoman System in the Balkan Peninsula (http://www.loyno.edu/history/journal/1998-9/Krummerich.htm)", Sean Krummerich, Loyola University New Orleans, *The Student Historical Journal*, volume 30 (1998–99)
[128] Turkish Toleration (http://www.globaled.org/nyworld/materials/ottoman/turkish.html), The American Forum for Global Education
[129] Belge, 389
[130] Moran, Berna. *Türk Romanına Eleştirel Bir Bakış Vol. 1*. p. 19. ISBN 975-470-054-0.
[131] "Eli Shah. The Ottoman Artistic Legacy" (http://www.mfa.gov.il/MFA/MFAArchive/1990_1999/1999/2/The Ottoman Artistic Legacy). Mfa.gov.il. . Retrieved 26 August 2010.
[132] Foroqhi, S. *Subjects of the Sultan: Culture and Daily Life in the Ottoman Empire*, I. B. Tauris, 2005, ISBN 1-85043-760-2, p. 152
[133] Foroqhi, p.153
[134] "Allaboutturkey" (http://www.allaboutturkey.com/karagoz.htm). Allaboutturkey. 2006-11-20. . Retrieved 2012-08-20.
[135] Emin Şenyer: Karagoz Traditional Turkish Shadow Theatre (http://www.karagoz.net/english/shadowplay.htm)
[136] Bert Fragner, "From the Caucasus to the Roof of the World: a culinary adventure", in Sami Zubaida and Richard Tapper, *A Taste of Thyme: Culinary Cultures of the Middle East*, London, Prague and New York, p. 52
[137] "Supply of Slaves" (http://coursesa.matrix.msu.edu/~fisher/hst373/readings/inalcik6.html). Coursesa.matrix.msu.edu. . Retrieved 26 August 2010.
[138] "Islam and slavery: Sexual slavery" (http://www.bbc.co.uk/religion/religions/islam/history/slavery_7.shtml). BBC. . Retrieved 26 August 2010.
[139] Université de Strasbourg. Institut de turcologie, Université de Strasbourg. Institut d'études turques, Association pour le développement des études turques. (1998). *Turcica*. Éditions Klincksieck. p. 198.

[140] (Ragep 2001a)
[141] F. Jamil Ragep (2001), "Freeing Astronomy from Philosophy: An Aspect of Islamic Influence on Science", *Osiris*, 2nd Series, Vol. 16, Science in Theistic Contexts: Cognitive Dimensions, p. 49-64, 66-71.
[142] Edith Dudley Sylla, "Creation and nature", in Arthur Stephen McGrade (2003), p. 178-179, Cambridge University Press, ISBN 0521000637.
[143] George Saliba, "Arabic planetary theories after the eleventh century AD", in Rushdī Rāshid and Régis Morelon (1996), *Encyclopedia of the History of Arabic Science*, p. 58-127 [123-124], Routledge, ISBN 0415124107.
[144] Sevim Tekeli, "Taqi al-Din", in Helaine Selin (1997), *Encyclopaedia of the History of Science, Technology, and Medicine in Non-Western Cultures*, Kluwer Academic Publishers, ISBN 0792340663.
[145] Zaken, Avner Ben (2004). "The heavens of the sky and the heavens of the heart: the Ottoman cultural context for the introduction of post-Copernican astronomy". *The British Journal for the History of Science* (Cambridge University Press) **37**: 1–28. doi:10.1017/S0007087403005302.
[146] G. Bademci (2006), "First illustrations of female Neurosurgeons in the fifteenth century by Serefeddin Sabuncuoglu", *Neurocirugía* **17**: 162-5
[147] Horton, Paul (1977). "Topkapi's Turkish Timepieces" (http://www.saudiaramcoworld.com/issue/197704/topkapi.s.turkish.timepieces.htm). *Saudi Aramco World, July–August 1977*: 10–13. . Retrieved 2008-07-12.

Notes

References

- Itzkowitz, Norman (1980). *Ottoman Empire and Islamic Tradition*. University of Chicago Press. ISBN 0-226-38806-9.
- Kinross, Patrick Balfour (1979). *The Ottoman Centuries: The Rise and Fall of the Turkish Empire*. William Morrow.

External links

- Ottoman Text Archive Project – University of Washington (http://courses.washington.edu/otap/)
- American Travelers to the Holy Land in the 19th Century (http://www.shapell.org/journeys.aspx?american-travelers-to-the-holy-land-in-the-19th-century) Shapell Manuscript Foundation
- Ottoman Empire in The Oxford Dictionary of Islam (http://www.oxfordislamicstudies.com/article/opr/t125/e1801?_hi=0&_pos=3)
- The Ottoman Empire: Resources – University of Michigan (http://www.umich.edu/~turkish/ottemp.html)
- The Ottoman Empire: A Chronological Outline (http://www.turizm.net/turkey/history/ottoman.html)
- Information about Ottomans (http://www.theottomans.org/)
- Ottoman Navy and Navigation (http://www.ottomannavy.com/)
- Detailed information about Ottoman Empire on Ansiklopedika.org (http://ansiklopedika.org/OsmanlÄ±_Devleti) **(Turkish)**

Iran

Islamic Republic of Iran *Jomhuri-ye Eslāmi-ye Irān*		

Motto: استقلال. آزادی. جمهوری اسلامی *Independence, Freedom, Islamic Republic*		
Anthem: جمهوری اسلامی ایران *National Anthem of the Islamic Republic of Iran*		
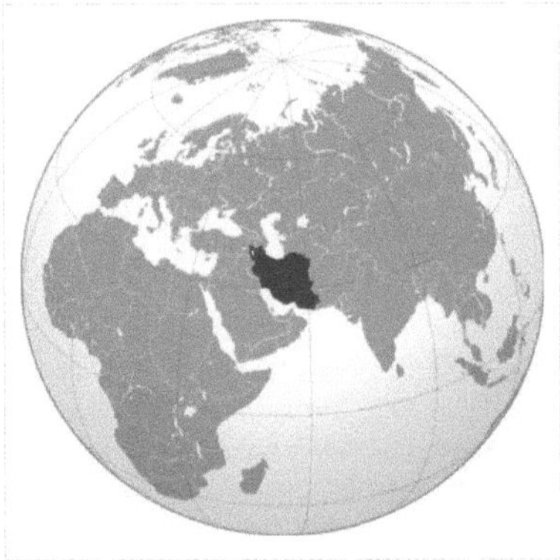		
Capital and largest city		Tehran 35°41′N 51°25′E
Official languages		Persian
Spoken languages		Persian, Azerbaijani, Kurdish, Lori, Gilaki, Mazandarani, Balochi, Arabic, Turkmen[1]
Demonym		Iranian
Government		Unitary state, Islamic republic
-	Supreme Leader	Ali Khamenei
-	President	Mahmoud Ahmadinejad
-	First Vice President	Mohammad-Reza Rahimi
-	Speaker of the Parliament	Ali Larijani
-	Chief Justice	Sadeq Larijani
Legislature		Islamic Consultative Assembly
Unification[2]		
-	Median Empire	625 BCE
-	Achaemenid Empire	550 BCE

-	Safavid Empire	1501[3]
-	Islamic Republic	24 October 1979
-	Current constitution	
Area		
-	Total	1,648,195 km^2 (18th) 636,372 sq mi
-	Water (%)	0.7
Population		
-	2012 census	75,149,669[4] (17th)
-	Density	48/km^2 (162rd) 124/sq mi
GDP (PPP)		2011 estimate
-	Total	$990.219 billion[5]
-	Per capita	$13,053[5]
GDP (nominal)		2011 estimate
-	Total	$482.445 billion[5]
-	Per capita	$6,359[5]
Gini (2008)		38[6] (medium
HDI (2011)		▲ 0.707[7] (high / 88th)
Currency		Rial (ریال) (IRR)
Time zone		IRST (UTC+3:30)
-	Summer (DST)	IRDT (UTC+4:30)
Drives on the		right
Calling code		98
ISO 3166 code		IR
Internet TLD		.ir, ایران.
1.	Bookrags.com [8]	
2.	Iranchamber.com [9]	
3.	Statistical Center of Iran. "جمعيت و متوسط رشد سالانه" [10] (in Persian). Retrieved 13 February 2009.	
4.	CIA Factbook [11]	

Iran (/ɪˈrɑːn/[12] or /aɪˈræn/;[13] Persian: ایران Persian pronunciation: [ʔiˈɾɒn] (listen)), officially the **Islamic Republic of Iran** (Persian: جمهوری اسلامی ایران, *Jomhuri-ye Eslāmi-ye Irān*), is a country in Western Asia.[14] [15] [16] The name "Iran", which in Persian means "Land of the Aryans", has been in native use since the Sassanian era, in antiquity. It came into use internationally in 1935, before which the country was known to the Western world as **Persia** (pron.: /ˈpɜːʒə/ or /ˈpɜːʃə/).[13] [17] Both "Persia" and "Iran" are used interchangeably in cultural contexts; however, "Iran" is the name used officially in political contexts.[18] [19]

The 18th-largest country in the world in terms of area at 1648195 km^2 (sq mi), Iran has a population of around 75 million.[14] [20] It is a country of particular geopolitical significance owing to its location in three spheres of Asia (West, Central, and South). Iran is bordered on the north by Armenia, Azerbaijan and Turkmenistan. As Iran is a littoral state of the Caspian Sea, which is an inland sea, Kazakhstan and Russia are also Iran's direct neighbors to the north. Iran is bordered on the east by Afghanistan and Pakistan, on the south by the Persian Gulf and the Gulf of Oman, on the west by Iraq and on the northwest by Turkey. Tehran is the capital, the country's largest city and the political, cultural, commercial and industrial center of the nation. Iran is a regional power,[21] [22] and holds an important position in international energy security and world economy as a result of its large reserves of petroleum and natural gas. Iran has the second largest proven natural gas reserves in the world and the fourth largest proven petroleum reserves.[23]

Iran is home to one of the world's oldest civilizations.[24] The first dynasty in Iran formed during the Elamite kingdom in 2800 BC. The Iranian Medes unified Iran into an empire in 625 BC.[2] They were succeeded by the Iranian Achaemenid Empire, the Hellenic Seleucid Empire and two subsequent Iranian empires, the Parthians and the Sassanids, before the Muslim conquest in 651 AD. Iranian post-Islamic dynasties and empires expanded the Persian language and culture throughout the Iranian plateau. Early Iranian dynasties which re-asserted Iranian independence included the Tahirids, Saffarids, Samanids and Buyids.

The blossoming of Persian literature, philosophy, medicine, astronomy, mathematics and art became major elements of Muslim civilization. Iranian identity continued despite foreign rule in the ensuing centuries[25] and Persian culture was adopted also by the Ghaznavid,[26] Seljuk,[27] [28] Ilkhanid[29] and Timurid[30] rulers. The emergence in 1501 of the Safavid dynasty,[3] which promoted Twelver Shia Islam[31] as the official religion of their empire, marked one of the most important turning points in Iranian and Muslim history.[32] The Persian Constitutional Revolution established the nation's first parliament in 1906, within a constitutional monarchy. Following a coup d'état instigated by the UK and US in 1953, Iran gradually became a more autocratic country. Growing dissent with foreign influence culminated during the Iranian Revolution which led to establishment of an Islamic republic on 1 April 1979.[20] [33]

Iran is a founding member of the UN, NAM, OIC and OPEC. The political system of Iran, based on the 1979 constitution, comprises several intricately connected governing bodies. The highest state authority is the Supreme Leader. Shia Islam is the official religion and Persian is the official language.[34]

Etymology

The name of Iran (ایران) is the Modern Persian derivative from the Proto-Iranian term *Aryānā*, meaning "Land of the Aryans", first attested in Zoroastrianism's Avesta tradition.[35] [36] [37] [38] The term *Ērān* is found to refer to Iran in a 3rd century Sassanid inscription, and the Parthian inscription that accompanies it uses the Parthian term "aryān" in reference to *Iranians*.[39] However historically Iran has been referred to as Persia or similar (*La Perse, Persien, Perzië, etc.*) by the Western world, mainly due to the writings of Greek historians who called Iran *Persis* (Περσίς), meaning land of the Persians. In 1935 Rezā Shāh requested that the international community should refer to the country as Iran. Opposition to the name change led to the reversal of the decision, and in 1959 both names were to be used interchangeably.[40] Since the Iranian Revolution in 1979 the official name of the country has been the "Islamic Republic of Iran".

The historical and cultural wider usage of 'Iran'[41] [42] [43] is not restricted to the modern state proper. Irānshahr[44] or Irānzamīn (Greater Iran)[45] corresponded to territories of Iranian cultural or linguistic zones. Besides modern Iran, it included portions of the Caucasus, Mesopotamia, the Indian subcontinent and Central Asia.

History

Early History

Further information: Archaeological sites in Iran, Tepe Sialk, Jiroft culture, and Shahr-e Sukhteh

The earliest archaeological artifacts in Iran, like those excavated at the Kashafrud and Ganj Par sites, attest to a human presence in Iran since the Lower Paleolithic era.[46] Neanderthal artefacts dating back to the Middle Paleolithic period have been found mainly in the Zagros region at sites such as Warwasi and Yafteh Cave.[47] [48] Early agricultural communities began to flourish in Iran at around 8000 BC,[49] [50] with settlements such as Chogha Bonut, Susa and Chogha Mish developing in the Zagros region.[51] [52] [53]

The emergence of Susa as a city is determined by C14 dating as early as 4395 BC.[54] There are dozens of pre-historic sites across the Iranian plateau pointing to the existence of ancient cultures and urban settlements in the fourth millennium BC.[53] [55] [56] During the Bronze age Iran was home to several civilisations such as Elam, Jiroft and Zayandeh Rud civilisations. Elam, the most prominent of these civilisations developed in the southwest of Iran alongside those in Mesopotamia. The development of writing in Elam in fourth millennium BC paralleled that in Sumer.[57] The Elamite kingdom continued its existence until the emergence of the Median and Achaemenid Empires.

Classical Era

During the second millenium BC, Proto-Iranian tribes arrived in Iran from the Eurasian steppes,[59] rivaling the native settlers of the country.[60] [61] As these tribes dispersed into the wider area of Greater Iran and beyond, the boundaries of modern Iran were dominated by the Persian, Parthian and Median tribes. Soon after Zoroastrianism emerged as the main religion of the Iranian tribes.[62] The unification of the Median tribes under a single ruler in 728 BC led to the creation of a Median empire, which by 612 BC controlled the whole of Iran as well as eastern Anatolia.[63]

Ruins of the Achaemenid palace of Persepolis, now a UNESCO World Heritage Site[58]

In 550 BC, Cyrus the Great from the state of Anshan took over the Median empire, and founded the Achaemenid empire by unifying other city states. The conquest of Media happened as a result of what is called the Persian revolt, which was initially triggered by the actions of the Median ruler Astyages, and quickly spread to other provinces, as they allied with the Persians. Later conquests under Cyrus and his successors expanded the empire to include Lydia, Babylon, Egypt and the lands to the west of the Indus an Oxus Rivers. Conflict on the western borders began with the famous Greco-Persian Wars which continued through the first half of the 5th century BC and ended with the Persian withdrawal from all of their European territories.[64] The empire had a centralised, bureaucratic administration under the Emperor and a large professional army and civil services, inspiring similar developments in later empires.[65]

In 334 BC, Alexander the Great invaded the Achaemenid Empire, defeating the last Achaemenid Emperor Darius III at the Battle of Issus in 333 BC. Following the premature death of Alexander, Iran came under the control of Hellenistic Seleucid Empire. In the middle of the 2nd century BC, Parthia rose to become the main power in Iran and continued as a feudal monarchy for nearly five centuries until 224 AD, when it was succeeded by the Sassanid Empire.[66] The Sassanids established an empire roughly within the frontiers achieved by the Achaemenids, with the capital at Ctesiphon.[67] Most of the period of the Parthian and Sassanid Empires were overshadowed by the

Roman-Persian Wars, which raged on their western borders for over 700 years. These wars exhausted both Romans and Sassanids,which arguably led to the defeat of both at the hands of the invading Muslim Arabs.

Middle Ages (652–1501)

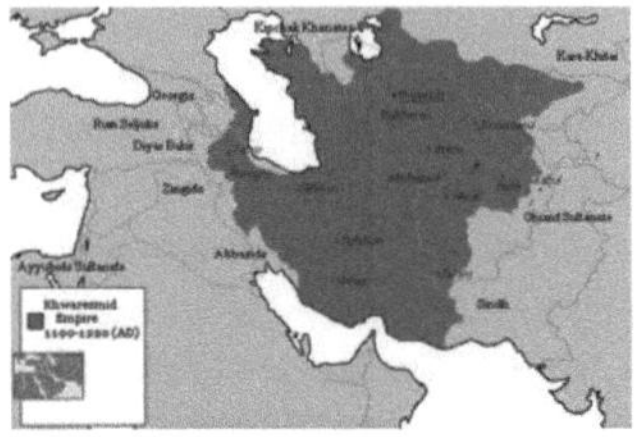

Khwarezmid Empire around 1200

The prolonged Roman-Persian wars, as well as social conflict within the Empire opened the way for an Islamic invasion of Iran in the 7th century.[68] [69] Initially defeated by the Rashidun Caliphate, Iran later came under the rule of their successors the Ummayad and Abbasid Caliphates. The process of conversion of Iranians to Islam which followed was a prolonged and gradual process. Under the new Arab elite of the Rashidun and later Ummayad Caliphates Iranians, both Muslim (mawali) and non-Muslim (Dhimmi), were discriminated, being excluded from government and military, and having to pay a special tax.[70] [71] In 750 the Abbasids succeeded in overthrowing the Ummayad Caliphate, mainly due to the support from dissatisfied Iranian mawali.[72] The mawali formed the majority of the rebel army, which was led by the Iranian general Abu Muslim.[73] [74] [75] After two centuries of Arab rule semi-independent and independent Iranian kingdoms (such as the Tahirids, Saffarids, Samanids and Buyids) began to appear on the fringes of the declining Abbasid Caliphate. By the Samanid era in the 9th and 10th centuries Iran's efforts to regain its independence had been well solidified.[76]

Persian miniature depicting Timur's campaign in India

The arrival of the Abbasid Caliphs saw a revival of Persian culture and influence, and a move away from Arabic culture. The role of the old Arab aristocracy was slowly replaced by a Persian bureaucracy.[77] The blossoming Persian literature, philosophy, medicine, and art became major elements in the forming of a Muslim civilization during the Islamic Golden Age.[78] [79] The Islamic Golden Age reached its peak in the 10th and 11th centuries, during which Persia was the main theatre of scientific activity.[80] After the 10th century, Persian, alongside Arabic, was used for scientific, philosophical, historical, mathematical, musical, and medical works, as important Iranian writers such as Nasir al-Din al-Tusi, Avicenna, Qotb al-Din Shirazi, Naser Khusraw and Biruni made contributions to Persian scientific writing.

The cultural revival that began in the Abbasid period led to a resurfacing of Iranian national identity, and so earlier attempts of Arabization never succeeded in Iran. The Iranian Shuubiyah movement became a catalyst for Iranians to regain their independence in their relations with the Arab invaders.[81] The most notable effect of the movement was the continuation of the Persian language attested to the epic poet Ferdowsi, now regarded as the most important figure in Persian literature.

The 10th century saw a mass migration of Turkic tribes from Central Asia into the Iranian plateau.[82] Turkic tribesmen were first used in the Abbasid army as slave-warriors (Mamluks), replacing Persian and Arab elements within the army.[73] As a result the Mamluks gained significant political power. In 999, Iran came under the rule of Ghaznavid dynasty, whose rulers were of Mamluk Turk origin, and later under the Turkish Seljuk and Khwarezmian

Empires. These Turks had been Persianised and had adopted Persian models of administration and rulership.[82] The result of the adoption and patronage of Persian culture by Turkish rulers was the development of a distinct Turko-Persian tradition.

In 1219-21 the Khwarezmian Empire suffered a devastating invasion by Genghis Khan's Mongol army. Mongol violence and depredations killed up to three-fourths of the population of the Iranian Plateau, possibly 10 to 15 million people. Some historians have estimated that Iran's population did not again reach its pre-Mongol levels until the mid-20th century.[83] Following the fracture of the Mongol Empire in 1256 Hulagu Khan, Genghis Khan's grandson, established the Ilkhanate dynasty in Iran. In 1370 yet another conqueror, Timur, commonly known as Tamerlane in the West, followed Hulagu's example, establishing the Timurid Dynasty which lasted for another 156 years. In 1387, Timur ordered the complete massacre of Isfahan, reportedly killing 70,000 citizens.[84] Hulagu, Timur and their successors soon came to adopt the ways and customs of that which they had conquered, choosing to surround themselves with a culture that was distinctively Persian.[85]

Modern era (1501–)

19th century drawing of Naqsh-e Jahan Square, Isfahan

At the start of the 16th century, Shah Ismail I (1500-1525), the founder of the Safavid Dynasty, established himself as ruler of western Persia and Azerbaijan.[82] He subsequently extended his authority over all of Persia. Ismail is also known for instigating a religious revolution in Iran, forcefully converting the predominantly Sunni population to the state religion of Shi'a Islam.[86] There was an intense rivalry between Safavid Persia and the Ottoman Empire, leading to the Ottoman–Persian Wars.[83] During the Safavid era Iran once again became a centre for high civilisation and wealth, peaking in the reign of the brilliant soldier, statesman and administrator Shah Abbas I (1587-1629).[32] [83] Under his rule the state became highly centralized, the army was reorganized and modernized, and a distinct style of architecture was developed in his new capital at Isfahan. However, following a slow decline in the late 17th and early 18th centuries, the Safavid dynasty was instead ended by Pashtun rebels who besieged Isfahan and defeated Soltan Hosein in 1722.

In 1729, a Khorasan chieftain, Nader Shah, successfully drove out, then conquered the Pashtuns invaders. He repeatedly defeated the Turks and, by 1735, had regained territory lost to the Ottomon and Russian Empires. In 1738-9 he made a very profitable incursion into the Mughal Empire. His military successes on all fronts earned him the nickname "Napoleon of Persia" or "the second Alexander". Following a brief period of civil war and turmoil, sparked by Nader Shah's assassination, Karim Khan came to power in 1750, giving himself the title *Vakil e-Ra'aayaa* (Representative of the People), and bringing a period of relative peace and prosperity.[83]

Mohammad Reza Pahlavi and Farah Pahlavi, the last Shah and Empress of Iran

Another civil war ensued after Karim Khan's death in 1779, out of which Aga Muhammad Khan emerged victorious, founding the Qajar Dynasty in 1794 and establishing Tehran as his capital. This cruel and brutal ruler was assassinated in1797. Qajar rule was marked by its inadequate response to change and its failure to maintain Iranian territorial integrity and sovereignty, and is consequently characterised by over a century of misrule.[82] The Great Persian Famine of 1870–1871 is believed to have caused the death of 1.5 million persons, or 20–25% of Persia's population.[87] Whilst resisting efforts to be colonised, Iran suffered in the 19th century as a result of Russian and British empire-building, known as 'The Great Game', losing much of its territory in the Russo-Persian and the Anglo-Persian Wars. A series of protests took place in response to the sale of concessions to foreigners by Nasser al-Din Shah and Mozaffar ad-Din Shah between 1872 and 1905, the last of which resulted in the Iranian Constitutional Revolution and establishment of Iran's first national parliament (*majles*) in 1906. The corrupt and despotic Mohammad Ali Shah, rescinded the constitution, bombed the majles building and abolished parliament in 1908. A Russian army helped him to suppress the revolt in 1909. However, the struggle continued until 1911, when Mohammad Ali was defeated and forced to abdicate. On the pretext of restoring order, the Russians occupied northern Iran in 1911. During World War I, the British occupied much of western Iran, not leaving until 1921.

In 1921 Reza Khan, Prime Minister of Iran and former general of the Persian Cossack Brigade, overthrew the incompetent and corrupt Qajar Dynasty and became Shah. An ardent nationalist, Reza Shah initiated policies of military, administrative and financial modernisation and centralization. He quickly persuaded the Russians to withdraw from Iran.[82] Industrialization, the construction of the Trans-Iranian Railway and the establishment of a national education system can be named as some of his reforms. However, in 1941 he had been forced to abdicate in favour of his son, Mohammad Reza Pahlavi, by Britain and the USSR, who were both fearful of Reza Shah's nascent [88] ties to Germany and in need of supply lines for the Allied war effort in the form of the new Trans-Iranian Railway.[89] During World War II, Iran was once again subject to British and Russian occupation.

In 1951, after the assassination of prime minister Ali Razmara, Dr. Mohammad Mosaddegh was elected prime minister by a parliamentary vote which was then ratified by the Shah. As prime minister, Mosaddegh became enormously popular in Iran after he nationalized Iran's petroleum industry and oil reserves. In response, the British government, headed by Winston Churchill, embargoed Iranian oil and successfully enlisted the United States to join in a plot to depose the democratically elected government of Mosaddegh. In 1953 US President Dwight D. Eisenhower authorized Operation Ajax. The operation, supported by the Shah, was successful, and Mosaddegh was arrested on 19 August 1953. The coup was the first time the US had openly overthrown an elected, civilian government of another sovereign state.[90]

After Operation Ajax, Shah Mohammad Reza Pahlavi favoured American and British oil interests and his rule became increasingly autocratic. With American support, the Shah was able to rapidly modernize the Iranian infrastructure and military. However, his rule was also corrupt and repressive. Arbitrary arrests and torture by his secret police, SAVAK, were used to crushed all forms of political opposition. Ayatollah Ruhollah Khomeini became an active critic of the Shah's White Revolution and publicly denounced the government.

The Shah visits the Kharg petrochemical complex, 1970

Khomeini was arrested and imprisoned for 18 months. After his release in 1964, Khomeini publicly criticized the United States government. The Shah sent him into exile. He went first to Turkey, then to Iraq and finally to France. While in exile, Khomeini continued to denounce the Shah.

By the mid-1970s, there was growing unrest with the Shah's repressive regime. The Iranian Revolution, also known as the Islamic Revolution,[91] [92] [93] began in January 1978 with the first major demonstrations against the Shah.[94] Minor political reforms and the release of some political prisoners in 1978 failed to satisfy the growing opposition. In November 1978, the Shah imposed martial law and implemented a new crackdown in an attempt to crush opposition. After strikes and demonstrations paralyzed the country and its economy, the Shah fled the country in January 1979 and Ayatollah Ruhollah Khomeini returned from exile to Tehran. His departure was tantamount to abdication. The Pahlavi dynasty collapsed ten days later, on 11 February, when Iran's military declared itself "neutral". Armed civilians and rebel troops overwhelmed troops loyal to the Shah in a climatic 3-days of street fighting. Iran officially became an Islamic Republic on 1 April 1979, when Iranians overwhelmingly approved a national referendum to make it so.[20] [33] In parallel nation wide uprisings against the new regime erupted in Kordestan, Khuzestan, Balochistan and other areas, though were eventually subdued, with some lasting until late 1980.

In December 1979, the country approved a theocratic constitution, whereby Khomeini became Supreme Leader of the country. The speed and success of the revolution surprised many throughout the world,[95] as it had not been precipitated by a military defeat, a financial crisis, or a peasant rebellion.[96] Although both nationalists and Marxists joined with Islamic traditionalists to overthrow the Shah, tens of thousands were killed and executed by the Islamic regime afterward.[97]

Ayatollah Khomeini, founder of the Islamic Republic of Iran

As a beneficiary of the exploitation of Iran's oil reserves, the USA had always been a strong supporter of the Shah's regime. Therefore, Iran – United States relations deteriorated rapidly as a result of the revolution. On 4 November 1979, a group of Iranian students seized US embassy personnel, labeling the embassy a "den of spies".[98] They accused its personnel of being CIA agents plotting to overthrow the revolutionary government, as the CIA had done to Mosaddegh in 1953. While the student ringleaders had not asked for permission from Khomeini to seize the embassy, Khomeini nonetheless supported the embassy takeover after hearing of its success.[99] Thirteen (being female and/or African American) of the 66 hostages were released after a couple of weeks.[99] The remaining hostages were held for 444 days. Attempts by the Jimmy Carter administration to negotiate for their release or rescue them were unsuccessful. In January 1981 the hostages were finally set free according to the Algiers Accords.

Iraqi leader Saddam Hussein decided to take advantage of what he perceived to be disorder in the wake of the Iranian Revolution and its unpopularity with Western governments. The once-strong Iranian military had been disbanded during the revolution. Saddam sought to expand Iraq's access to the Persian Gulf by acquiring territories that Iraq had claimed earlier from Iran during the Shah's rule. Of chief importance to Iraq was

Khuzestan, which not only has a substantial Arab population, but boasted rich oil fields as well. On the unilateral behalf of the United Arab Emirates, the islands of Abu Musa and the Greater and Lesser Tunbs became objectives as well. At this stage Saddam Hussein enjoyed the support of the USA, the Gulf States and many leading states in the West who were hostile to the Iranian Revolution. On 22 September 1980 the Iraqi army invaded Iran at Khuzestan, precipitating the Iran–Iraq War.

Although Saddam Hussein's forces made several early advances, by 1982, and after bitter fighting, the Iranian forces managed to drive the Iraqi army back into Iraq. Despite receiving large amounts of foreign financial and military aid (including chemical weapons), all of Saddam's subsequent offensives were thrown back. The war continued until 1988, when Khomeini, in his words, "drank the cup of poison" and accepted a truce mediated by the UN. The total Iranian casualties in the war were estimated to be between 500,000 and 1,000,000;[100] with more than 100,000 Iranians being victims of Iraq's chemical weapons.[101] Almost all relevant international agencies have confirmed that Saddam engaged in chemical warfare to blunt the Iranian counter-attacks. These agencies also unanimously confirmed that Iran never used chemical weapons during the war.[102] [103] [104] An estimated 95,000 Iranian child soldiers were killed during the Iran–Iraq War.[105]

June 2009 Iranian election protests in Tehran

Following the Iran–Iraq War, President Akbar Hashemi Rafsanjani and his administration concentrated on a pragmatic pro-business policy of rebuilding and strengthening the economy without making any dramatic break with the ideology of the revolution. Rafsanjani served until 1997 when he was succeeded by the moderate reformist Mohammad Khatami. During his two terms as president, Khatami advocated freedom of expression, tolerance and civil society, constructive diplomatic relations with other states including European Union and Asian governments, and an economic policy that supported free market and foreign investment. However, Khatami is widely regarded as having been unsuccessful in achieving his goal of making Iran more free and democratic.[106]

In the 2005 presidential elections, Iran made yet another change in political direction, when the conservative populist candidate Mahmoud Ahmadinejad was elected over Akbar Hashemi Rafsanjani.[107] A significant challenge to Ahmadinejad's political power, and the foundations of the Islamic Republic itself occurred during the 2009 Iranian presidential election that was held on 12 June 2009,[108] the tenth presidential election to be held in the country.[109] The Interior Ministry, announced incumbent president Mahmoud Ahmadinejad had won the election receiving 24.5 million votes, amounting to 62.63% of the vote, while Mir-Hossein Mousavi had come in second place with 13.2 million votes, being 33.75% of the vote.[110] [111] However, there were allegations of large irregularities in the results, provoking the 2009–2010 Iranian election protests both within Iran and in major capitals in the West.[112] [113]

Geography

Iran is the eighteenth largest country in the world, with an area of 1648195 km^2 (sq mi).[23] Its area roughly equals that of the United Kingdom, France, Spain, and Germany combined, or somewhat more than the US state of Alaska.[114] Iran lies between latitudes 24° and 40° N, and longitudes 44° and 64° E. Its borders are with Azerbaijan (611 km/380 mi (with Azerbaijan-Naxcivan exclave (179 km/111 mi))[115] and Armenia (35 km/22 mi) to the north-west; the Caspian Sea to the north; Turkmenistan (992 km/616 mi) to the north-east; Pakistan (909 km/565 mi) and Afghanistan (936 km/582 mi) to the east; Turkey (499 km/310 mi) and Iraq (1458 km/906 mi) to the west; and finally the waters of the Persian Gulf and the Gulf of Oman to the south.

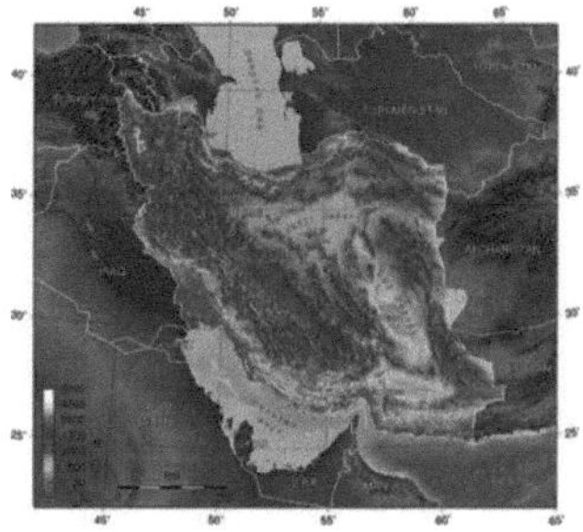

Topography of Iran

Mount Damavand, Iran's highest point, is located in Amol County, Mazanderan.

Iran consists of the Iranian Plateau with the exception of the coasts of the Caspian Sea and Khuzestan Province. It is one of the world's most mountainous countries, its landscape dominated by rugged mountain ranges that separate various basins or plateaux from one another. The populous western part is the most mountainous, with ranges such as the Caucasus, Zagros and Alborz Mountains; the last contains Iran's highest point, Mount Damavand at 5610 m (18406 ft), which is also the highest mountain on the Eurasian landmass west of the Hindu Kush.[116]

The northern part of Iran is covered by dense rain forests called Shomal or the Jungles of Iran. The eastern part consists mostly of desert basins such as the Dasht-e Kavir, Iran's largest desert, in the north-central portion of the country, and the Dasht-e Lut, in the east, as well as some salt lakes. This is because the mountain ranges are too high for rain clouds to reach these regions. The only large plains are found along the coast of the Caspian Sea and at the northern end of the Persian Gulf, where Iran borders the mouth of the Shatt al-Arab (or the Arvand Rūd) river. Smaller, discontinuous plains are found along the remaining coast of the Persian Gulf, the Strait of Hormuz and the Gulf of Oman.

Climate

Iran's climate ranges from arid or semiarid, to subtropical along the Caspian coast and the northern forests. On the northern edge of the country (the Caspian coastal plain) temperatures rarely fall below freezing and the area remains humid for the rest of the year. Summer temperatures rarely exceed 29 °C (84.2 °F).[117] [118] Annual precipitation is 680 mm (26.8 in) in the eastern part of the plain and more than 1700 mm (66.9 in) in the western part.

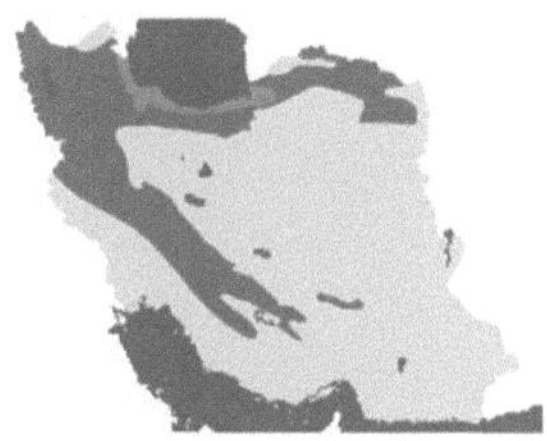

Simplified climatic map of Iran Caspian Mild Mountains Arid and Semi-Arid

To the west, settlements in the Zagros basin experience lower temperatures, severe winters with below zero average daily temperatures and heavy snowfall. The eastern and central basins are arid, with less than 200 mm (7.9 in) of rain, and have occasional deserts.[118] Average summer temperatures exceed 38 °C (100.4 °F). The coastal plains of the Persian Gulf and Gulf of Oman in southern Iran have mild winters, and very humid and hot summers. The annual precipitation ranges from

135 to 355 mm (5.3 to 14.0 in).[118]

Fauna

Persian Leopard

Iran's wildlife is composed of several animal species including bears, gazelles, wild pigs, wolves, jackals, panthers, Eurasian Lynx, and foxes. Domestic animals include, sheep, goats, cattle, horses, water buffalo, donkeys, and camels. The pheasant, partridge, stork, eagles and falcon are also native to Iran.

One of the most famous members of Iranian wildlife is the critically endangered Asiatic cheetah, also known as the Iranian Cheetah, whose numbers were greatly reduced after the Iranian Revolution. Today there are ongoing efforts to increase its population and introduce it back in India. Iran had lost all its Asiatic Lion and the now extinct Caspian Tigers by the earlier part of the 20th century.[119]

Provinces and cities

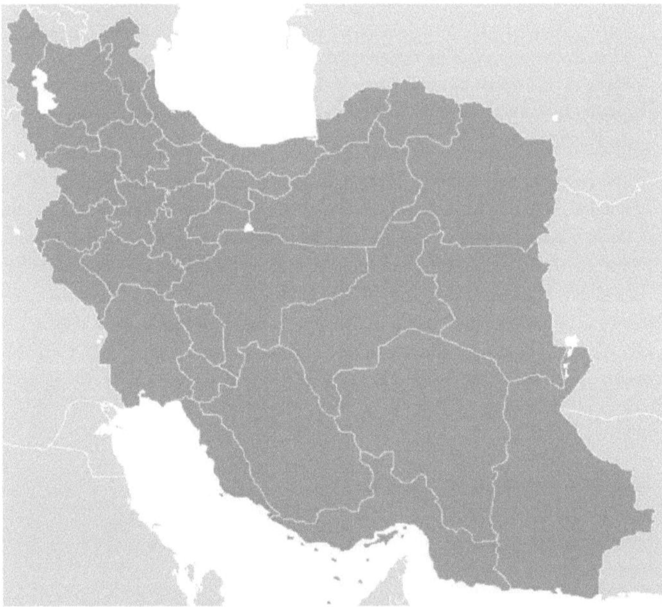

Alborz

Ardabil

Bushehr

Chaharmahal and Bakhtiari

Isfahan

Fars

Gilan

Golestan

Hamadan

Hormozgan

Ilam

Kerman

Kermanshah

Khuzestan

Kohgiluyeh and Boyer-Ahmad

Kurdistan

Loristan

Markazi

Mazandaran

Qazvin

Qom

Razavi Khorasan

Semnan

Sistan and Baluchestan

Tehran

Yazd

Zanjan

North Khorasan

South Khorasan

West Azerbaijan

East Azerbaijan

Caspian Sea

Persian Gulf

Turkmenistan

Afghanistan

Pakistan

Azerbaijan

Armenia

T

u

r

k
e
y
Iraq
Kuwait
Saudi Arabia

Iran is divided into thirty one provinces (*ostān*), each governed by an appointed governor (استاندار, ostāndār). The provinces are divided into counties (*shahrestān*), and subdivided into districts (*bakhsh*) and sub-districts (*dehestān*).

Iran has one of the highest urban growth rates in the world. From 1950 to 2002, the urban proportion of the population increased from 27% to 60%.[120] The United Nations predicts that by 2030, 80% of the population will be urban.[121] Most internal migrants have settled near the cities of Tehran, Isfahan, Ahvaz, and Qom. The listed populations are from the 2006/07 (1385 AP) census.[122] Tehran, with a population of 7,705,036, is the largest city in Iran and is the capital. Tehran, like many big cities, suffers from severe air pollution. It is the hub of the country's communication and transport network.

Mashhad, with a population of 2,410,800, is the second largest Iranian city and the centre of the Razavi Khorasan Province. Mashhad is one of the holiest Shia cities in the world as it is the site of the Imam Reza shrine. It is the centre of tourism in Iran, and between 15 and 20 million pilgrims go to the Imam Reza's shrine every year.[123] [124]

Another major Iranian city is Isfahan (population 1,583,609), which is the capital of Isfahan Province. The Naqsh-e Jahan Square in Isfahan has been designated by UNESCO as a World Heritage Site. The city contains a wide variety of Islamic architectural sites ranging from the 11th to the 19th century. The growth of the suburban area around the city has turned Isfahan into Iran's second most populous metropolitan area (3,430,353).[125]

The fourth major city of Iran is Tabriz (population 1,378,935), the capital of the East Azerbaijan Province. It is also the second industrial city of Iran after Tehran. Tabriz had been the second largest city in Iran until the late 1960s and one of its former capitals and residence of the crown prince under the Qajar dynasty. The city has proven extremely influential in the country's recent history.

The fifth major city is Karaj (population 1,377,450), located in Alborz Province and situated 20 km west of Tehran, at the foot of the Alborz mountains; however, the city is increasingly becoming an extension of metropolitan Tehran.

The sixth major Iranian city is Shiraz (population 1,214,808); it is the capital of Fars Province. The Elamite civilization to the west greatly influenced the area, which soon came to be known as Persis. The ancient Persians were present in the region from about the 9th century BC, and became rulers of a large empire under the Achaemenid dynasty in the 6th century BC. The ruins of Persepolis and Pasargadae, two of the four capitals of the Achaemenid Empire, are located in or near Shiraz. Persepolis was the ceremonial capital of the Achaemenid Empire and is situated 70 kilometres (mi) northeast of modern Shiraz. UNESCO declared the citadel of Persepolis a World Heritage Site in 1979.

Largest cities or towns of Iran

Statistical Center of Iran: Results of national census, 2012 [126]

	Rank	City name	Province	Pop.	Rank	City name	Province	Pop.	
	1	**Tehran**	Tehran	8,778,535[127]	11	**Urmia**	West Azerbaijan	662,734	
Tehran	2	**Mashhad**	Razavi Khorasan	2,567,243	12	**Kerman**	Kerman	621,374	Isfahan
	3	**Isfahan**	Esfahan	1,755,382	13	**Zahedan**	Sistan and Baluchestan	594,683	
Mashhad	4	**Tabriz**	East Azerbaijan	1,657,967	14	**Arak**	Markazi	480,560	Tabriz
	5	**Karaj**	Alborz	1,459,755	15	**Hamadan**	Hamadan	479,640	
	6	**Shiraz**	Fars	1,287,423	16	**Qazvin**	Qazvin	446,760 [128]	
	7	**Qom**	Qom	1,071,503	17	**Yazd**	Yazd	432,194	
	8	**Ahwaz**	Khuzestan	1,068,908	18	**Ardabil**	Ardabil	418,262	
	9	**Rasht**	Gilan	863,470	19	**Bandar Abbas**	Hormozgan	379,301	
	10	**Kermanshah**	Kermanshah	828,313	20	**Zanjan**	Zanjan	376,284	

Government and politics

The political system of the Islamic Republic is based on the 1979 Constitution. Accordingly, it is the duty of the Islamic government to furnish all citizens with equal and appropriate opportunities, to provide them with work, and to satisfy their essential needs, so that the course of their progress may be assured.[129]

Ali Khamenei, Supreme Leader of Iran

The system comprises several intricately connected governing bodies. The Leader of the Revolution (commonly called "Supreme Leader" in the US and the UK) is responsible for delineation and supervision of the general policies of the Islamic Republic of Iran.[130] The Supreme Leader is Commander-in-Chief of the armed forces, controls the military intelligence and security operations; and has sole power to declare war or peace.[130]

The heads of the judiciary, state radio and television networks, the commanders of the police and military forces and six of the twelve members of the Guardian Council are appointed by the Supreme Leader.[130] The Assembly of Experts elects and dismisses the Supreme Leader on the basis of qualifications and popular esteem.[131] The Assembly of Experts is responsible for supervising the Supreme Leader in the performance of legal duties.

After the Supreme Leader, the Constitution defines the President of Iran as the highest state authority.[130] [132] The President is elected by universal suffrage for a term of four years and can only be re-elected for one term.[132] Presidential candidates must be approved by the Guardian Council prior to running in order to ensure their allegiance to the ideals of the Islamic revolution.[133]

The President is responsible for the implementation of the Constitution and for the exercise of executive powers, except for matters directly related to the Supreme Leader, who has the final say in all matters.[130] The President appoints and supervises the Council of Ministers, coordinates government decisions, and selects government policies to be placed before the legislature.[134] Eight Vice-Presidents serve under the President, as well as a cabinet of twenty two ministers, who must all be approved by the legislature.[135] Unlike many other states, the executive branch in Iran does not control the armed forces. Although the President appoints the Ministers of Intelligence and

Defense, it is customary for the President to obtain explicit approval from the Supreme Leader for these two ministers before presenting them to the legislature for a vote of confidence. Iran's current president, Mahmoud Ahmadinejad, was first elected in a run-off poll in the 2005 presidential elections and re-elected in the 2009 presidential elections.

As of 2012, the legislature of Iran (known in English as the Islamic Consultative Assembly) is a unicameral body.[136] Before the Iranian Revolution, the legislature was bicameral, but the upper house was removed under the new constitution. The Majlis of Iran comprises 290 members elected for four-year terms.[136] The Majlis drafts legislation, ratifies international treaties, and approves the national budget. All Majlis candidates and all legislation from the assembly must be approved by the Guardian Council.[137]

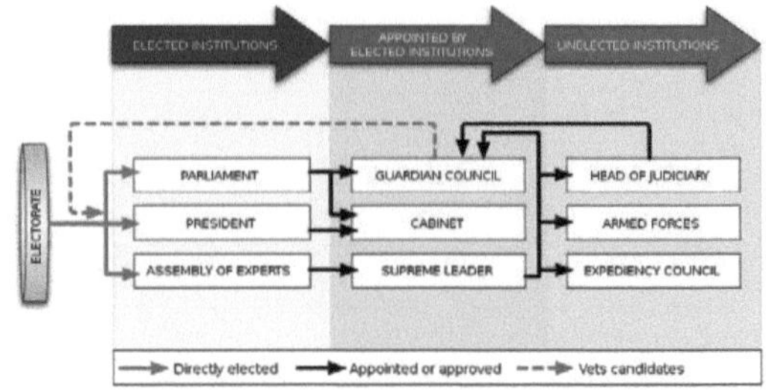

Iran's Syncretic political system combines elements of a modern Islamic theocracy with democracy.

The Guardian Council comprises twelve jurists including six appointed by the Supreme Leader. The others are elected by the Parliament from among the jurists nominated by the Head of the Judiciary.[138] [139] The Council interprets the constitution and may veto Parliament. If a law is deemed incompatible with the constitution or Sharia (Islamic law), it is referred back to Parliament for revision.[132] In a controversial exercise of its authority, the Council has drawn upon a narrow interpretation of Iran's constitution to veto parliamentary candidates. The Expediency Council has the authority to mediate disputes between Parliament and the Guardian Council, and serves as an advisory body to the Supreme Leader, making it one of the most powerful governing bodies in the country.[140]

President Mahmoud Ahmadinejad at Columbia University

The Supreme Leader appoints the head of Iran's judiciary, who in turn appoints the head of the Supreme Court and the chief public prosecutor.[141] There are several types of courts including public courts that deal with civil and criminal cases, and "revolutionary courts" which deal with certain categories of offenses, including crimes against national security. The decisions of the revolutionary courts are final and cannot be appealed.[141] The Special Clerical Court handles crimes allegedly committed by clerics, although it has also taken on cases involving lay people. The Special Clerical Court functions independently of the regular judicial framework and is accountable only to the Supreme Leader. The Court's rulings are final and cannot be appealed.[141]

The Assembly of Experts, which meets for one week annually, comprises 86 "virtuous and learned" clerics elected by adult suffrage for eight-year terms. As with the presidential and parliamentary elections, the Guardian Council determines candidates' eligibility.[141] The Assembly elects the Supreme Leader and has the constitutional authority to remove the Supreme Leader from power at any time.[141] It has not challenged any of the Supreme Leader's decisions.[141]

Local city councils are elected by public vote to four-year terms in all cities and villages of Iran. According to article seven of Iran's Constitution, these local councils together with the Parliament are "decision-making and administrative organs of the State". This section of the constitution was not implemented until 1999 when the first local council elections were held across the country. Councils have many different responsibilities including electing mayors, supervising the activities of municipalities; studying, planning, co-ordinating and implementing of social, cultural, educational, health, economic, and welfare requirements of their constituencies.

Economy

There is an estimated population of 1.2 million weavers in Iran producing carpets for domestic markets and international export.[142]

The Central Bank of the Islamic Republic of Iran is responsible for developing and maintaining the Iranian rial, which serves as the country's currency. The economy of Iran is the twenty-fifth largest in the world by GDP (nominal) and the eighteenth largest economy in the world by purchasing power parity (PPP). Iran's economy is a mixture of central planning, state ownership of oil and other large enterprises, village agriculture, and small-scale private trading and service ventures.[143] Its economic infrastructure has been improving steadily over the past two decades but continues to be affected by inflation and unemployment.[144] In the early 21st century the service sector contributed the largest percentage of the GDP, followed by industry (mining and manufacturing) and agriculture. In 2006, about 45% of the government's budget came from oil and natural gas revenues, and 31% came from taxes and fees.[145]

Government spending contributed to an average annual inflation rate of 14% in the period 2000–2004. As at 2007, Iran had earned $70 billion in foreign exchange reserves mostly (80%) from crude oil exports.[146] In 2011 GDP was $482.4 billion ($1.003 trillion at PPP), or $13,200 at PPP per capita, signifying a 2% growth in GDP.[23] Because of these figures and the country's diversified but small industrial base, the United Nations classifies Iran's economy as semi-developed (1998).

Iran's automobile production crossed the 1 million mark in 2005. Iran Khodro is the largest car manufacturer in the Middle-East. It has established joint-ventures with foreign partners from 4 continents.

About 1,659,000 foreign tourists visited Iran in 2004; most came from Asian countries, including the republics of Central Asia, while a small share came from the countries of the European Union and North America. Iran currently ranks 89th in tourist income, but is rated among the "10 most touristic countries" in the world in terms of its history.[147] [148] Weak advertising, unstable regional conditions, a poor public image in some parts of the world, and absence of efficient planning schemes in the tourism sector have all hindered the growth of tourism.

The administration continues to follow the market reform plans of the previous one and indicated that it will diversify Iran's oil-reliant economy. Iran has also developed a biotechnology, nanotechnology, and pharmaceuticals industry.[149] The strong oil market since 1996 helped ease financial pressures on Iran and allowed for Tehran's timely debt service payments.

Iranian budget deficits have been a chronic problem, mostly due to large-scale state subsidies, that include foodstuffs and especially gasoline, totaling more than $84 billion in 2008 for the energy sector alone.[150] [151] In 2010, the economic reform plan was approved by parliament to cut subsidies gradually and replace them with targeted social assistance. The objective is to move towards free market prices in a 5-year period and increase productivity and social justice.[152]

Over the past 15 years, the authorities have placed an emphasis on the local production of domestic-consumption oriented goods such as home appliances, cars, agricultural products, pharmaceutical, etc. Today, Iran possesses a good manufacturing industry, despite restrictions imposed by foreign countries. However, nationalized industries such as the bonyads have often been managed badly, making them ineffective and uncompetitive with years. Currently, the government is trying to privatize these industries, and, despite successes, there are still several problems to be overcome, such as the lagging corruption in the public sector and lack of competitiveness. Iran ranks

69th out of 139 in Global Competitiveness Report.[153]

Iran has leading manufacturing industries in the fields of car-manufacture and transportation, construction materials, home appliances, food and agricultural goods, armaments, pharmaceuticals, information technology, power and petrochemicals in the Middle East.[154]

TSE has been one of the world's best performing stock exchanges in recent years.[155] [156]

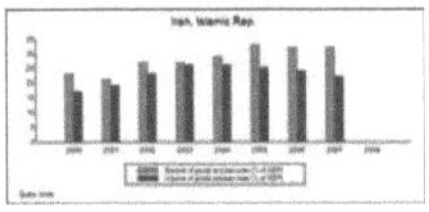

Iran's trade balance (2000–2007). For the first time, the value of Iran's non-oil exports is expected to reach the value of imports by 2012.[157]

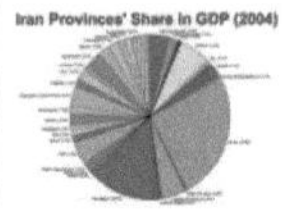

Iranian provinces' contribution to GDP. Tehran host 45% of Iran's industries.[158]

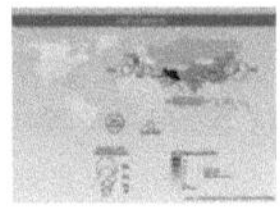

Map of Iran's non-oil exports

Foreign relations and military

Iranian Air Force MiG-29 fighter aircraft

Iran's stated goal is to establish a new world order based on world peace and justice.[159] [160] Iran's foreign relations are based on two strategic principles: eliminating outside influences in the region and pursuing extensive diplomatic contacts with developing and non-aligned countries. Iran maintains diplomatic relations with almost every member of the United Nations, except for Israel, which Iran does not recognize, and the United States since the Iranian Revolution.[161] Since 2005, Iran's nuclear program has become the subject of contention with the Western world due to suspicions that Iran could divert the civilian nuclear technology to a weapons program. This has led the UN Security Council to impose sanctions against Iran on select companies linked to this program, thus furthering its economic isolation on the international scene. The US Director of National Intelligence said in February 2009 that Iran would not realistically be able to a get a nuclear weapon until 2013, if it chose to develop one.[162]

Iranian Navy *Kilo* -class attack submarine

The Islamic Republic of Iran has two types of armed forces: the regular forces Islamic Republic of Iran Army, Islamic Republic of Iran Air Force, Islamic Republic of Iran Navy and the Islamic Revolutionary Guards Corps (IRGC), totaling about 545,000 active troops. Iran also has around 350,000 Reserve Force totaling around 900,000 trained troops.[163] Iran has a paramilitary, volunteer militia force within the IRGC, called the Basij, which includes about 90,000 full-time, active-duty uniformed members. Up to 11 million men and women are members of the Basij who could potentially be called up for service; GlobalSecurity.org estimates Iran could mobilize "up to one million men". This would be among the largest troop mobilizations in the world.[164] In 2007, Iran's military spending represented 2.6% of the GDP or $102 per capita, the lowest figure of the Persian Gulf nations.[165] Iran's military doctrine is based on deterrence.[166]

Since the Iranian Revolution, to overcome foreign embargo, Iran has developed its own military industry, produced its own tanks, armored personnel carriers, guided missiles, submarines, military vessels, guided missile destroyer, radar systems, helicopters and fighter planes.[167] [168] [169] In recent years, official announcements have highlighted the development of weapons such as the Hoot, Kowsar, Zelzal, Fateh-110, Shahab-3 and Sajjil missiles, and a variety of unmanned aerial vehicles (UAVs).[170] The Fajr-3 (MIRV) is currently Iran's most advanced ballistic missile, it is a liquid fuel missile with an undisclosed range which was developed and produced domestically.

Infrastructure

Science and technology

An 18th century Persian astrolabe

Ancient Iranians built Qanats and Yakhchal to provide and keep water. The first windmill appeared in Iran in the 9th century.[171] Iranians contributed significantly to the current understanding of astronomy, natural science, medicine, mathematics, and philosophy. Khwarizmi is widely hailed as the father of algebra. Ethanol (alcohol) was first identified by Persian alchemists such as Muhammad ibn Zakarīya Rāzi. Throughout the Middle Ages, the philosophy and mathematics of the Ancient Greeks and Persians were furthered and preserved within Persia. The Academy of Gundishapur was a renowned centre of learning in the city of Gundeshapur during late antiquity and was the most important medical centre of the ancient world during the 6th and 7th centuries.[172] During this period, Persia became a centre for the manufacture of scientific instruments, retaining its reputation for quality well into the 19th century.

Iran strives to revive the golden age of Persian science. The country has increased its publication output nearly tenfold from 1996 through 2004, and has been ranked first in terms of output growth rate followed by China.[173] According to SCImago, Iran could rank fourth in the world in terms of research output by 2018.[174] Despite the limitations in funds, facilities, and international collaborations, Iranian scientists remain highly productive in several experimental fields, such as pharmacology, pharmaceutical chemistry, organic chemistry, and polymer chemistry. Iranian scientists are also helping construct the Compact Muon Solenoid, a detector for CERN's Large Hadron Collider. In 2009, a SUSE Linux-based HPC system made by the Aerospace Research Institute of Iran (ARI) was launched with 32 cores and now runs 96 cores. Its performance was pegged at 192 GFLOPS.[175] Sorena 2 Robot, which was designed by engineers at University of Tehran, was unveiled in 2010. the Institute of Electrical and Electronics Engineers (IEEE) has placed the name of Surena among the five prominent robots of the world after analyzing its performance.[176]

In the biomedical sciences, Iran's Institute of Biochemistry and Biophysics is a UNESCO chair in biology.[177] In late 2006, Iranian scientists successfully cloned a sheep by somatic cell nuclear transfer, at the Rouyan research centre in Tehran.[178] According to a study by David Morrison and Ali Khademhosseini (Harvard-MIT and Cambridge), stem cell research in Iran is amongst the top 10 in the world.[179] Iran ranks 15th in the world in nanotechnologies.[180] [181] [182]

Omid satellite. Iran is the 9th country to put a domestically built satellite into orbit.

The Iranian nuclear program was launched in the 1950s. Iran is the 7th country in production of uranium hexafluoride[183] and controls the entire cycle for producing nuclear fuel.[184] Iran's current facilities includes several research reactors, a uranium mine, an almost complete commercial nuclear reactor, and uranium processing facilities that include a uranium enrichment plant.

The Iranian Space Agency launched its first reconnaissance satellite named Sina-1 in 2006, and a space rocket in 2007,[185] which aimed at improving science and research for university students.[186] Iran placed its domestically built satellite, Omid into orbit on the 30th anniversary of the Iranian Revolution, on 2 February 2009,[187] through Safir rocket, becoming the ninth country in the world capable of both producing a satellite and sending it into space from a domestically made launcher.[188] In July 2012, Iranian Supereme leader Ali Khamanei stated on a State TV broadcast that US-led sanctions have not succeeded in impeding the technological advancement of his nation.[189]

Iranian scientists outside Iran have also made some major contributions to science. In 1960, Ali Javan co-invented the first gas laser and fuzzy set theory was introduced by Lotfi Zadeh.[190] Iranian cardiologist, Tofy Mussivand invented and developed the first artificial cardiac pump, the precursor of the artificial heart. Furthering research and treatment of diabetes, HbA1c was discovered by Samuel Rahbar. Iranian physics is especially strong in string theory, with many papers being published in Iran.[191] Iranian-American string theorist Kamran Vafa proposed the Vafa-Witten theorem together with Edward Witten.

Telecommunications

In Iran, the state-owned Telecommunication Company of Iran handles telecommunications.

Internet

Iran originally received access to the internet in 1993. In Iran,

Energy

Iran holds 10% of the world's proven oil reserves and 15% of its gas. It is OPEC's second largest exporter and the world's fourth oil producer.

Iran ranks second in the world in natural gas reserves and third in oil reserves.[192] [193] It is OPEC's 2nd largest oil exporter and is an energy superpower.[194] [195] In 2005, Iran spent US$4 billion on fuel imports, because of contraband and inefficient domestic use.[196] Oil industry output averaged 4 million barrels per day (m^3/d) in 2005, compared with the peak of six million barrels per day reached in 1974. In the early years of the 2000s (decade), industry infrastructure was increasingly inefficient because of technological lags. Few exploratory wells were drilled in 2005.

In 2004, a large share of natural gas reserves in Iran were untapped. The addition of new hydroelectric stations and the streamlining of conventional coal and oil-fired stations increased installed capacity to 33,000 megawatts. Of that amount, about 75% was based on natural gas, 18% on oil, and 7% on hydroelectric power. In 2004, Iran opened its first wind-powered and geothermal plants, and the first solar thermal plant is to come online in 2009. Iran is the third country in the world to have developed GTL technology.[197]

Demographic trends and intensified industrialization have caused electric power demand to grow by 8% per year. The government's goal of 53,000 megawatts of installed capacity by 2010 is to be reached by bringing on line new gas-fired plants and by adding hydroelectric, and nuclear power generating capacity. Iran's first nuclear power plant at Bushehr went online in 2011.[198] [199]

Culture

City Theater of Tehran, the largest auditorium in Western Asia

The culture of Iran is a mix of ancient pre-Islamic culture and Islamic culture. Iranian culture has long been a predominant culture of the region, with Persian considered the language of intellectuals during much of the 2nd millennium, and the language of religion and the populace before that.

The Sassanid era was an important and influential historical period in Iran as Iranian culture influenced China, India and Roman civilization considerably,[200] and so influenced as far as Western Europe and Africa.[201] This influence played a prominent role in the formation of both Asiatic and European medieval art.[202] This influence carried forward to the Islamic world. Much of what later became known as Islamic learning, such as philology, literature, jurisprudence, philosophy, medicine, architecture and the sciences were based on some of the practises taken from the Sassanid Persians to the broader Muslim world.[203] [204] [205]

Since the Islamization of Iran, Islamic rituals have penetrated the Iranian culture. The most noticeable of them is the commemoration of Husayn ibn Ali: every year on the Day of Ashura, most Iranians, including Armenians and Zoroastrians, participate in mourning for the martyrs of the battle of Karbala. Daily life in modern Iran is closely interwoven with Shia Islam and the country's art, literature, and architecture are an ever-present reminder of its deep national tradition and of a broader literary culture.[205] [206]

The Iranian New Year (Nowruz) is an ancient tradition celebrated on 21 March to mark the beginning of spring in Iran. It is also celebrated in Afghanistan, Republic of Azerbaijan, Uzbekistan, Turkmenistan, Tajikistan, Kazakhstan and previously also in Georgia and Armenia. It is also celebrated by the Iraqi and Anatolian Kurds.[207] Nowruz was registered on the list of Masterpieces of the Oral and Intangible Heritage of Humanity[208] and described as the *Persian New Year*[209] [210] [211] [212] by the United Nations Educational, Scientific and Cultural Organization (UNESCO) in 2009.

Language and literature

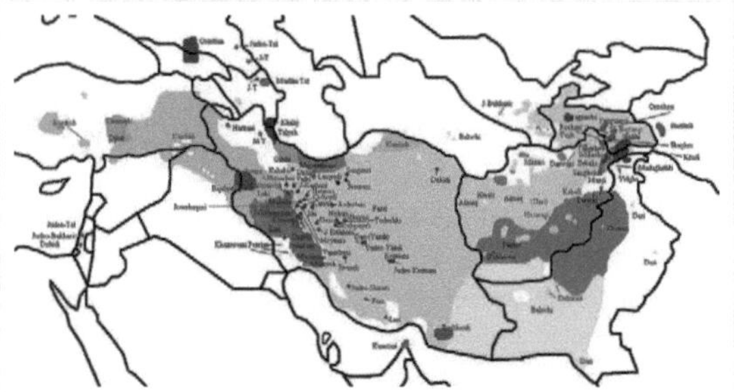

Geographic distribution of the Modern Iranian languages: Persian (green), Pashto (purple) and Kurdish (turquoise), Lurish (red), Baloch (Yellow), as well as smaller communities of other Iranian languages

Article 15 of the Iranian constitution states that the "Official language (of Iran)... is Persian...[and]... the use of regional and tribal languages in the press and mass media, as well as for teaching of their literature in schools, is allowed in addition to Persian." Persian serves as a *lingua franca* in Iran and most publications and broadcastings are in this language.

Next to Persian, there are many publications and broadcastings in other relatively popular languages of Iran such as Azerbaijani, Kurdish and even in less popular ones such as Arabic and Armenian. Many languages originated in Iran, but Persian is the most used language. Persian belongs to the Iranian branch of the family of Indo-European languages. The oldest records in Old Persian date to the Achaemenid Empire,[213] and examples of Old Persian have been found in present-day Iran, Iraq, Turkey and Egypt.

In the late 8th century, Persian was highly Arabized and written in a modified Arabic script. This caused a movement supporting the revival of Persian. An important event of this revival was the writing of the *Shahname* by Ferdowsi (Persian: *Epic of Kings*), Iran's national epic, which is said to have been written entirely in native Persian. This gave rise to a strong reassertion of Iranian national identity, and is in part credited for the continued existence of Persian as a separate language.

بسی رنج بردم در این سال سی
عجم زنده ردم بدین پارسی
For thirty years, I suffered much pain and strife
with Persian I gave the Ajam verve and life
—*Ferdowsi* (935–1020)

Kelileh va Demneh Persian manuscript copy dated 1429

Persian beside Arabic has been a medium for literary and scientific contributions to the Islamic world especially in Anatolia, central Asia and Indian subcontinent. Poetry is a very important part of Persian culture. Poetry is used in many Persian classical works, whether from literature, science, or metaphysics. Persian literature has been considered by such thinkers as Goethe as one of the four main bodies of world literature.[214]

The Persian language has produced a number of famous poets; however, only a few poets as Rumi and Omar Khayyám have surfaced among western popular readership, even though the likes of Hafiz, Saadi, Nizami[215] Attar, Sanai, Nasir Khusraw, Jami, Taleb Amoli are considered by many Iranians to be just as influential. The books of famous poets have been translated into western languages since 1634. An example of Persian poetic influence is the poem below which is widely popular:

دنر یپ ي یاضعا مدآ ینب
دنرهوگ ي زشنيرفآ رد ه
راگزور دروآ درد هب یوضع وچ
رارق دنامن ار اهوضع رگد

Of one Essence is the human race
thus has Creation put the base
One Limb impacted is sufficient
For all Others to feel the Mace
—*Saadi* (1184–1283)

Hafezi'eh, tomb of Hafiz in Shiraz

Philosophy

While there are ancient relations between the Indian Vedas and the Iranian Avesta, the two main families of the ancient Indo-Iranian philosophical traditions were characterized by fundamental differences in their implications for the human being's position in society and their view on the role of man in the universe.[216] [217]

Iranian philosophy after the acceptance of Islam in Persia, is characterized by different interactions with the Ancient Iranian Philosophy, the Ancient Greek philosophy and with the development of Islamic philosophy. Illuminationism and transcendent theosophy are regarded as two of the main philosophical traditions of that era in Persia. Among important contributors to philosophy in Iran are Zoroaster, Jamasp, Mardan-Farrux Ohrmazddadan, Adurfarnbag Farroxzadan, Adurbad Emedan, Iranshahri, Farabi, Avicenna, Ali ibn Sahl Rabban al-Tabari, Suhrawardi, Nasir Khusraw, Biruni, Muhammad ibn Zakariya al-Razi, Abu Yaqub al-Sijistani, Nasir al-Din Tusi, Qutb al-Din Shirazi, Mir Damad, Mulla Sadra, Mir Fendereski and Hadi Sabzevari.

Music

17th century painting from Hasht Behesht palace, Isfahan

The musical culture of Persia, while distinct, is closely related to other musical systems of the West and Central Asia. It has also affinities to the music cultures of the Indian subcontinent, to a certain degree even to those of Africa, and, in the period after 1850 particularly, to that of Europe. Its history can be traced to some extent through these relationships. Like that of most of the world's cultures, the music of Persia has depended on oral/aural transmission and learning.[218]

Cinema

Iranian cinema has thrived in modern Iran, and many Iranian directors have garnered worldwide recognition for their work. Iranian movies have won over three hundred awards in the past twenty-five years including Oscars. One of the best-known directors is Abbas Kiarostami. The media of Iran is a mixture of private and state-owned, but books and movies must be approved by the Ministry of Culture and Islamic Guidance before being released to the public. The Internet has become enormously popular among the Iranian youth. Iran is now the world's fourth largest country of bloggers.[219]

Art and architecture

Iran is home to one of the richest artistic traditions in world history and encompasses many disciplines, including architecture, painting, weaving, pottery, calligraphy, metalworking and stonemasonry. Carpet-weaving is one of the most distinguished manifestations of Persian culture and art, and dates back to ancient Persia. Persians were among the first to use mathematics, geometry, and astronomy in architecture and also have extraordinary skills in making massive domes which can be seen frequently in the structure of bazaars and mosques. The main building types of classical Iranian architecture are the mosque and the palace. Besides being home to a large number of art houses and galleries, Iran also holds one of the largest and most valuable jewel collections in the world.

Iran ranks seventh among countries in the world with the most archeological architectural ruins and attractions from antiquity as recognized by UNESCO.[220] Fifteen of UNESCO's World Heritage Sites are creations of Iranian architecture.

Panoramic view of Naqsh-e Jahan Square

Cuisine

The cuisine of Iran is diverse, with each province featuring dishes, as well as culinary traditions and styles, distinct to their regions. The main Persian cuisines are combinations of rice with meat, chicken or fish and some onion, vegetables, nuts, and herbs. Herbs are frequently used along with fruits such as plums, pomegranates, quince, prunes, apricots, and raisins. Iranians also usually eat plain yogurt (Persian: ماست, māst) with lunch and dinner; it is a staple of the diet in Iran. To achieve a balanced taste, characteristic flavourings such as saffron, dried limes, cinnamon, and parsley are mixed delicately and used in some special dishes. Onions and garlic are normally used in the preparation of the accompanying course, but are also served separately during meals, either in raw or pickled form. Iran is also famous for its caviar.[221] Iranian food is not piquant.

Sports

Azadi Football Stadium

With two thirds of Iran's population under the age of 25, many sports are practised in Iran, both traditional and modern. Iran is the birthplace of polo,[222] and Varzesh-e Pahlavani. Freestyle wrestling has been traditionally regarded as Iran's national sport, however today, the most popular sport in Iran is soccer with the country having won the Asian Cup on three occasions. Basketball is also very popular in Iran where the national team won two of the last three Asian Championships.[223] In 1974, Iran became the first country in West Asia to host the Asian Games. Iran is home to several unique skiing resorts,[224] with the Tochal resort being the world's fifth-highest ski resort (3730 m/12238 ft at its highest station), and located only fifteen minutes away from Tehran. Being a mountainous country, Iran is a venue for hiking, rock climbing,[225] and mountain climbing.[226] [227] [228]

Demographics

Iran is a diverse country, consisting of people of many religious and ethnic backgrounds, which is cemented by the Persian language and culture.[229] The majority of the population speaks the Persian language, which is also the official language of the country, as well as other Iranian languages or dialects. Turkic languages and dialects, most importantly Azerbaijani language, are spoken in different areas in Iran. Additionally, Arabic is spoken in the southwestern parts of the country, although Arabs constitute a minority in these regions. The exact ethnic breakdown of Iran is unknown as there are no official numbers, however some organizations have made estimates. The World Factbook released the estimate: Persians (61%), Azerbaijanis (16%), Kurds (10%), Lurs (6%), Arabs (2%), Balochs (2%), Turkmens and Turkic tribes (2%),[230] Laks, Qashqai, Armenians, Georgians, Persian Jews, Assyrians, Circassians, Tats, Mandaeans, Gypsies, Brahuis, Kazakhs and others (1%).[23] However according to them Persian is spoken as first language by 53%, while Azerbaijani and other Turkic dialects is spoken by 18%, Kurdish by 10%, Gilaki and Mazandarani by 7%, Luri by 6%, Balochi by 2%, Arabic by 2% and that some 2% have other languages as first language.[23]

Iran's ethnoreligious distribution

The Library of Congress estimates are as following: Persians (65%), Azerbaijanis (16%), Kurds (7%), Lurs (6%), Arabs (2%), Baluchi (2%), Turkmens (1%), Turkic tribal groups such as the Qashqai (1%), and non-Persian, non-Turkic groups such as Armenians, Assyrians and Georgians (less than 1%). According to it Persian is spoken as a mother tongue by at least 65% of the population and as a second language by a large proportion of the remaining 35%.[231]

Population of Iran

Iran's population increased dramatically during the latter half of the 20th century, reaching about 75 million by 2009.[232] According to the 1956 census the population of Iran was about 19 million.[233] In recent years, however, Iran's birth rate has dropped significantly, and is 1.29% in July 2012.[4] Studies project that Iran's rate of population growth will continue to slow until it stabilizes above 105 million by 2050.[234] [235] More than two-thirds of the population is under the age of 30, and the literacy rate is 82%.[236] Women today compose more than half of the incoming classes for universities around the country and increasingly continue to play pivotal roles in society.

Iran hosts one of the largest refugee populations in the world, with more than one million refugees, mostly from Afghanistan and Iraq.[237] Since 2006, Iranian officials have been working with the UNHCR and Afghan officials for their repatriation.[238] According to estimates, about five million Iranian citizens have emigrated to other countries, mostly since the Iranian Revolution in 1979.[239] [240]

According to the Iranian Constitution, the government is required to provide every citizen of the country with access to social security that covers retirement, unemployment, old age, disability, accidents, calamities, health and medical treatment and care services. This is covered by public revenues and income derived from public contributions. The World Health Organization in on health systems ranked Iran's performance on health level 58th, and its overall health system performance 93rd among the world's nations in 2000.[241]

Religion

Iran religiosity	
Religion	**Percent**
Islam	98%
Other	2%

Religion in Iran is dominated by the Twelver Shia branch of Islam, which is the official state religion and to which about 90% to 95%[242] [243] [244] of Iranians belong. About 4% to 8% of Iranians belong to the Sunni branch of Islam, mainly Kurds and Iran's Balochi Sunni. The remaining 2% are non-Muslim religious minorities, including Bahá'ís, Mandeans, Hindus, Yezidis, Yarsanis, Zoroastrians, Jews, and Christians.[23]

The latter three minority religions are officially recognized and protected, and have reserved seats in the *Majlis* (Parliament). However the Bahá'í Faith, Iran's largest religious minority,[245] is not officially recognized, and has been persecuted during its existence in Iran. Since the 1979 revolution the persecution of Bahá'ís has increased with executions, the denial of civil rights and liberties, and the denial of access to higher education and employment.[246] [247]

Hinduism in Iran has a history stretching back to the Middle Ages. Presently, Hindus are known to travel to Iran, but the vast majority consist of migrant workers from India. Out of Iran's population of 68,017,860, there are 68,017 recorded Hindus, making them 0.1% of the total population.[248] Some of the poetry of Sohrab Sepehri shows Buddhist influence, and another major contemporary poet, Ahmad Shamlou, translated a book of Japanese haiku poetry into Persian.

See also

- Outline of Iran
- Index of Iran-related articles

References

[1] http://www.ethnologue.org/show_country.asp?name=IR

[2] Encyclopædia Britannica. "Encyclopædia Britannica Encyclopedia Article: Media ancient region, Iran" (http://www.britannica.com/EBchecked/topic/372125/Media). Britannica.com. . Retrieved 25 August 2010.

[3] Andrew J. Newman, Safavid Iran: Rebirth of a Persian Empire, I. B. Tauris (30 March 2006)

[4] "Iran – اهش غیرمنتظره نرخ رشد جمعیت در ایران" (http://www.dw.de/dw/article/0,,16112833,00.html?maca=per-rss-per-all_volltext-5613-xml-mrss). DW Persian. . Retrieved 19 July 2012.

[5] "Islamic Republic of Iran" (http://www.imf.org/external/pubs/ft/weo/2012/01/weodata/weorept.aspx?pr.x=47&pr.y=6&sy=2009&ey=2012&scsm=1&ssd=1&sort=country&ds=.&br=1&c=429&s=NGDPD,NGDPDPC,PPPGDP,PPPPC,LP&grp=0&a=). International Monetary Fund. . Retrieved 18 April 2012.

[6] "Central Bank of Iran: Economic Trends 2008/2009" (http://www.cbi.ir/page/5881.aspx). . Retrieved 18 June 2011.

[7] "Human Development Report 2010" (http://hdr.undp.org/en/media/HDR_2011_EN_Table1.pdf). United Nations. 2010. . Retrieved 5 November 2010.

[8] http://www.bookrags.com/browse/Encyclopedia%20of%20Religious%20Practices/50

[9] http://www.iranchamber.com/geography/articles/flag_anthem.php

[10] http://www.sci.org.ir/content/userfiles/_sci/sci/SEL/f02/2.1.html

[11] https://www.cia.gov/library/publications/the-world-factbook/geos/ir.html#Econ

[12] "Definition for Iran – Oxford Dictionaries Online (World English)" (http://oxforddictionaries.com/definition/Iran?q=iran). Oxforddictionaries.com. . Retrieved 7 April 2012.

[13] "Iran" (http://education.yahoo.com/reference/dictionary/entry/Iran). *The American Heritage Dictionary of the English Language*. Boston: Houghton Mifflin. 2000. .

[14] "Iran Country Profile" (http://www.bbc.co.uk/news/world-middle-east-14541327). BBC NEWS. . Retrieved 8 August 2012.

[15] ""CESWW" – Definition of Central Eurasia" (http://cesww.fas.harvard.edu/ces_definition.html). Cesww.fas.harvard.edu. . Retrieved 1 August 2010.

[16] [[National Geographic Society|National Geographic (http://travel.nationalgeographic.com/places/countries/country_iran.html)].]

[17] D. N. Mackenzie (15 December 1998). "Ērān, Ērānšahr" (http://www.iranicaonline.org/articles/eran-eransah). *Encyclopaedia Iranica*. iranicaonline.org. . Retrieved 25 August 2010.

[18] Iransaga (http://www.art-arena.com/history.html), "Persia or Iran, a brief history".

[19] Iranian.ws (http://www.iranian.ws/art.htm), Iranian & Persian Art.

[20] "Iran" (http://www.britannica.com/EBchecked/topic/293359/Iran). *Encyclopædia Britannica*. Encyclopædia Britannica. 2012. . Retrieved 8 August 2012.

[21] The Committee Office, House of Commons. "Select Committee on Foreign Affairs, Eighth Report, Iran" (http://www.publications.parliament.uk/pa/cm200607/cmselect/cmfaff/363/36310.htm). Publications.parliament.uk. . Retrieved 18 June 2011.

[22] Iran @ 2000 and Beyond lecture series, opening address, W. Herbert Hunt, 18 May 2000 (http://www.petro-hunt.com/lectures/LectureOpen.htm). Retrieved 1 October 2007.

[23] CIA World Factbook. "Iran" (https://www.cia.gov/library/publications/the-world-factbook/geos/ir.html). . Retrieved 7 August 2012.

[24] Lowell Barrington, Michael J. Bosia, Kathleen Bruhn, "Comparative Politics: Structures and Choices", Cengage Learning, 2009. Excerpt from page 34: "Like China, Iran is home to one of the world's oldest civilizations"

[25] "Ahmad Ashraf, "Iranian Identity" in Encyclopaedia Iranica" (http://www.iranicaonline.org/articles/iranian-identity-iii-medieval-islamic-period). Iranicaonline.org. 15 December 2006. . Retrieved 14 September 2011.

[26] B. Spuler, "The Disintegration of the Caliphate in the East", in the Cambridge History of Islam, Vol. IA: The Central islamic Lands from Pre-Islamic Times to the First World War, ed. by P.M. Holt, Ann K.S. Lambton, and Bernard Lewis (Cambridge: Cambridge University Press, 1970). pg 147: One of the effects of the renaissance of the Persian spirit evoked by this work was that the Ghaznavids were also Persianized and thereby became a Persian dynasty.

[27] Grousset, Rene, The Empire of the Steppes, (Rutgers University Press, 1991), 161,164; "..renewed the Seljuk attempt to found a great Turko-Persian empire in eastern Iran..", "It is to be noted that the Seljuks, those Turkomans who became sultans of Persia, did not Turkify Persia-no doubt because they did not wish to do so. On the contrary, it was they who voluntarily became Persians and who, in the manner of the great old Sassanid kings, strove to protect the Iranian populations from the plundering of Ghuzz bands and save Iranian culture from the Turkoman menace."

[28] Bosworth, C.E.; Hillenbrand, R.; Rogers, J.M.; Blois, F.C. de; Bosworth, C.E.; Darley-Doran, R.E., Saldjukids, Encyclopaedia of Islam. Edited by: P. Bearman, Th. Bianquis, C.E. Bosworth, E. van Donzel and W.P. Heinrichs. Brill, 2009. Brill Online: "Culturally, the constituting of the Seljuq Empire marked a further step in the dethronement of Arabic from being the sole lingua franca of educated and polite society in the Middle East. Coming as they did through a Transoxania which was still substantially Iranian and into Persia proper, the Seljuqs with no high-level Turkish cultural or literary heritage of their own – took over that of Persia, so that the Persian language became the administration and culture in their land of Persia and Anatolia. The Persian culture of the Rum Seljuqs was particularly splendid, and it was only gradually that Turkish emerged there as a parallel language in the field of government and adab; the Persian imprint in Ottoman civilization was to remain strong until the 19th century. "

[29] Ross E. Dunn, "The adventures of Ibn Battuta, a Muslim traveler of the fourteenth century", University of California Press, 1986. pg 144: "Indeed, under Ilkhanid sovereignty the high culutre of eastern and central Anatolia became more Persianized than ever before"

[30] Maria Subtelny, "Timurids in Transition", BRILL; illustrated edition (30 September 2007). pg 40: "Nevertheless, in the complex process of transition, members of the Timurid dynasty and their Turko-Mongolian supporters became acculturate by the surrounding Persinate millieu adopting Persian cultural models and tastes and acting as patrons of Persian culture, painting, architecture and music." pg 41: "The last members of the dynasty, notably Sultan-Abu Sa'id and Sultan-Husain, in fact came to be regarded as ideal Perso-Islamic rulers who develoted as much attention to agricultural development as they did to fostering Persianate court culture."

[31] R.M. Savory, *Safavids*, Encyclopaedia of Islam, 2nd edition

[32] "The Islamic World to 1600", The Applied History Research Group, The University of Calgary, 1998 (http://www.ucalgary.ca/applied_history/tutor/islam/empires/safavid/abbas.html). Retrieved 1 October 2007

[33] Encyclopædia Britannica (http://www.britannica.com/eb/article-32981) 23 January 2008

[34] "قانون اساسی جمهوری اسلامی ایران" (http://fa.wikisource.org/wiki/قانون_اساسی_جمهوری_اسلامی_ایران) (in Persian). . Retrieved 23 January 2008.

[35] V.Jayaram (9 January 2007). "The Concepts of Hinduism – Arya" (http://www.hinduwebsite.com/hinduism/concepts/arya.asp). Hinduwebsite.com. . Retrieved 18 June 2011.

[36] LSS.wis.edu (http://imp.lss.wisc.edu/~aoliai/languagepage/iranianlanguages.htm), "Iranian Languages", Political, Social, Scientific, Literary & Artistic (Monthly) October 2000, No. 171, Dr. Suzan Kaviri, pp. 26–7retrieved 1 October 2007

[37] ""Iran – The Ancient Name of Iran", N.S. Gill" (http://ancienthistory.about.com/od/persianempiremaps/qt/Iran.htm). Ancienthistory.about.com. 4 September 2010. . Retrieved 18 June 2011.

[38] Bailey, Harold Walter (1987). "Arya" (http://www.iranicaonline.org/articles/arya-an-ethnic-epithet). *Encyclopaedia Iranica*. **2**. New York: Routledge & Kegan Paul. pp. 681–683. .

[39] MacKenzie, David Niel (1998). "Ērān, Ērānšahr" (http://www.iranicaonline.org/articles/eran-eransah). *Encyclopedia Iranica*. **8**. Costa Mesa: Mazda. .

[40] "Renaming Persia" (http://www.persiansarenotarabs.com/renaming-persia/). *persiansarenotarabs.com*. 2007. . Retrieved 26 April 2011.

[41] Richard N. Frye, interview by Asieh Namdar, CNN, 20 October 2007. "I spent all my life working in Iran. and as you know I don't mean Iran of today, I mean Greater Iran, the Iran which in the past, extended all the way from China to borders of Hungary and from other Mongolia

to Mesopotamia". (http://azadegan.info/files/Dr.Frye-discusses-greater-Iran-on-CNN.mp4)

[42] Richard Nelson Frye, The Harvard Theological Review, Vol. 55, No. 4 (Oct., 1962), pp. 261-268 http://www.jstor.org/pss/1508723 I use the term Iran in an historical context[...]Persia would be used for the modern state, more or less equivalent to "western Iran". I use the term "Greater Iran" to mean what I suspect most Classicists and ancient historians really mean by their use of Persia - that which was within the political boundaries of States ruled by Iranians.

[43] Marcinkowski, Christoph (2010). Shi'ite Identities: Community and Culture in Changing Social Contexts. LIT Verlag Münster. p. 83. ISBN 978-3-643-80049-7. "The 'historical lands of Iran' – 'Greater Iran' – were always known in the Persian language as Irānshahr or Irānzamīn. Both terms refer to the Iranian plateau in addition to the Persianate world at large, those regions that had been historically under significant Persian cultural influence, roughly corresponding to the territories ruled over by the ancient Parthians and Sasanids – i.e., in addition to 'Iran proper', also the Caucasus, Mesopotamia (Iraq), Central Asia, and large parts of what is now Pakistan and Afghanistan and conforming to the Persian 'historical understanding' of the 'full territorial extent' of Iran. The capital of this entity was, at times, situated in what is now Iraq"

[44] Mīr Khvānd, Muḥammad ibn Khāvandshāh, Tārīkh-i rawẓat al-ṣafā. Taṣnīf Mīr Muḥammad ibn Sayyid Burhān al-Dīn Khāvand Shāh al-shahīr bi-Mīr Khvānd. Az rū-yi nusakh-i muta'addadah-i muqābilah gardīdah va fihrist-i asāmī va a'lām va qabāyil va kutub bā chāphā-yi digar mutamāyiz mībāshad.[Tehrān] Markazī-i Khayyām Pīrūz [1959-60]. ایرانشهر از ران فرات تا جیحون است و وسط آبادانی عالم است. Iranshahr streches from the Euphrates to the Oxus, and it is the center of the prosperity of the World

[45] Frye, R. (2011). Persia. Routledge. p. 13. ISBN 978-0-415-57033-6. "This 'greater Iran' included and still includes part of the Caucasus Mountains, Central Asia, Afghanistan and Iraq; for Kurds, Baluchis, Afghans, Tajiks, Ossetes, and other smaller groups are Iranians" (http://books.google.com/books?id=9QOfAvCP1jkC&pg=PA13&dq=This+'greater+Iran'+included+and+still+includes+part+of+the+Caucasus+Mountains,+Central+Asia,+Afghanistan+and+Iraq;+for+Kurds,+Baluchis,+Afghans,+Tajiks,+Ossetes,+and+other+smaller+groups+are+Iranians.&hl=en&sa=X&ei=2lmhUM2GIMWJiwLZsIH4BA&ved=0CDAQ6AEwAA#v=onepage&q=This 'greater Iran' included and still includes part of the Caucasus Mountains, Central Asia, Afghanistan and Iraq; for Kurds, Baluchis, Afghans, Tajiks, Ossetes, and other smaller groups are Iranians&f=false)

[46] Biglari, Fereidoun; Saman Heydari, Sonia Shidrang. "Ganj Par: The first evidence for Lower Paleolithic occupation in the Southern Caspian Basin, Iran" (http://www.antiquity.ac.uk/projgall/biglari302/). Antiquity. . Retrieved 27 April 2011.

[47] Pbase.com (http://www.pbase.com/k_amj/tehran_museum). Retrieved 27 March 2008

[48] J. D. Vigne, J. Peters and D. Helmer, *First Steps of Animal Domestication*, Proceedings of the 9th Conference of the International Council of Archaeozoology, Durham, August 2002, ISBN 1-84217-121-6

[49] *Excavations at Chogha Bonut: The earliest village in Susiana, Irant* (http://oi.uchicago.edu/research/pubs/nn/spr97_alizadeh.html), by Abbas Alizadeh - The Oriental Institute and the Department of Near Eastern Languages and Civilizations The University of Chicago

[50] Hole, Frank (20 July 2004). "NEOLITHIC AGE IN IRAN" (http://www.iranicaonline.org/articles/neolithic-age-in-iran). *Encyclopedia Iranica*. Encyclopaedia Iranica Foundation. . Retrieved 9 August 2012.

[51] *Ancient Near Eastern art* (http://books.google.com.au/books?id=RTGc9YH-C38C&source=gbs_navlinks_s) by Dominique Collon

[52] K. Kris Hirst. "Chogha Mish (Iran)" (http://archaeology.about.com/od/cterms/g/choghamish.htm). .

[53] Xinhua, "New evidence: modern civilization began in Iran", 10 Aug 2007 (http://news.xinhuanet.com/english/2007-08/10/content_6508609.htm). Retrieved 1 October 2007.

[54] The Archaeology of Elam: Formation and Transformation of an Ancient Iranian State - by D. T. Potts, Cambridge University Press, 29/07/1999 - pp. 45-46 - ISBN 0521563585 hardback

[55] Iran Daily, "Panorama", 3 March 2007 (http://www.iran-daily.com/1385/2795/html/panorama.htm). Retrieved 1 October 2007. Archived (http://web.archive.org/20080124101129/http://www.iran-daily.com/1385/2795/html/panorama.htm) January 24, 2008 at the Wayback Machine

[56] Iranian.ws, "Archaeologists: Modern civilization began in Iran based on new evidence", 12 August 2007 (http://www.iranian.ws/iran_news/publish/article_22427.shtml). Retrieved 1 October 2007.

[57] "Ancient Scripts:Elamite" (http://www.ancientscripts.com/elamite.html). 1996. . Retrieved 28 April 2011.

[58] "Persepolis" (http://whc.unesco.org/en/list/114). UNESCO. . Retrieved 29 April 2011.

[59] http://www.ivarta.com/columns/OL_051212.htm

[60] "The Palaeolithic Indo-Europeans" (http://www.panshin.com/trogholm/wonder/indoeuropean/indoeuropean3.html)—Panshin.com. Retrieved 4 June 2006.

[61] Afary, Janet; Peter William Avery,Khosrow Mostofi. "Iran (Ethnic Groups)" (http://www.britannica.com/EBchecked/topic/293359/Iran). *Encyclopedia Britannica*. . Retrieved 28 April 2011.

[62] http://www.bestirantravel.com/culture/zoroastrian.html

[63] "Median Empire" (http://www.iranchamber.com/history/median/median.php). Iran Chamber Society. 2001. . Retrieved 29 April 2011.

[64] "Greco-Persian Wars" (http://www.britannica.com/EBchecked/topic/244117/Greco-Persian-Wars). *Encyclopedia Britannica*. . Retrieved 29 April 2011.

[65] Schmitt, Rüdiger. "Achaemenid dynasty" (http://www.iranica.com/newsite/articles/unicode/v1f4/v1f4a109.html). *Encyclopaedia Iranica*. **vol. 3**. Routledge & Kegan Paul. .

[66] Jakobsson, Jens (2004). "Seleucid Empire" (http://www.iranchamber.com/history/seleucids/seleucids.php). Iran Chamber Society. . Retrieved 29 April 2011.

[67] "Sassanid Empire" (http://www.iranchamber.com/history/sassanids/sassanids.php). Iran Chamber Society. . Retrieved 29 April 2011.

[68] Liska, George (1998). "Projection contra Prediction: Alternative Futures and Options =" (http://books.google.com/?id=oaPlJgAACAAJ&dq=Liska,+Expanding+Realism). *Expanding Realism: The Historical Dimension of World Politics*. Rowman & Littlefield. p. 170. ISBN 0-847-68680-9. .

[69] Student Resources, Chapter 12: The First Global Civilization: The Rise and Spread of Islam, The Arab Empire of the Umayyads -Weakness of the Adversary Empires (http://occawlonline.pearsoned.com/bookbind/pubbooks/stearns_awl/chapter12/objectives/deluxe-content.html)

[70] H. Patrick Glenn, *Legal Traditions of the World*. Oxford University Press, 2007, pg. 218–219.

[71] Student Resources, Chapter 12: The First Global Civilization: The Rise and Spread of Islam, The Arab Empire of the Umayyads - Converts and "People of the Book" (http://occawlonline.pearsoned.com/bookbind/pubbooks/stearns_awl/chapter12/objectives/deluxe-content.html)

[72] "Islamic History: The Abbasid Dynasty" (http://www.religionfacts.com/islam/history/abbasid.htm). Religion Facts. . Retrieved 30 April 2011.

[73] Hooker, Richard (1996). "The Abbasid Dynasty" (http://public.wsu.edu/~dee/ISLAM/ABASSID.HTM). Washington State University. . Retrieved 17 June 2011.

[74] Joel Carmichael, "The Shaping of the Arabs: A Study in Ethnic Identity", Published by Macmillan, 1967. pg 235. Excerpt: "Abu Muslim, the Persian general and popular leader".

[75] Richard Nelson Frye, "Iran", Edition: 2, revised Published by G. Allen & Unwin, 1960. p. 47: "A Persian Muslim called Abu Muslim.

[76] Bosworth C. E., Cambridge History of Iran, vol. 4, p. 90.

[77] Applied History Research Group , University of Calgary, " The Islamic World to 1600", Last accessed 30 October 2008 (http://www.ucalgary.ca/applied_history/tutor/islam/fractured/)

- Ehsan Yarshater, "The Persian Presence in the Islamic World" in Richard G. Hovannisian, Georges Sabagh, "The Persian Presence in the Islamic World", Published by Cambridge University Press, 1997. pp. 6–7: "The Golden age of Islam, as the early Abbassid period has been labeled, was distinguished by intellectual advances, literary innovations, and cultural exuberance attributable, in no small measure, to the vital participation of Persian men of letters, philosophers, theologians, grammarians, mathematicians, musicians, astronomers, geographers, and physicians"
- Bernard Lewis, "Iran in History", excerpt: "Culturally, politically, and most remarkable of all even religiously, the Iranian contribution to this new Islamic civilization is of immense importance. The work of Iranians can be seen in every field of cultural endeavor, including Arabic poetry, to which poets of Iranian origin composing their poems in Arabic made a very significant contribution." TAU.ac.il (http://www.tau.ac.il/dayancenter/mel/lewis.pdf)

[79] Bernard Lewis, "Iran in History", TAU.ac.il (http://www.tau.ac.il/dayancenter/mel/lewis.pdf) excerpt: "Iran was indeed Islamized, but it was not Arabized. Persians remained Persians. And after an interval of silence, Iran reemerged as a separate, different and distinctive element within Islam, eventually adding a new element even to Islam itself. Culturally, politically, and most remarkable of all even religiously, the Iranian contribution to this new Islamic civilization is of immense importance. The work of Iranians can be seen in every field of cultural endeavor, including Arabic poetry, to which poets of Iranian origin composing their poems in Arabic made a very significant contribution. In a sense, Iranian Islam is a second advent of Islam itself, a new Islam sometimes referred to as Islam-i Ajam. It was this Persian Islam, rather than the original Arab Islam, that was brought to new areas and new peoples: to the Turks, first in Central Asia and then in the Middle East in the country which came to be called Turkey, and of course to India. The Ottoman Turks brought a form of Iranian civilization to the walls of Vienna."

[80] William Bayne Fisher, et al., *The Cambridge History of Iran 4* Published by Cambridge University Press, 1975, ISBN 0-521-20093-8, p. 396.

[81] C.E. Bosworth, "Ajam" in Encyclopedia Iranica

[82] Garthwaite, Gene R (2007). *the Persians* (http://books.google.co.uk/books?id=FqzrZ3b8aakC&printsec=frontcover&dq=the+persians&hl=en&ei=yAn9Tc7LF9Ge-Qah87ngAw&sa=X&oi=book_result&ct=result#v=onepage&q&f=false). Singapore: Blackwell. ISBN 1557868603. .

[83] R. Ward, Steven (2009). *Immortal: a military history of Iran and its armed forces* (http://books.google.com/books?id=8eUTLaaVOOQC&pg=PA39&dq&hl=en#v=onepage&q=&f=false). Georgetown University Press. p. 39. ISBN 1589012585. .

[84] Isfahan: Iran's Hidden Jewel (http://www.smithsonianmag.com/people-places/Irans-Hidden-Jewel.html?c=y&page=2). *Smithsonian*.

[85] Bertold Spuler. *The Muslim World. Vol. I The Age of the Caliphs*. Leiden. E.J. Brill. 1960 ISBN 0-685-23328-6 p. 29.

[86] "Safavid Empire (1501-1722)" (http://www.bbc.co.uk/religion/religions/islam/history/safavidempire_1.shtml). *BBC Religion*. BBC. 7 September 2009. . Retrieved 20 June 2011.

[87] Yeroushalmi, David (2009). *The Jews of Iran in the nineteenth century: aspects of history, community* (http://books.google.com/books?id=XYIGS3s3zTQC&pg=PA327&dq#v=onepage&q=&f=false). BRILL. p. 327. ISBN 90-04-15288-1. .

[88] http://www.thefreedictionary.com/nascent

[89] *Historic Personalities of Iran: Reza Shah Pahlavi* (http://www.iranchamber.com/history/reza_shah/reza_shah.php), Iran Chamber Society, , retrieved 2 July 2011

[90] Stephen Kinzer: *All the Shah's Men. An American Coup and the Roots of Middle East Terror*, John Wiley and Sons, 2003, p.x

[91] "Islamic Revolution of 1979" (http://www.iranchamber.com/history/islamic_revolution/islamic_revolution.php). Iranchamber.com. . Retrieved 18 June 2011.

[92] "Islamic Revolution of Iran" (http://www.webcitation.org/query?id=1257012838707054). *Islamic Revolution of Iran*. Encarta. Archived from the original (http://encarta.msn.com/encyclopedia_761588431/Islamic_Revolution_of_Iran.html) on 31 October 2009. . Retrieved 19 June 2011.

[93] Fereydoun Hoveyda, *The Shah and the Ayatollah: Iranian Mythology and Islamic Revolution* ISBN 0-275-97858-3, Praeger Publishers

[94] "The Iranian Revolution" (http://www.fsmitha.com/h2/ch29ir.html). Fsmitha.com. 22 March 1963. . Retrieved 18 June 2011.

[95] Jahangir Amuzegar, *The Dynamics of the Iranian Revolution*, (1991), p.4, 9–12 ISBN 0-7914-0731-4

[96] Arjomand, *Turban* (1988), p. 191.

[97] Cheryl Benard, Zalmay Khalilzad, *"The Government of God"* ISBN 0-231-05376-2, Columbia University Press (1984), p. 18.

[98] "American Experience, Jimmy Carter, "444 Days: America Reacts"" (http://www.pbs.org/wgbh/amex/carter/sfeature/sf_hostage.html). Pbs.org. . Retrieved 18 June 2011.

[99] Guests of the Ayatollah: The Iran Hostage Crisis: The First Battle in America's War with Militant Islam, Mark Bowden, p. 127 ISBN 0-8021-4303-2, Grove Press

[100] Ian Black (23 September 2010). "Iran and Iraq remember war that cost more than a million lives" (http://www.guardian.co.uk/world/2010/sep/23/iran-iraq-war-anniversary). *The Guardian* (London). .

[101] Centre for Documents of The Imposed War, Tehran. (گنج تاقیقحت و تاعلاطم ز رم)

[102] "News" (http://www.fas.org/news/iran/1997/970205-480132.htm). FAS. . Retrieved 23 January 2008.

[103] Federation of American Scientists (http://www.fas.org/cw/intro.htm) 23 January 2008

[104] NTI Chemical profile of Iran (http://www.nti.org/e_research/profiles/Iran/Chemical/2340_2965.html) 23 January 2008 (Archive (http://web.archive.org/web/20080725072420/http://nti.org/e_research/profiles/Iran/Chemical/2340_2965.html))

[105] " Child-Soldier Treaty Has Wide Support (http://www.nytimes.com/1988/12/19/opinion/l-child-soldier-treaty-has-wide-support-697888.html?src=pm)". The New York Times. 19 December 1988.

[106] Dan De Luce in Tehran (4 May 2004). "The Guardian, Tuesday 4 May 2004, Khatami blames clerics for failure" (http://www.guardian.co.uk/world/2004/may/04/iran). *The Guardian* (London). . Retrieved 25 August 2010.

[107] "Iran hardliner becomes president" (http://news.bbc.co.uk/2/hi/middle_east/4740441.stm). BBC. 3 August 2005. . Retrieved 6 December 2006.

[108] "Iran To Hold Presidential Election In June 2009" (http://www.rferl.org/content/Iran_To_Hold_Presidential_Election_In_June_2009/1196953.html) (Reuters). *Radio Free Europe/Radio Liberty*. 7 September 2008. . Retrieved 2 December 2008.

[109] "Ahmadinejad Wins Landslide" (http://www.webcitation.org/5hWcBsp9V). *Iran Daily* (http://www.iran-daily.com/). 13 June 2009. Archived from the original (http://www.iran-daily.com/1388/3423/html/) on 14 June 2009. . Retrieved 13 June 2009.

[110] "یروهمج تسایر تاباختنا ُهرود نیمهد یاهن جیاتن" (http://moi.ir/Portal/Home/ShowPage.aspx?Object=News&CategoryID=832a711b-95fe-4505-8aa3-38f5e17309c9&LayoutID=dd8faff4-f71b-4c65-9aef-a1b6d0160be3&ID=5e30ab89-e376-434b-813f-8c22255158e1) (in Persian). Ministry of Interior of Iran. 13 June 2009. . Retrieved 27 June 2009.

[111] Press TV Leader addresses nation on election results (http://www.presstv.ir/detail.aspx?id=98016&sectionid=351070101), 13 June 2009

[112] BBC report http://www.youtube.com/watch?v=FjQLC4w1XMY

[113] Tait, Robert; Black, Ian; Tran, Mark (17 June 2009). "Iran protests: Fifth day of unrest as regime cracks down on critics" (http://www.guardian.co.uk/world/2009/jun/17/iran-protests-day-five). *The Guardian* (London). .

[114] "Iran-Location, size, and extent" (http://www.nationsencyclopedia.com/Asia-and-Oceania/Iran-LOCATION-SIZE-AND-EXTENT.html). Nationsencyclopedia.com. 7 February 2011. . Retrieved 18 June 2011.

[115] "CIA – The World Factbook" (https://www.cia.gov/library/publications/the-world-factbook/geos/aj.html). Cia.gov. . Retrieved 7 April 2012.

[116] "SurfWax: News, Reviews and Articles On Hindu Kush" (http://news.surfwax.com/geography/files/Hindu_Kush_Mountains.html). News.surfwax.com. . Retrieved 18 June 2011.

[117] Nature & Mountains of Iran (http://www.weneedtrees.com/iran/nature.asp). Retrieved 25 February 2008. Archived (http://web.archive.org/20071122191112/http://www.weneedtrees.com/iran/nature.asp) November 22, 2007 at the Wayback Machine

[118] "Iran- Current Information" (http://my.simmons.edu/lcis/zoom/2005/10/iran-_current_i.html). My.simmons.edu. . Retrieved 18 June 2011.

[119] Guggisberg, C.A.W. (1961). *Simba: The Life of the Lion*. Howard Timmins, Cape Town.

[120] Payvand. "Iran: Focus on reverse migration" (http://www.payvand.com/news/03/nov/1135.html). . Retrieved 17 April 2006.

[121] "Islamic Azad University" (http://www.sci.org.ir/portal/faces/public/census85/census85.natayej/census85.abadipage). Retrieved 28 January 2008. Archived (http://web.archive.org/20071110154704/http://www.sci.org.ir/portal/faces/public/census85/census85.natayej/census85.abadipage) November 10, 2007 at the Wayback Machine

[122] Iranian National Portal of Statistics (http://www.sci.org.ir/portal/faces/public/census85/census85.natayej/census85.abadipage). Retrieved 27 February 2008. Archived (http://web.archive.org/20071110154704/http://www.sci.org.ir/portal/faces/public/census85/census85.natayej/census85.abadipage) November 10, 2007 at the Wayback Machine

[123] Religious Tourism Potentials Rich (http://www.iran-daily.com/1383/2211/html/panorama.htm). Retrieved 28 February 2008.

[124] "Mashhad, Iran" (http://www.sacredsites.com/middle_east/iran/mashad.htm). Sacredsites.com. . Retrieved 18 June 2011.

[125] Iran – Statistical Centre (http://www.sci.org.ir/content/userfiles/_census85/census85/natayej/township/Os10.xls). Retrieved 27 February 2008.

[126] Statistical Center of Iran: Results of national census, 2007 (http://amar.sci.org.ir/Detail.aspx?Ln=F&no=258263&S=GW)

[127] World Gezatteer Bevölkerungsdaten der Stadt Tehran 2012 (http://bevoelkerungsstatistik.de/wg.php?x=&men=gpro&lng=de&des=wg&srt=pnan&col=adhoq&msz=1500&geo=452435143)

[128] http://world-gazetteer.com/wg.php?x=&men=gcis&lng=en&des=wg&geo=-106&srt=npan&col=abcdefghinoq&msz=1500&pt=c&va=&srt=1npan

[129] University of Bern; Iranian Constitution summary (http://www.servat.unibe.ch/law/icl/ir00000_.html). Retrieved 30 November 2009.

[130] Leader.ir (http://www.leader.ir/langs/en/index.php?p=leader_law). Retrieved 13 May 2008

[131] Federal Research Division, Library of Congress. "Iran – The Constitution" (http://countrystudies.us/iran/81.htm). . Retrieved 14 April 2006.

[132] "Iran The Presidency" (http://www.photius.com/countries/iran/government/iran_government_the_presidency.html). Photius.com. . Retrieved 18 June 2011.

[133] Chibli Mallat, The Renewal of Islamic Law: Muhammad Baqer As-Sadr, Najaf and the Shi'i international, ISBN 0-521-53122-5, Cambridge University Press

[134] "Iran – The Prime Minister and the Council of Ministers" (http://countrystudies.us/iran/84.htm). Countrystudies.us. . Retrieved 18 June 2011.

[135] "The Structure of Power in Iran" (http://www.iranchamber.com/government/articles/structure_of_power.php). Iranchamber.com. 24 June 2005. . Retrieved 18 June 2011.

[136] "IFES Election Guide" (http://www.electionguide.org/country.php?ID=103). Electionguide.org. . Retrieved 18 June 2011.

[137] "Iran – The Council of Guardians" (http://countrystudies.us/iran/86.htm). Countrystudies.us. . Retrieved 18 June 2011.

[138] "Iran The Council of Guardians" (http://www.photius.com/countries/iran/government/iran_government_the_council_of_guard~276.html). Photius.com. . Retrieved 18 June 2011.

[139] Manou & Associates Inc.. "Iranian Government Constitution, English Text" (http://www.iranonline.com/iran/iran-info/Government/constitution-6-2.html). Iranonline.com. . Retrieved 18 June 2011.

[140] "Expediency council" (http://news.bbc.co.uk/1/shared/spl/hi/middle_east/03/iran_power/html/expediency_council.stm). *BBC News*. . Retrieved 3 February 2008.

[141] "Iran Chamber Society: The Structure of Power in Iran" (http://www.iranchamber.com/government/articles/structure_of_power.php). Iranchamber.com. 24 June 2005. . Retrieved 18 June 2011.

[142] "Iran to form carpet export consortium" (http://www.presstv.com/Detail.aspx?id=74845&sectionid=351020102). PressTV. 9 November 2008. . Retrieved 1 March 2009.

[143] "Iran economy" (http://www.traveldocs.com/ir/economy.htm). Traveldocs.com. . Retrieved 18 June 2011.

[144] "World Bank: Iran's Economic Indices Improving" (http://web.archive.org/web/20070927231116/http://iran-daily.com/1386/2887/html). Iran Daily. 8 July 2007. Archived from the original (http://iran-daily.com/1386/2887/html) on 27 September 2007. . Retrieved 8 July 2007.

[145] "IRNA: Crude price pegged at dlrs 39.6 a barrel under next year's budget" (http://www.payvand.com/news/08/jan/1250.html). Payvand.com. 22 November 2006. . Retrieved 18 June 2011.

[146] Iran Daily (http://iran-daily.com/1386/3068/html/economy.htm) Forex Reserves Put at $70b Retrieved 24 February 2008

[147] "Iran-daily.com" (http://web.archive.org/web/20080513074502/http://www.iran-daily.com/1384/2241/html/focus.htm). Web.archive.org. Archived from the original (http://www.iran-daily.com/1384/2241/html/focus.htm) on 13 May 2008. . Retrieved 7 November 2010.

[148] "Iranian Hospitality Industry" (http://presstv.com/Program/142594.html). PressTV. 6 December 2010. . Retrieved 18 June 2011.

[149] List of Iranian Nanotechnology companies (http://www.nanovip.com/nanotechnology-companies/iran). Retrieved 23 January 2008. Archived (http://web.archive.org/20071013044517/http://www.nanovip.com/nanotechnology-companies/iran) October 13, 2007 at the Wayback Machine

[150] "Ahmadinejad's Achilles Heel: The Iranian Economy" (http://www.payvand.com/news/07/jan/1295.html). Payvand.com. . Retrieved 18 June 2011.

[151] "Energy subsidies reach $84b" (http://web.archive.org/web/20080506122659/http://www.iran-daily.com/1387/3111/html/economy.htm). Iran-Daily. 8 January 2007. Archived from the original (http://www.iran-daily.com/1387/3111/html/economy.htm) on 6 May 2008. . Retrieved 27 April 2008.

[152] "Iran – Country Brief" (http://go.worldbank.org/KQD2RP3RX0). Go.worldbank.org. . Retrieved 30 January 2010.

[153] "World Economic Forum: Iran ranks 69th out of 139 in global competitiveness" (http://payvand.com/blog/blog/2010/09/13/iran-ranks-69th-out-of-139-in-global-competitiveness-world-economic-forum/). Payvand.com. 13 September 2010. . Retrieved 18 June 2011.

[154] UK Trade & Investment (https://www.uktradeinvest.gov.uk/ukti/appmanager/ukti/countries?_nfls=false&_nfpb=true&_pageLabel=CountryType1&navigationPageId=/iran). Retrieved 26 February 2008.

[155] "Iran blocks share price gains" (http://news.bbc.co.uk/1/hi/business/3129995.stm). *BBC News*. 6 August 2003. . Retrieved 15 February 2010.

[156] "Tehran Exchange Trades Futures to Attract Investors" (http://www.businessweek.com/news/2010-07-26/tehran-exchange-trades-futures-to-attract-investors.html). *BusinessWeek*. 2009. . Retrieved 29 July 2010.

[157] http://www.mehrnews.com/en/newsdetail.aspx?NewsID=1497809

[158] CSIS: The US, Israel, the Arab States and a Nuclear Iran (http://csis.org/files/media/csis/pubs/081006_iran_nuclear.pdf)

[159] Iran urges NAM to make collective bids to establish global peace (http://www.presstv.ir/detail/2012/08/26/258180/iran-urges-joint-bids-to-promote-peace/). PressTV, August 26, 2012. Retrieved November 20, 2012.
[160] Ahmadinejad calls for new world order based on justice (http://presstv.com/detail/2012/05/26/243242/ahmadinejad-new-world-order/). PressTV May 26, 2012. Retrieved November 20, 2012.
[161] "Key Events in Iran Since 1921" (http://www.pbs.org/newshour/indepth_coverage/middle_east/iran/timeline.html). Pbs.org. . Retrieved 18 June 2011.
[162] Charbonneau, Louis (26 October 2009). "RPT-EXCLUSIVE-Iran would need 18 months for atom bomb-diplomats" (http://www.reuters.com/article/idUSN25158068). Reuters. . Retrieved 1 August 2010.
[163] IISS Military Balance 2006, Routledge for the IISS, London, 2006, p.187
[164] John Pike. "Niruyeh Moghavemat Basij Mobilisation Resistance Force" (http://www.globalsecurity.org/intell/world/iran/basij.htm). Globalsecurity.org. . Retrieved 18 June 2011.
[165] "Iran's defense spending 'a fraction of Persian Gulf neighbors'" (http://www.payvand.com/news/06/jun/1011.html). Payvand.com. 22 November 2006. . Retrieved 18 June 2011.
[166] "Iran's doctrine based on deterrence" (http://www2.irna.com/en/news/view/line-24/0804185731142306.htm). IRNA. . Retrieved 18 June 2011.
[167] "Iran Launches Production of Stealth Sub" (http://www.foxnews.com/story/0,2933,156160,00.html). *Fox News*. 10 May 2005. . Retrieved 27 February 2008.
[168] PressTv: Advanced attack chopper joins Iran fleet (http://www.presstv.com/detail.aspx?id=95824&sectionid=351020101) Retrieved 24 May 2009
[169] "Iran launches advanced Jamaran destroyer" (http://www.presstv.com/detail.aspx?id=118982&sectionid=351020101). Presstv.com. 19 February 2010. . Retrieved 1 August 2010.
[170] "Iran tests new long-range missile" (http://news.bbc.co.uk/2/hi/middle_east/7725951.stm). *BBC*. 12 November 2008. . Retrieved 12 November 2008.
[171] Ahmad Y Hassan, Donald Routledge Hill (1986). *Islamic Technology: An illustrated history*, p. 54. Cambridge University Press. ISBN 0-521-42239-6.
[172] The Cambridge History of Iran Vol 4, p396. ISBN 0-521-20093-8
[173] Expert: Dr.VSR.Subramaniam (18 October 2006). "Economics: economic, medical uses of alcohol, uses of alcohol" (http://experts.about.com/q/Economics-2301/economic.htm). Experts.about.com. . Retrieved 18 June 2011.
[174] Iran could rank fourth in the world in terms of research output in 2018: SCImago (http://tehrantimes.com/politics/101672-iran-would-rank-fourth-in-the-world-in-terms-of-research-output-in-2018-scimago). *Tehran Times*, September 22, 2012. Retrieved November 21, 2012.
[175] Patrick Thibodeau (22 June 2009). "AMD Chips Used in Iranian HPC for Rocket Research" (http://www.computerworld.com/action/article.do?command=viewArticleBasic&taxonomyName=knowledge_center&articleId=340338&taxonomyId=1&intsrc=kc_top). Computerworld.com. . Retrieved 7 April 2012.
[176] "No. 3817 | Front page | Page 1" (http://www.iran-daily.com/1389/8/18/MainPaper/3817/Page/1/Index.htm#). Irandaily. . Retrieved 21 October 2011.
[177] "Institute of Biochemistry and Biophysics" (http://www.ibb.ut.ac.ir/). Ibb.ut.ac.ir. 2 February 2011. . Retrieved 18 June 2011.
[178] Middle East Online ن اي ر ان ا ول ا ست ي س ا ل ديم (http://www.middle-east-online.com/english/?id=17674) The first successfully cloned animal in Iran retrieved 7 August 2008
[179] "Iranian Studies Group at MIT" (http://isg-mit.org/projects-storage/StemCell/stem_cell_iran.pdf) (PDF). . Retrieved 25 August 2010.
[180] "INIC – News – 73% of Tehran's Students Acquainted with Nanotechnology" (http://en.nano.ir/index.php/news/show/1477). En.nano.ir. 18 January 2010. . Retrieved 1 August 2010.
[181] "Iran Ranks 15th In Nanotech Articles" (http://www.bernama.com/bernama/v5/bm/newsworld.php?id=453647). Bernama. 9 November 2009. . Retrieved 1 August 2010.
[182] Iran daily: Iranian Technology From Foreign Perspective (http://www.iran-daily.com/1388/3372/html/science.htm)
[183] "Iran, 7th in UF6 production – IAEO official" (http://www.payvand.com/news/08/dec/1156.html). Payvand.com. 22 November 2006. . Retrieved 1 August 2010.
[184] (http://news.yahoo.com/s/ap/20090411/ap_on_re_mi_ea/ml_iran_nuclear_4)
[185] Rocket launch (http://asia.news.yahoo.com/070225/afp/070225120805top.html). Retrieved 23 January 2008. Archived (http://web.archive.org/20070830060558/http://asia.news.yahoo.com/070225/afp/070225120805top.html) August 30, 2007 at the Wayback Machine
[186] http://newsmax.com/archives/articles/2007/2/26/90124.shtml
[187] "Iran scientific savvy 'amazes world'" (http://www.presstv.ir/Detail.aspx?id=84914&sectionid=351020101). Presstv.ir. . Retrieved 18 June 2011.
[188] "West 'shocked by Iran spaceshot'" (http://www.presstv.ir/Detail.aspx?id=84906&sectionid=351020101). Presstv.ir. . Retrieved 18 June 2011.
[189] "Iran immune to sanctions: Khamenei" (http://www.thehindu.com/news/international/article3627775.ece). 12 July 2012. .
[190] http://www-bisc.cs.berkeley.edu/Zadeh-1965.pdf

[191] "CERN Press Release – CERN signs draft Memorandum of Understanding with Iran" (http://press.web.cern.ch/press/PressReleases/Releases2001/PR09.01EIran.html). Press.web.cern.ch. . Retrieved 18 June 2011.

[192] "CIA.gov" (https://www.cia.gov/library/publications/the-world-factbook/rankorder/2178rank.html). CIA.gov. . Retrieved 7 April 2012.

[193] "Iran – U.S. Energy Information Administration (EIA)" (http://www.eia.doe.gov/emeu/cabs/Iran/Background.html). Eia.doe.gov. . Retrieved 7 April 2012.

[194] "American Journal of Scientific Research ISSN 1450-223X Issue 45 (2012), pp. 76–84" (http://www.eurojournals.com/AJSR_45_10.pdf) (PDF). . Retrieved 7 February 2012.

[195] The EU should be playing Iran and Russia off against each other (http://www.eurasianhome.org/xml/t/opinion.xml?lang=en&nic=opinion&pid=558), by Julian Evans, Eurasian Home, 8 November 2006

[196] Kim Murphy – Los Angeles Times (7 January 2007). "U.S. targets Iran's vulnerable oil" (http://www.heraldextra.com/content/view/205986/3/). Heraldextra.com. . Retrieved 18 June 2011.

[197] . "Iran, Besieged by Gasoline Sanctions, Develops GTL to Extract Gasoline from Natural Gas" (http://oilprice.com/Energy/Energy-General/Iran-Besieged-by-Gasoline-Sanctions-Develops-GTL-to-Extract-Gasoline-from-Natural-Gas.html). Oilprice.com. . Retrieved 7 February 2012.

[198] "Iran" (http://lcweb2.loc.gov/frd/cs/profiles/Iran.pdf) (PDF). . Retrieved 18 June 2011.

[199] "No Operation" (http://www.presstv.com/detail.aspx?id=111980). Presstv.com. . Retrieved 14 September 2011.

[200] J. B. Bury, History of the Later Roman Empire: From the Death of Theodosius I to the Death of Justinian Volume 1, p.109 ISBN 0-486-20398-0, Dover Publications

[201] "Sassanids in Africa" (http://www.transoxiana.com.ar/0104/sasanians.html). Transoxiana.com.ar. . Retrieved 18 June 2011.

[202] "Iransaga: The art of Sassanids" (http://www.artarena.force9.co.uk/sass2.htm). Artarena.force9.co.uk. . Retrieved 18 June 2011.

[203] "Iran – A country study" (http://www.parstimes.com/history/title.html). Parstimes.com. . Retrieved 18 June 2011.

[204] "History of Islamic Science 5" (http://www.levity.com/alchemy/islam16.html). Levity.com. . Retrieved 18 June 2011.

[205] Afary, Janet (2006). "Iran" (http://www.britannica.com/eb/article-9106324/Iran). *Encyclopædia Britannica Online*. . Retrieved 29 October 2007.

[206] "گزارش عزاداری ، ارامنه و زرتشتیان ایران" (http://www.tebyan.net/Literature_Art/GodlyPeople/Occasions/2005/2/22/10530.html). Tebyan.net. . Retrieved 7 April 2012.

[207] "The Zarathushtrian Assembly" (http://www.zoroastrian.org/articles/nowruz.htm). Zoroastrian.org. . Retrieved 18 June 2011.

[208] "Leaders of regional countries to mark Nowruz" (http://www.presstv.com/detail.aspx?id=121693&sectionid=351020101). PressTV. . Retrieved 18 June 2011.

[209] "Norouz Persian New Year" (http://www.britishmuseum.org/whats_on/events_calendar/march_2010/norouz_persian_new_year.aspx). British Museum. 25 March 2010. . Retrieved 6 April 2010.

[210] "General Assembly Fifty-fifth session 94th plenary meeting Friday, 9 March 2001, 10 a.m. New York" (http://www.un.org/spanish/aboutun/organs/ga/55/verbatim/a55pv94e.pdf). United Nations General Assembly. 9 March 2001. . Retrieved 6 April 2010.

[211] "Nowrooz, a Persian New Year Celebration, Erupts in Iran – Yahoo!News" (http://news.yahoo.com/s/time/20100317/wl_time/08599197278600). News.yahoo.com. 16 March 2010. . Retrieved 6 April 2010.

[212] "US mulls Persian New Year outreach" (http://washingtontimes.com/news/2010/mar/19/us-mulls-persian-new-year-outreach-to-iran/). *Washington Times*. 19 March 2010. . Retrieved 6 April 2010.

[213] Katzner, Kenneth (2002). *The Languages of the World*. Routledge. p. 163. ISBN 0-415-25004-8.

[214] Von David Levinson, Karen Christensen, Encyclopedia of Modern Asia, Charles Scribner's Sons. 2002 pg 48

[215] C. A. (Charles Ambrose) Storey and Franço de Blois (2004), "Persian Literature – A Biobibliographical Survey: Volume V Poetry of the Pre-Mongol Period.", RoutledgeCurzon; 2nd revised edition (21 June 2004). ISBN 0-947593-47-0. Pg 363: "Nizami Ganja'i, whose personal name was Ilyas, is the most celebrated native poet of the Persians after Firdausi. His nisbah designates him as a native of Ganja (Elizavetpol, Kirovabad) in Azerbaijan, then still a country with an Iranian population

[216] Philip G. Kreyenbroek: "Morals and Society in Zoroastrian Philosophy" in "Persian Philosophy". Companion Encyclopedia of Asian Philosophy: Brian Carr and Indira Mahalingam. Routledge, 2009.

[217] Mary Boyce: "The Origins of Zoroastrian Philosophy" in "Persian Philosophy". Companion Encyclopedia of Asian Philosophy: Brian Carr and Indira Mahalingam. Routledge, 2009.

[218] Bruno Nettl, "Iran: Music" in Encyclopaedia Iranica. Retrieved November 2010 at iranicaonline.org (http://www.iranicaonline.org/articles/iran-xi-persian-music)

[219] Alavi, N (20 December 2004). "Freedom in Farsi blogs" (http://www.guardian.co.uk/technology/2004/dec/20/iran.blogging). *The Guardian* (London). . Retrieved 23 January 2008.

[220] "Virtual Conference" (http://www.american.edu/ted/iran-tour.htm). American.edu. . Retrieved 18 June 2011.

[221] Iran-daily.com (http://www.iran-daily.com/1383/2228/html/focus.htm)

[222] Harrison, Frances (22 September 2005). "Polo comes back home to Iran" (http://news.bbc.co.uk/2/hi/middle_east/4272210.stm). *BBC News*. . Retrieved 23 January 2008.

[223] Alipour, Sam (21 April 2012). "Mission Improbable" (http://sports.espn.go.com/espnmag/story?id=3671265). *ESPN*. . Retrieved 21 April 2012.

[224] Wolfensberger, Marc (13 April 2006). "Iran's Sun, Snow Lure European Skiers to Nuclear Pariah's `Fun'" (http://www.bloomberg.com/apps/news?pid=10000088&sid=aGpcjac1ZTF8&refer=culture). Bloomberg. . Retrieved 18 June 2011.

[225] "Rock Climbing Routes, Gear, Photos, Videos & Articles" (http://www.rockclimbing.com/). Rockclimbing.com. 27 October 2009. . Retrieved 18 June 2011.

[226] "Iran Mountain Zone (IMZ)" (http://www.mountainzone.ir/). Mountainzone.ir. 11 June 1966. . Retrieved 18 June 2011.

[227] "Mountaineering in Iran" (http://www.abc-of-mountaineering.com/middle-east/iran/). Abc-of-mountaineering.com. . Retrieved 18 June 2011.

[228] http://abclocal.go.com/kgo/story?section=News&id=2062458

[229] "Encyclopaedia Iranica. R. N. Frye. Peoples of Iran" (http://www.iranicaonline.org/articles/iran-v1-peoples-survey). Iranicaonline.org. . Retrieved 14 September 2011.

[230] "Central Intelligence Agency" (https://www.cia.gov/library/publications/the-world-factbook/fields/2075.html). Cia.gov. . Retrieved 7 April 2012.

[231] Library of Congress, Library of Congress – Federal Research Division. "Ethnic Groups and Languages of Iran" (http://lcweb2.loc.gov/frd/cs/profiles/Iran.pdf). . Retrieved 2 December 2009.

[232] Asia-Pacific Population Journal, United Nations. "A New Direction in Population Policy and Family Planning in the Islamic Republic of Iran" (http://www.un.org/Depts/escap/pop/journal/v10n1a1.htm). . Retrieved 14 April 2006.

[233] "Iran – population" (http://countrystudies.us/iran/32.htm). Countrystudies.us. . Retrieved 18 June 2011.

[234] Census Bureau, Government of the U.S.A.. "IDB Summary Demographic Data for Iran" (http://www.census.gov/cgi-bin/ipc/idbsum.pl?cty=IR). . Retrieved 14 April 2006.

[235] Iran News, Payvand.com. "Iran's population growth rate falls to 1.5 percent: UNFP" (http://www.payvand.com/news/04/aug/1017.html). . Retrieved 18 October 2006.

[236] (http://www.thebusinessyear.com/tby_demo//publications.php?art_id=113&type=sector&sector=healtheducation)

[237] " Afghanistan-Iran: Iran says it will deport over one million Afghans (http://www.irinnews.org/Report.aspx?ReportId=77107)

[238] United Nations, UNHCR. "Tripartite meeting on returns to Afghanistan" (http://www.unhcr.org/news/NEWS/452b78394.html). . Retrieved 14 April 2006.

[239] "Iran to start bank for expat-investors" (http://www.presstv.com/detail.aspx?id=123686&sectionid=351020102). Presstv.com. 18 April 2010. . Retrieved 1 August 2010.

[240] "Migration Information Institute: Characteristics of the Iranian Diaspora" (http://www.migrationinformation.org/Profiles/display.cfm?ID=424). Migrationinformation.org. . Retrieved 18 June 2011.

[241] WHO, World Health Organisation. "The World Health Report 2000" (http://www.who.int/whr/2000/en/annex10_en.pdf) (PDF). . Retrieved 12 October 2006.

[242] Walter Ralston Martin, Ravi K. Zacharias, *The Kingdom of the Cults* (2003), p.421; Excerpt: Ninety-five percent of Iran's Muslims are Shi'ites.

[243] Bhabani Sen Gupta, *The Persian Gulf and South Asia:: prospects and problems of inter-regional cooperation*. p.158; Excerpt: Shias constitute seventy-five percent of the population of the Gulf. Of this, ninety-five percent of Iranians and sixty of Iraqis are Shias.

[244] Daily report, Issues 64–73 -'United States. Foreign Broadcast Information Service- Page 50, Excerpt: Ninety -five percent of Iranians are Shiite Moslems.

[245] International Federation for Human Rights (1 August 2003). "Discrimination against religious minorities in Iran" (http://www.fidh.org/IMG/pdf/ir0108a.pdf) (PDF). fdih.org. p. 6. . Retrieved 17 January 2009.

[246] International Federation for Human Rights (1 August 2003). "Discrimination against religious minorities in Iran" (http://www.fidh.org/IMG/pdf/ir0108a.pdf) (PDF). fdih.org. . Retrieved 19 March 2007.

[247] Iran Human Rights Documentation Center (2007). "A Faith Denied: The Persecution of the Bahá'ís of Iran" (http://www.iranhrdc.org/english/pdfs/Reports/bahai_report.pdf) (PDF). Iran Human Rights Documentation Center. . Retrieved 19 March 2007.

[248] (The percentage of Hindu population of Iran was taken from the United States Department of State's International Religious Freedom Report 2004)

External links

- The e-office of the Supreme Leader of Iran (http://www.leader.ir/langs/en/)
- The President of Iran (http://www.president.ir/en/)
- Iran.ir (http://en.iran.ir/)
- Iran (https://www.cia.gov/library/publications/the-world-factbook/geos/ir.html) entry at *The World Factbook*
- Iran (http://ucblibraries.colorado.edu/govpubs/for/iran.htm) web resources provided by GovPubs at the University of Colorado–Boulder Libraries
- Iran (http://www.dmoz.org/Regional/Middle_East/Iran/) at the Open Directory Project
- Wikimedia Atlas of Iran

Ismail I

Ismail I	
Shahanshah of Persia	
 Shah Ismail I.	
Reign	1501–1524
Coronation	1501
Born	July 17, 1487
Birthplace	Ardabil
Died	May 23, 1524
Place of death	Tabriz
Buried	Ardabil
Successor	Tahmasp I
Consort	daughter of Shirvanshah II Khalilullah[1] [2] [3] [4]
Royal House	Safavid dynasty
Father	Haydar Safavi
Mother	Martha, daughter of Uzun Hasan

Ismail I (July 17, 1487 – May 23, 1524), known in Persian as **Shāh Ismāʿīl** (Persian: شاه اسماعیل; full name: *Abū l-Muzaffar bin Haydar as-Safavī*), was a Shah of Iran (1502) [5] [6] and the founder of the Safavid dynasty which survived until 1736. Isma'il started his campaign in Iranian Azerbaijan in 1500 as the leader of the Safaviyya, an Twelver Shia militant religious order and unified all of Iran by 1509.[7] Born in Ardabil in Northwestern Iran, he reigned as Shah Ismail I of Iran from 1502 to 1524.

Ismail played a key role in the rise of Twelver Islam; he converted Iran from Sunni to Ismaili Shi'i Islam, importing religious authorities from the Levant.[8] In Alevism, Shah Ismail remains revered as a spiritual guide.

Ismail was also a prolific poet who, under the pen name **Khatā'ī** ("Sinner") contributed greatly to the literary development of the Azerbaijani language.[9]

Life and political history

Ismail was born to Martha and Shaykh Haydar on July 17, 1487 in Ardabil. His father, Haydar, was the sheikh of the Safaviyya Sufi order and a direct descendant of its Kurdish[10] [11] [12] founder, Safi-ad-din Ardabili (1252–1334). Ismail was the last in line of hereditary Grand Masters of the Safaviyah Sufi order, prior to his ascent to a ruling dynasty. As a boy only a year old, he lost his father, the leader of a growing Qizilbash Shi'i community in Iranian Azerbaijan, who died in battle. His mother, Martha, was the daughter of Uzun Hasan by his Pontic Greek wife Theodora, better known as Despina Khatun.[13] Theodora was the daughter of Emperor John IV of Trebizond. (She had married Uzun Hassan in a deal to protect Trebizond from the Ottomans.[14]) Ismail grew up bilingual, speaking Persian and Azeri.[15] [16]

The battle between the young Ismail and Shah Farrukh Yassar of Shirvan

As legend has it, the infant Ismail went into hiding for several years. With his followers, he finally returned to Tabriz, vowing to make Shi'i Islam the official religion of Iran. Ismail found significant support among the people of Azerbaijan, as well as some parts of the Ottoman Empire, mainly in eastern Anatolia. Ismail's advent to power was due to Turkoman tribes of Anatolia and Azerbaijan, who formed the most important part of the Qizilbash movement.[17] Centuries of Sunni rule followed by non-Muslim Mongol hegemony lent fertile ground for new teachings. In the summer of 1500, about 7000 Qizilbash forces, consisted of Ustaclu, Shamlu, Rumlu, Tekelu, Zhulkadir, Afshar, Qajar and Varsak tribes, responded to the invitation of Ismail in Erzincan.[18] Qizilbash forces passed over the Kura River in Novermber 1500, and marched towards the Shirvanshah's state. They defeated the forces under the Shirvanshah Farrukh Yassar near Cabanı (present-day Shamakhi Rayon, Azerbaijan), and conquered Baku.[19] In July 1501, Ismail was enthroned as Shah of Azerbaijan,[20] choosing Tabriz, Azerbaijan, as his capital. When the Safavids came to power in 1501, Shah Ismail was 14 or 15 years old; by 1510 he had conquered the whole of Iran.[21] After defeating the Aq Qoyunlu in 1502, he took the title of Shah of Iran.[6]

In 1510, Ismail I moved against the Sunni Uzbeg tribe. In battle near the city of Merv, some 17,000 Qizilbash warriors ambushed and defeated a superior Uzbek force numbering 28,000. The Uzbek ruler, Muhammad Shaybani, was caught and killed trying to escape the battle and the shah had his skull made into a jeweled drinking goblet.

The Battle of Marv in 1510

In 1514, Selim I, the Sunni Sultan of the Ottoman Empire, attacked Ismail's kingdom to stop the spread of Shiism into Ottoman dominions. Selim and Ismail had been exchanging a series of belligerent letters prior to the attack.

Selim I defeated Ismail at the battle of Chaldiran in 1514.[22] Ismail's army was more mobile and their soldiers were better prepared but the Ottomans prevailed due in large part to their efficient modern army, and possession of artillery, black powder and muskets. Ismail was wounded and almost captured in battle. Selim I entered the Iranian capital of Tabriz in triumph on September 5,[23] but did not linger. A mutiny among his troops fearing a counter attack and entrapment by the fresh Safavid forces called in from the interior, forced the triumphant Ottomans to withdraw prematurely. This allowed Ismail to recover quickly. Among the booties from Tabriz was Ismail's favorite wife, for whose release the Sultan demanded huge concessions, which were refused. Despite his defeat at Chaldiran, Ismail quickly recovered most of his kingdom, from east of the Lake Van to the shores of the Indian Ocean.

After Chaldiran, however, Ismail lost his supernatural air and the aura of invincibility, gradually falling into heavy drinking of alcohol.[24] Ismail retired to his palace and withdrew from active participation in the affairs of the state, leaving these to his minister, Mirza Shah-Hussayn.[25] He died on 23 May 1524 at the relatively early age of thirty-six.

Ismail's reign featured extensive conquests, shaping the map of Iran up to the present day. Iraq, including Baghdad and the holy Shi'a shrines of Najaf and Karbala, had been already seized from the Jalayirids and remained a part of Ismail's kingdom until its end. Despite the defeat at Chaldiran, where he lost western Armenia and northwestern Kurdistan, he managed to pass on to his son, Tahmasp I, his entire vast empire. Ismail destroyed Sunni sites in Baghdad including tombs of Abbasid Caliphs, tombs of Sunni Imam Abū Ḥanīfah and Abdul Qadir Gilani.[26]

His son Tahmasp I succeeded him as Shah.

Ismail's poetry

The battle between Shah Ismail I and the Borjigin Abu'l-Khayr Khan.

Ismail is also known for his poetry using the pen-name *Khatā'ī* (Arabic: یئاطخ "Sinner").[27] According to Encyclopædia Iranica, "Ismail was a skillful poet who used prevalent themes and images in lyric and didactic-religious poetry with ease and some degree of originality". He was also deeply influenced by the Persian literary tradition of Iran, particularly by the *Shahnameh* of Ferdowsi, which probably explains the fact that he named all of his sons after *Shahnameh*-characters. Dickson and Welch suggest that Ismail's "Shāhnāmaye Shāhī" was intended as a present to the young Tahmasp.[28] After defeating Muhammad Shaybani's Uzbeks, Ismail asked Hatefi, a famous poet from Jam (Khorasan), to write a Shahnameh-like epic about his victories and his newly established dynasty. Although the epic was left unfinished, it was an example of *mathnawis* in the heroic style of the Shahnameh written later on for the Safavid kings.[29]

He wrote in the Azerbaijani language,[30] and in the Persian language. He is considered an important figure in the literary history of Azerbaijani language and has left approximately 1400 verses in this language, which he chose to use for political reasons.[30] Approximately 50 verses of his Persian poetry have also survived.

Most of the poems are concerned with love — particularly of the mystical Sufi kind — though there are also poems propagating Shi'i doctrine and Safavi politics. His other serious works include the Nasihatnāme, a book of advice, and the unfinished *Dahnāme*, a book which extols the virtues of love.

As Ismail believed in his own divinity and in his descent from Ali, in his poems he tended to strongly emphasize these claims.

Along with the poet Imadaddin Nasimi, Khatā'ī is considered to be among the first proponents of using a simpler Azeri language in verse that would thereby appeal to a broader audience. His work is most popular in Azerbaijan, as well as among the Bektashis of Turkey. There is a large body of Alevi and Bektashi poetry that has been attributed to him. The major impact of his religious propaganda, in the long run, was the conversion of Persia from Sunni to Shia Islam.[31]

The following anecdote demonstrates the status of vernacular Turkish and Persian in the Ottoman Empire and in the incipient Safavid state. Khatā'ī sent a poem in Turkish to the Ottoman Sultan Selim I before going to war in 1514. In a reply the Ottoman Sultan answered in Persian to indicate his contempt.

Issue

- Sons
  - Tahmasp I
  - Prince 'Abul Ghazi Sultan Alqas Mirza (15 March 1515 - 9 April 1550) Governor of Shirvan 1538-1547. He rebelled against his brother Tahmasp, captured and imprisoned at the Fortress of Qahqahan. m. Khadija Sultan Khanum, having had issue, two sons, Sultan Ahmad Mirza (died 1568) and Sultan Farrukh Mirza (died 1568)
  - Prince Sultan Rustam Mirza (born 13 September 1517)
  - Prince 'Abul Naser Sultan Sam Mirza (28 August 1518 - December 1567) Governor-General of Khorasan 1521-1529 and 1532–1534, and of Ardabil 1549-1571. He rebelled against his brother Tahmasp, captured and imprisoned at the Fortress of Qahqahan. He had issue, two sons and one daughter. His daughter, married Prince Jésé (d. 1583) Governor of Sakki the third son of Levan King of Kakheti in Georgia.

Ismail I's Statue in Ardabil, Iran.

  - Prince 'Abu'l Fat'h Sultan Moez od-din Bahram Mirza (7 September 1518 - 16 September 1550) Governor of Khorasan 1529-1532, Gilan 1536-1537 and Hamadan 1546-1549. m. Zainab Sultan Khanum. She had issue, four sons and one daughter: Sultan Hassan Mirza died in his youth, Sultan Husain Mirza (died 1567) 'Abu'l Fat'h Sultan Ibrahim Mirza (1541–1577), Sultan Badi uz-Zaman Mirza (*k.*1577)
  - Prince Soltan Hossein Mirza (born 11 December 1520)
- Daughters
  - Princess Fulaneh Begum, m. as his second wife, before 14 May 1513, Prince Murad Effendi, elder son of Şehzade Ahmet, Crown Princess of Ottoman Empire, son of Bayezid II.
  - Princess Gunish Khanum (26 February 1507 – 2 March 1533) m. (first) at Hamadan, 24 August 1518, Sultan Mozaffar Amir-i-Dibaj (*k.* at Tabriz, 23 September 1536), Governor of Rasht and Fooman 1516-1535, son of Amir Hisam od-din Amir-i-Dibaj.
  - Princess Pari Khan Khanum m. in 4 October 1521, Shirvanshah Khalil II Governor of Shirvan 1523-1536, son of Shirvanshah Ibrahim II.
  - Princess Khair un-nisa Khanish Khanum (died 12 March 1564) m. 1537, Seyyed Nur od-din Nimatu'llah Baqi Yazdi (*d.* 21 July 1564), son of Mir Nezam od-din 'Abdu'l Baqi Yazdi.
  - Princess Shah Zainab Khanum (born 1519)
  - Princess Farangis Khanum (born 1519)
  - Princess Mahin Banu Khanum (1519- 20 January 1562)[32]

Legacy

Ismail's greatest legacy was establishing an enduring empire which lasted over 200 years. Even after the fall of Safavids in 1736, their cultural and political influence endured through the era of Afsharid, Zand, Qajar, and Pahlavi dynasties into the modern Islamic Republic of Iran, where Shi'a Islam is still the official religion as it was during the Safavids.

Defeat of Shah Ismail I during the Battle of Chaldiran.

Memory

In the name of Ismail I mentioned:

- Metro, District and Facility[33] in Azerbaijan.
- The street in Ganja and Prospect in Baku.
- In 1993, in Baku was erected a monument to Ismail I.
- The sculpture was erected in Khachmaz (city) to Ismail I.

Alevism

In Alevism, Shah Ismail is seen as a religious figure, and a moral spiritual leader. His teachings are in the Buyruk.

See also

- Safavid dynasty family tree
- List of Turkic Languages poets
- Safavid conversion of Iran from Sunnism to Shiism

Footnotes

[1] Мюнеджжим-баши, с.173

[2] Шараф-хан Бидлиси, т.ц с.169

[3] Хондемир, 'т.III, ч.4, с.570-571, 599-601

[4] Дорн, с.593-595 Эфендиев. Некотырые сведения, с.90.

[5] Ismāʿīl I (http://www.britannica.com/EBchecked/topic/296081/Ismail-I), in Encyclopædia Britannica, online ed., 2011

[6] Woodbridge Bingham, Hilary Conroy, Frank William Iklé, *A History of Asia: Formations of Civilizations, From Antiquity to 1600*, and Bacon, 1974, p. 116. (http://books.google.com/books?id=ogkQAQAAMAAJ&q="Shah+of+Iran"+1502&dq="Shah+of+Iran"+1502&hl=en&ei=9Hg5TsrhCY7zmAXJ1P25Bw&sa=X&oi=book_result&ct=result&resnum=3&ved=0CDYQ6AEwAg)

[7] Encyclopedia Iranica. R.M. Savory. Esmail Safawi (http://www.iranica.com/newsite/articles/v8f6/v8f665.html)

[8] Ismāʿīl I (http://www.britannica.com/EBchecked/topic/296081/Ismail-I) at *Encyclopædia Britannica*

[9] G. Doerfer, "Azeri Turkish", Encyclopaedia Iranica, viii, Online Edition, p. 246. (http://www.iranicaonline.org/articles/azerbaijan-viii)

[10] Richard Tapper, *Frontier nomads of Iran: a political and social history of the Shahsevan*, Cambridge University Press, 1997, ISBN 978-0-521-58336-7, p. 39. (http://books.google.com.tr/books?id=uAzGTtWlp7gC&printsec=frontcover&dq=Richard+Tapper,+Frontier+nomads+of+Iran:+a+political+and+social&hl=tr&ei=-FkrTKfvFIGlcezT8JUD&sa=X&oi=book_result&ct=result&resnum=1&ved=0CCcQ6AEwAA#v=onephttp://en.wikipedia.org/w/index.php?title=Safi-ad-din_Ardabili&action=editage&q&f=false)

[11] EBN BAZZAZ Encyclopedia Iranica (http://www.iranica.com/newsite/articles/v8f1/v8f1010.html)

[12] Muḥammad Kamāl, *Mulla Sadra's Transcendent Philosophy*, Ashgate Publishing Inc, 2006, ISBN 0-7546-5271-8, p. 24. (http://books.google.de/books?id=EwB7Zo7lVp0C&pg=PA12&dq="According+to+Mulla+Salih+Ibrahimi,+Shahab+al-Din+Suhrawardi+was+originally"&hl=en&redir_esc=y#v=snippet&q=Kurdish&f=false)

[13] Peter Charanis. "Review of Emile Janssens' *Trébizonde en Colchide*", *Speculum, Vol. 45, No. 3,*, (Jul., 1970), p. 476

[14] Anthony Bryer, *open citation*, p. 136

[15] Roger M. Savory. „Safavids“ in Peter Burke, Irfan Habib, Halil Inalci:»History of Humanity-Scientific and Cultural Development: From the Sixteenth to the Eighteenth Century", Taylor & Francis. 1999. Excerpt from pg 259:"Доказательства, имеющиеся в настоящее время, приводят к уверенности, что семья Сефевидов имеет местное иранское происхождение, а не тюркское, как это иногда утверждают. Скорее всего, семья возникла в Персидском Курдистане, а затем перебралась в Азербайджан, где ассимилировалась с говорящими по-тюркски азерийцами, и в конечном итоге поселились в маленьком городе Ардебиль где-то в одиннадцатом веке [Evidence available at the present time leads to the conviction that the Safavid family came from indigenous Iranian stock, and not from Turkish ancestry as it is sometimes claimed. It is probable that the family originated in Persian Kurdistan, and later moved to Azerbaijan, where it became assimilated to Turkic-speaking Azeris and eventually settled in the small town of Ardabil sometime during the eleventh century.]".

[16] Вопрос о языке, на котором говорил шах Исмаил, не идентичен вопросу о его «расе» или «национальности». Его происхождение было смешанным: одна из его бабушек была греческая принцесса Комнина. Хинц приходит к выводу, что кровь в его жилах была главным образом, не тюркской. Уже его сын шах Тахмасп начал избавляться от своих туркменских преторианцев. [The question of the language used by Shah Ismail is not identical with that of his race or of his "nationality". His ancestry was mixed: one of his grandmothers was a Greek Comnena princess. Hinz, *Aufstieg*, 74, comes to the conclusion that the blood in his veins was chiefly non-Turkish. Already, his son Shah Tahmasp began to get rid of his Turcoman praetorians.] — V. Minorsky, "The Poetry of Shah Ismail I," *Bulletin of the School of Oriental and African Studies*, University of London 10/4 (1942): 1006–53.

[17] Encyclopaedia Iranica. R. N. Frye. Peoples of Iran. (http://www.iranica.com/newsite/articles/v13f3/v13f3004a.html)

[18] Faruk Sümer, *Safevi Devletinin Kuruluşu ve Gelişmesinde Anadolu Türklerinin Rolü*, Türk Tarih Kurumu Yayınları, Ankara, 1992, p. 15. (**Turkish**)

[19] Nesib Nesibli, "Osmanlı-Safevî Savaşları, Mezhep Meselesi ve Azerbaucan", *Türkler*, Cilt 6, Yeni Türkiye Yayınları, Ankara, 2002, ISBN 975-6782-39-0, p. 895. (**Turkish**)

[20] *The New Encyclopædia Britannica: Micropædia*, Encyclopædia Britannica, 1991, ISBN 978-0-85229-529-8, p. 295. (http://books.google.com/books?id=mXsoAQAAIAAJ&q="Shah+of+Azerbaijan"+1501+July&dq="Shah+of+Azerbaijan"+1501+July&hl=en&ei=THE5TvOOHMqdmQWHmpW6Bw&sa=X&oi=book_result&ct=result&resnum=1&ved=0CCwQ6AEwAA)

[21] BBC, (LINK (http://www.bbc.co.uk/religion/religions/islam/history/safavidempire_1.shtml))

[22] Michael Axworthy *Iran: Empire of the Mind* (Penguin, 2008) p.133

[23] The later Crusades, 1274-1580: from Lyons to Alcazar Door Norman Housley, page 120, 1992

[24] *The Cambridge History of Islam*, Part 1, By Peter Malcolm Holt, Ann K. S. Lambton, Bernard Lewis, p. 401.

[25] Momen (1985), p. 107.

[26] History of the Ottoman Empire and modern Turkey (http://books.google.ca/books?id=Xd422lS6ezgC&pg=PA95&lpg=PA95&dq=hanifa+tomb+destroyed+safavids&source=bl&ots=hZR3iDt0fv&sig=UdZDakp2nwFdDD7GOyzEsyaD1g8&hl=en&ei=5gCuTeLqKIjagAfx6ITsCw&sa=X&oi=book_result&ct=result&resnum=1&ved=0CBkQ6AEwAA#v=onepage&q&f=false)

[27] Encyclopedia Iranica. *Ismail Safavi* (http://www.iranica.com/newsite/articles/v8f6/v8f665.html)

[28] M.B. Dickson and S.C. Welch, *The Houghton Shahnameh*, 2 vols. (Cambridge, Mass., and London, 1981. See p. 34 of vol. I).

[29] R.M. Savory, "Safavids", *Encyclopedia of Islam*, 2nd edition

[30] V. Minorsky, "The Poetry of Shah Ismail I," *Bulletin of the School of Oriental and African Studies*, University of London 10/4 (1942): 1006–53.

[31] http://www.iranica.com/newsite/articles/v8f6/v8f665.html

[32] The Royal Ark (http://www.royalark.net/Persia/safawi2.htm)

[33] Отмечен день рождения Шаха Исмаила Хатаи (http://www5.day.az/news/showbiz/10296.html)

References

- Momen, Moojan (1985). *An Introduction to Shi`i Islam: The History and Doctrines of Twelve.* Yale University Press. ISBN 0-300-03531-4.
- R.M. Savory, *Esmā'il Safawī*, Encyclopaedia Iranica, Online Edition, (LINK (http://www.iranicaonline.org/articles/esmail-i-safawi))
- Mirză Răsul İsmailzadä, *Şah İsmail Säfävi (Xätai) küllüyyatı: qäzällär, qäsidälär, näsihätnamä, dähnamä, qoşmalar / Xätai ; mätnin elmi-tänqidi täktibatçısı*; Alhoda Publishers, Iran, 2004 (in Azeri), ISBN 964-8121-09-5, OCLC 62561234
- M. Momen, *An Introduction to Shi'i Islam*, Yale Univ. Press, 1985, pp. 397, ISBN 0-300-03499-7

External links

- Sheikh Safiaddin Ardabili's Mausoleum Virtual Tour (Ismail I tomb) (http://www.virtualtourengine.com/tour.aspx?id=5)

Safavid dynasty

Safavid Dynasty of Iran[1] [2]	
← ← 1501–1736 ↓ ← ←	
 Flag of the Safavids since 17th century.[3]	 One of several state emblems since the early Safavid period.[4]
 A map of the Safavid empire-Imperii Persici (Persian Empire) drawn by Johann Homann.	

Capital		Tabriz (1501–1555) Qazvin (1555–1598) Isfahan (1598–1722)
Languages		• Persian (official,[5] coinage,[6] [7] civil administration,[8] court (since Isfahan became capital),[9] high literature,[8] literary,[6] [10] theological discourse,[6] diplomatic correspondence, belles-lettres (adab), historiography,[11] court-based religious posts[12]) • Azerbaijani (court, religious dignitaries, military)[11] [13] [14] [15]
Religion		Twelver Shia Islam (state religion)[16]
Government		Monarchy
Shah		
-	1501–24	Ismail I
-	1524–76	Tahmasp I
-	1587–1629	Abbas I
-	1694–1722	Sultan Husayn
-	1729–32	Tahmasp II
-	1732–36	Abbas III
-	1732–36	Nader Afshar
History		
-	Establishment of the Safaviyya by Safi-ad-din Ardabili	1301
-	Established	1501
-	Hotaki Invasion	1722
-	Reconquest under Nader Afshar	1726–29
-	Disestablished	1736
-	Nader Shah crowned	October 1, 1736
Area		2850000 km² (1100391 sq mi)
Currency		Tuman, Abbasi, Shahi.[17] • 1 Tuman = 50 Abbasi. • 1 Tuman = 50 French Livre. • 1 Tuman = £3 6s 8d.

Preceded by	Succeeded by
Timurid dynasty	Hotaki dynasty
Ak Koyunlu	Afsharid dynasty
Marashiyan	Mosul Eyalet
Paduspanis	Baghdad Eyalet
	Basra Eyalet

Today part of	

The **Safavid dynasty** (Persian: سلسلهٔ صفویان; Azerbaijani:) was one of the most significant ruling dynasties of Iran. They ruled one of the greatest Persian empires after the Muslim conquest of Persia[18] [19] [20] [21] and established the Twelver school of Shi'a Islam[22] as the official religion of their empire, marking one of the most important turning points in Muslim history. The Safavids ruled from 1501 to 1722 (experiencing a brief restoration from 1729 to 1736) and at their height, they controlled all of modern Iran, Azerbaijan and Armenia, most of Iraq, Georgia, Afghanistan, and the Caucasus, as well as parts of Pakistan, Turkmenistan and Turkey. Safavid Iran was one of the Islamic "gunpowder empires", along with its neighbours, the Ottoman and Mughal empires.

The Safavid dynasty had its origin in the Safaviyya Sufi order, which was established in the city of Ardabil in the Azerbaijan region. It was of mixed ancestry (Azerbaijani,[23] Kurdish[24] and Turkmen,[25] which included intermarriages with Georgian[26] and Pontic Greek[27] dignitaries). From their base in Ardabil, the Safavids established control over all of Greater Iran and reasserted the Iranian identity of the region,[28] thus becoming the first native dynasty since the Sassanid Empire to establish a unified Iranian state.[29]

Despite their demise in 1736, the legacy that they left behind was the revival of Persia as an economic stronghold between East and West, the establishment of an efficient state and bureaucracy based upon "checks and balances", their architectural innovations and their patronage for fine arts. The Safavids have also left their mark down to the present era by spreading Shi'a Islam in Iran, as well as major parts of the Caucasus, South Asia, Central Asia, and Anatolia.

Genealogy – The Ancestors of The Safavids and its multi-cultural identity

The Safavid Kings themselves claimed to be Seyyeds,[30] family descendants of the prophet Muhammad, although many scholars have cast doubt on this claim.[31] There seems now to be a consensus among scholars that the Safavid family hailed from Persian Kurdistan,[22] and later moved to Azerbaijan, finally settling in the 5th/11th century at Ardabil. Traditional pre-1501 Safavid manuscripts trace the lineage of the Safavids to Kurdish dignitary, Firuz Shah Zarin-Kulah.[24] [32]

According to some historians,[33] [34] including Richard Frye, the Safavids were of Azeri (Turkish) origin:[23]

> The Turkish speakers of Azerbaijan are mainly descended from the earlier Iranian speakers, several pockets of whom still exist in the region. A massive migration of Oghuz Turks in the 11th and 12th centuries not only Turkified Azerbaijan but also Anatolia. Azeri Turks were the founders of Safavid dynasty.

Other historians, such as Vladimir Minorsky[35] and Roger Savory, refute this idea:[36]

> From the evidence available at the present time, it is certain that the Safavid family was of indigineous Iranian stock, and not of Turkish ancestry as it is sometimes claimed. It is probable that the family originated in Persian Kurdistan, and later moved to Azerbaijan, where they adopted the Azari form of Turkish spoken there, and eventually settled in the small town of Ardabil sometimes during the eleventh century.

By the time of the establishment of the Safavid empire, the members of the family were native Turkish-speaking and Turkicized,[13] [37] and some of the Shahs composed poems in their native Turkish language. Concurrently, the Shahs themselves also supported Persian literature, poetry and art projects including the grand Shahnama of Shah Tahmasp,[38] [39] while members of the family and some Shahs composed Persian poetry as well.[40] [41] In terms of identity, it should be noted that the authority of the Safavids were religiously based and they based their legitimacy on being direct male descendants of the Ali,[42] the cousin of the Prophet Muhammad, and the first Shi'ite Imam.

Background – The Safavid Sufi Order

Safavid history begins with the establishment of the Safaviyya by its eponymous founder Safi-ad-din Ardabili (1252–1334). In 700/1301, Safi al-Din assumed the leadership of the Zahediyeh, a significant Sufi order in Gilan, from his spiritual master and father-in-law Zahed Gilani. Due to the great spiritual charisma of Safi al-Din, the order was later known as the Safaviyya. The Safavid order soon gained great influence in the city of Ardabil and Hamdullah Mustaufi noted that most of the people of Ardabil were followers of Safi al-Din.

Extant religious poetry from him, written in the Old Azari language[43] [44] – a now-extinct Northwestern Iranian language – and accompanied by a paraphrase in Persian which helps their understanding, has survived to this day and has linguistic importance.[43]

After Safī al-Dīn, the leadership of the Safaviyya passed onto Sadr al-Dīn Mūsā († 794/1391–92). The order at this time was transformed into a religious movement which conducted religious propaganda throughout Persia, Syria and Asia Minor, and most likely had maintained its Sunni Shafi'ite origin at that time. The leadership of the order passed on from Sadr ud-Dīn Mūsā to his son Khwādja Ali († 1429) and in turn to his son Ibrāhīm († 1429–47).

When Shaykh Junayd, the son of Ibrāhim, assumed the leadership of the Safaviyya in 1447, the history of the Safavid movement was radically changed. According to R.M. Savory, "'Sheikh Junayd was not content with spiritual authority and he sought material power'". At that time, the most powerful dynasty in Persia was that of the Kara Koyunlu, the "Black Sheep", whose ruler Jahan Shah ordered Junāyd to leave Ardabil or else he would bring destruction and ruin upon the city.[22] Junayd sought refuge with the rival of Kara Koyunlu Jahan Shah, the Aq Qoyunlu (White Sheep Turkomans) Khan Uzun Hassan, and cemented his relationship by marrying Uzun Hassan's sister, Khadija Begum. Junayd was killed during an incursion into the territories of the Shirvanshah and was succeeded by his son Haydar Safavi. Haydar married Martha 'Alamshah Begom,[27] Uzun Hassan's daughter, who gave birth to Ismail I, founder of the Safavid dynasty. Martha's mother Theodora – better known as Despina Khatun[45] – was a Pontic Greek princess, the daughter of the Grand Komnenos John IV of Trebizond. She had been married to Uzun Hassan[46] in exchange for protection of the Grand Komnenos from the Ottomans.

After Uzun Hassan's death, his son Ya'qub felt threatened by the growing Safavid religious influence. Ya'qub allied himself with the Shirvanshah and killed Haydar in 1488. By this time, the bulk of the Safaviyya were nomadic Oghuz Turkic-speaking clans from Asia Minor and Azerbaijan and were known as Qizilbash "Red Heads" because of their distinct red headgear. The Qizilbash were warriors, spiritual followers of Haydar, and a source of the Safavid military and political power.

After the death of Haydar, the Safaviyya gathered around his son Ali Mirza Safavi, who was also pursued and subsequently killed by Ya'qub. According to official Safavid history, before passing away, Ali had designated his young brother Ismail as the spiritual leader of the Safaviyya.[22]

History

Founding of the dynasty by Shāh Ismāil I

Persia prior to Ismāil's rule

After the decline of the Timurid Empire (1370–1506), Persia was politically splintered, giving rise to a number of religious movements. The demise of Tamerlane's political authority created a space in which several religious communities, particularly Shi'i ones, could now come to the fore and gain prominence. Among these were a number of Sufi brotherhoods, the Hurufis, Nuqtawis and Musha'sha'. Of these various movements, the Safawid Qizilbash was the most politically resilient, and it was on account of its success that Shah Isma'il I gained political prominence in 1501 CE.[47] There were many local states prior to the Iranian state established by Ismāil.[48] The most important local rulers about 1500 were:

- Huṣayn Bāyqarā, the Timurid ruler of Herāt

- Alwand Mīrzā, the Aq Qoyunlu Khan of Tabrīz
- Murad Beg, Aq Qoyunlu ruler of Irāq al-Ajam
- Farrokh Yaşar, the Shah of Širvan
- Badi Alzamān Mīrzā, local ruler of Balkh
- Huşayn Kīā Chalavī, the local ruler of Semnān
- Murād Beg Bayandar, local ruler of Yazd

Ismāil was able to unite all these lands under the Iranian Empire he created.

Rise of Shāh Ismāil I

Shah Ismail I

The Safavid dynasty was founded about 1501 by Shāh Ismāil I.[49] Shah Ismail's background is disputed: the language he used is not identical with that of his "race" or "nationality" and he was bilingual from birth.[50] Some scholars argue that Ismāil was of mixed Azeri, Kurdish, and Pontic Greek descent,[23] although others argue that he was non-Azeri[50] and was a direct descendant of Sheikh Safi al-Din. As such, he was the last in the line of hereditary Grand Masters of the Safaviyeh order, prior to its ascent to a ruling dynasty. Ismāil was known as a brave and charismatic youth, zealous with regards to his Shi'a faith, and believed himself to be of divine descent—practically worshipped by his Qizilbāsh followers. In 1500, Ismāil invaded neighboring Shirvan to avenge the death of his father, Sheik Haydar, who had been murdered in 1488 by the ruling Shirvanshah, Farrukh Yassar. Afterwards, Ismail went on a conquest campaign, capturing Tabriz in July 1501, where he enthroned himself the Shāh of Azerbaijan,[51] [52] [53] proclaimed himself Shahanshah of Iran[54] [55] [56] and minted coins in his name, proclaiming Shi'ism the official religion of his domain.[22] The establishment of Shi'ism as the state religion led to various Sufi orders openly declaring their Shi'i position, and others, to promptly assume Shi'ism. Among these, the founder of one of the most successful Sufi orders, Ni'matullah (d. 1431) traced his descent from the Ismaili Imam Muhammad b. Ismail, as evidenced in a poem as well as another unpublished literary composition. Though Nimatullah was apparently Sunni, the Ni'matullahi order soon declared his order to be Shi'I after the rise of the Safavid dynasty.[57]

Although Ismail I initially gained mastery over Azerbaijan alone, the Safavids ultimately won the struggle for power in all of Persia which had been going on for nearly a century between various dynasties and political forces. A year after his victory in Tabriz, Ismāil claimed most of Persia as part of his territory,[22] and within 10 years established a complete control over all of it. Ismail followed the line of Turkmen rulers prior to him by assumption of the title "Padishah-i-Iran", previously held by Uzun Hasan.[58] The Ottoman sultans addressed him as *the king of Persian lands and the heir to Jamshid and Kaykhusraw.*[59] Hamadan fell under his power in 1503, Shiraz and Kerman in 1504, Najaf and Karbala in 1507, Van in 1508, Baghdad in 1509, and Herat, as well as other parts of Khorasan, in 1510. By 1511, the Uzbeks in the north-east, led by their Khan Muhammad Shaybāni, were driven far to the north, across the Oxus River where they continued to attack the Safavids. Ismail's decisive victory over the Uzbeks, who had occupied most of Khorasan, ensured Iran's eastern borders, and the Uzbeks never since expanded beyond the Hindukush. Although the Uzbeks continued to make occasional raids to Khorasan, the Safavid empire was able to keep them at bay throughout its reign.

Clashes with the Ottomans

More problematic for the Safavids was the powerful Ottoman Empire. The Ottomans, a Sunni dynasty, considered the active recruitment of Turkmen tribes of Anatolia for the Safavid cause as a major threat. To counter the rising Safavid power, in 1502, Sultan Bayezid II forcefully deported many Shi'as from Anatolia to other parts of the Ottoman realm. In 1514, Bayezid's son, Sultan Selim I marched through Anatolia and reached the plain of Chaldiran near the city of Khoy, and a decisive war was fought there (Battle of Chaldiran). Most sources agree that the Ottoman army was at least double the size of that of Ismāil,[49] however, what gave the Ottomans the advantage was the artillery which the Safavid army lacked. According to R. M. Savory, *"Salim's plan was to winter at Tabriz and complete the conquest of Persia the following spring. However, a mutiny among his officers who refused to spend the winter at Tabriz forced him to withdraw across territory laid waste by the Safavid forces, eight days later"*.[49] Although Ismāil was defeated and his capital was captured, the Safavid empire survived. The war between the two powers continued under Ismāil's son, Shāh Tahmāsp I (q.v.), and the Ottoman Sultan Suleiman I, until Shāh Abbās (q.v.) retook the area lost to the Ottomans by 1602.

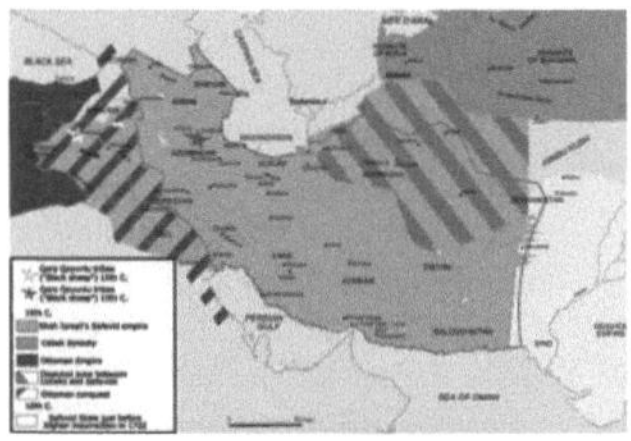

Shāh Ismāil's empire

The consequences of the defeat at Chaldiran were also psychological for Ismāil: the defeat destroyed Ismāil's belief in his invincibility, based on his claimed divine status.[22] His relationships with his Qizilbāsh followers were also fundamentally altered. The tribal rivalries between the Qizilbāsh, which temporarily ceased before the defeat at Chaldiran, resurfaced in intense form immediately after the death of Ismāil, and led to ten years of civil war (930-40/1524-33) until Shāh Tahmāsp regained control of the affairs of the state.

Early Safavid power in Iran was based on the military power of the Qizilbāsh. Ismāil exploited the first element to seize power in Iran. But eschewing politics after his defeat in Chaldiran, he left the affairs of the government to the office of the Wakīl (q.v.). Ismāil's successors, and most ostensibly Shāh Abbās I successfully diminished the Qizilbāsh's influence on the affairs of the state.

Shāh Tahmāsp

Shah Tahmasp, fresco on the walls of the Chehel Sotoun Palace

Shāh Tahmāsp, the young *governor of Herat*, succeeded his father Ismāil in 1524, when he was ten years and three months old. He was the ward of the powerful Qizilbash *amir* Ali Beg Rūmlū (titled *"Div Soltān"*) who saw himself as the de facto ruler of the state. The qizilbash, which still suffered under the legacy of the battle of Chaldiran, was engulfed in internal rivalries. The low morale within the military, and the decentralized structure of the government, with much power in the hands of local governors, eventually led to 10 years of civil war. Rival Qizilbāsh factions fought amongst themselves for the control of the empire until Shāh Tahmāsp came of age and reasserted his authority. Tahmasp reigned for 52 years, the longest reign in Safavid history.[22]

The Uzbeks, during the reign of Tahmāsp, attacked the eastern provinces of the kingdom five times and the Ottomans under Soleymān I initiated four invasions of Persia.[60] Losing territory in Iraq and the north-west, Tahmāsp realized that his capital was not secure, and he was forced to move the capital from Tabriz to Qazvin. Tahmasp made the Peace of Amasya with the Ottomans in 1555, ending the war during his life.[22]

Safavid Persian Empire, in 1598

Alliances to the East – The Mughal Emperor at the Shah's court

Almost simultaneously with the emergence of the Safavid Empire, another Muslim society was developing in South-Asia. The Mughal Empire, which ruled a largely Hindu population, adhered to Sunni Islam. But a common foe, in the Uzbeks, would eventually lead the two empires closer together. During the reign of Tahmasp, Shah Humayun of Mughal Hindustan found himself in a desperate situation, with devastating wars being fought against the Afghans and the Uzbeks and Humayuns brother, Kamran, attempting a coup d'état.[61] Having to flee from city to city, Humayun eventually sought refuge at the court of Tahmasp. Tahmasp, who refused to hand him over to his brother, greeted Humayun at his court in Qazvin as the true emperor of the Mughal dynasty, despite the fact that Humayun had been living in exile for more than fifteen years.[61] [62] After converting to Shia Islam,[63] Tahmasp offered him military assistans to fight off the revolts in return for Kandahar, which had for long been a battle ground between the two empires, and a combined Persian-Mughal force managed to seize Kandahar and occupy Kabul.[64] This eventually led to strong ties between the Safavids and the Mughals, and they persisted, almost unabated, throughout the history of the Safavid dynasty.

Map of Safavid Persia, in 1610

Legacy of Shah Tahmasp

When Shah Tahmasp entered the throne at a young age, Persia was in a dire state. But despite of a weak economy, a civil war and wars being fought on two fronts, Tahmasp had managed to maintain his position as the shah. During the first 30 years of his long reign, he had managed to suppress the internal divisions, slowly elevate the strength of the military to a level that finally led to the retreat of the Ottomans during the fourth war in 1533, and, in 1553, even wage a campaign against the Ottomans. This resulted in the peace treaty of Amasya, a treaty that favoured the Persians and secured Tabriz and the North-Western borders.[65] Some years before, in 1528, he had also converted an unfavorable war against the Uzbeks, at the battle of Jam, into a victory by the Persians.[66] When Shah Tahmasp's throne was overtaken by his successor, Persia was in a calm state, with secure borders and cordial relations with the neighbours to both east and west. What remained unchanged, was the decentralized power structure of the government, and that would not change until the throne was overtaken by his grandson, Shah Abbas.

Shah Tahmasp greets the exiled Humayun

After the death of Tahmāsp in 984/1576, the struggle for a dominant position in the state flared up again and was complicated by rival groups and factions. Dominant political factions vied for power and support three different candidates. The mentally unstable Ismāil, the son of Tahmāsp and the purblind Muhammad Khudābanda were some of the candidates but did not get the support of all the Qizilbāsh chiefs. The Turkmen Ustājlū tribe, one of the most powerful tribes among the Qizilbāsh, threw its support behind Haydar, who was of a Georgian mother, but the majority of the Qizilbāsh chiefs saw this as a threat to their own, Turkmen-dominated power. Instead, they first placed Ismāil II. on the throne (1576–77) and after him Muhammad Shāh Khudābanda (1578–88).[22]

In addition, Tahmasp must be credited for the revival of the fine arts, which flourished under his patronage and were brought to the pitch of perfection. Safavid culture is often admired for the large-scale city planning and architecture, achievements made during the reign of later shahs, but the arts of persian miniature, book-binding and calligraphy, in fact, never received as much attention as they did during his time.[67]

Shah Abbas

Shah 'Abbās King of the Persians. Copper engraving by Dominicus Custos, *Atrium heroicum Caesarum* (1600–2).

The greatest of the Safavid monarchs, Shah Abbas I (1587–1629) came to power in 1587 aged 16 following the forced abdication of his father, Shah Muhammad Khudābanda, having survived Qizilbashi court intrigues and murders. He recognized the ineffectualness of his army which was consistently being defeated by the Ottomans who had captured Georgia and Armenia and by Uzbeks who had captured Mashhad and Sistan in the east. First he sued for peace in 1590 with the Ottomans giving away territory in the north-west. Then two Englishmen, Robert Sherley and his brother Anthony, helped Abbas I to reorganize the Shah's soldiers into an officer-paid and well-trained standing army similar to a European model (which the Ottomans had already adopted). He wholeheartedly adopted the use of gunpowder (See Military history of Iran). The army divisions were: Ghulams غلام (crown servants,[68] usually conscripted from Georgians and Circassians), Tofangchis (تفنگچی, musketeers), and Topchis (Tupchis, توپچی, artillery-men).

Abbas moved the capital to Isfahan, deeper into central Iran. Abbas I built a new city next to the ancient Persian one. From this time the state began to take on a more Persian character. The Safavids ultimately succeeded in establishing a new Persian national monarchy.

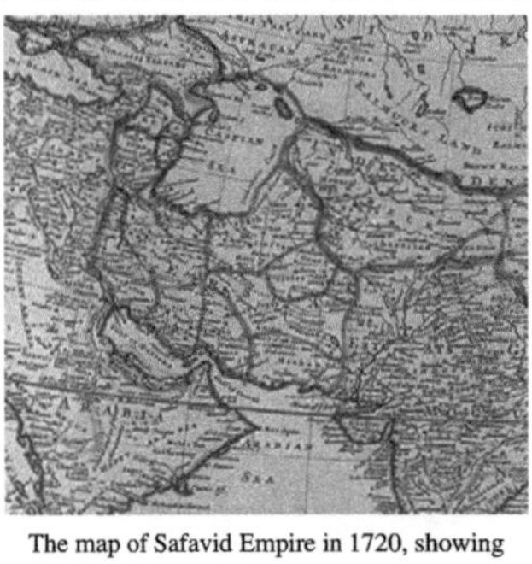

The map of Safavid Empire in 1720, showing different states of Persia.

Abbas I first fought the Uzbeks, recapturing Herat and Mashhad in 1598. Then he turned against the Ottomans recapturing Baghdad, eastern Iraq and the Caucasian provinces by 1622. He also used his new force to dislodge the Portuguese from Bahrain (1602) and, with English help, from Hormuz (1622), in the Persian Gulf (a vital link in Portuguese trade with India). He expanded commercial links with the English East India Company and the Dutch East India Company. Thus Abbas I was able to break the dependence on the Qizilbash for military might and therefore was able to centralize control.

The Ottoman Turks and Safavids fought over the fertile plains of Iraq for more than 150 years. The capture of Baghdad by Ismail I in 1509 was only followed by its loss to the Ottoman Sultan Suleiman I in 1534. After subsequent campaigns, the Safavids recaptured Baghdad in 1623 yet lost it again to Murad IV in 1638. Henceforth a treaty, signed in Qasr-e Shirin, was established delineating a border between Iran and Turkey in 1639, a border which still stands in northwest Iran/southeast Turkey. The 150-year tug-of-war accentuated the Sunni and Shi'a rift in Iraq.

Part of Safavid Persian Empire (on right) and middle east. Emanuel Bowen, 1744–52.

In 1609–10, a war broke out between Kurdish tribes and the Safavid Empire. After a long and bloody siege led by the Safavid grand vizier Hatem Beg, which lasted from November 1609 to the summer of 1610, the Kurdish stronghold of Dimdim was captured. Shah Abbas ordered a general massacre in Beradost and Mukriyan (Mahabad, reported by Eskandar Beg Monshi, Safavid Historian (1557–1642), in "Alam Ara Abbasi") and resettled the Turkic Afshar tribe in the region while deporting many Kurdish tribes to Khorasan.[69] [70] Nowadays, there is a community of nearly 1.7 million people who are descendants of the tribes deported from Kurdistan to Khurasan (Northeastern Iran) by the Safavids.[71]

Due to his obsessive fear of assassination, Shah Abbas either put to death or blinded any member of his family who aroused his suspicion. One of his sons was executed and two blinded. Since two other sons had predeceased him, the result was personal tragedy for Shah Abbas. When he died on 19 January 1629, he had no son capable of succeeding him.[72]

The beginning of the 17th century saw the power of the Qizilbash decline, the original militia that had helped Ismail I capture Tabriz and which had gained many administrative powers over the centuries. Power was shifting to a new class of merchants, many of them ethnic Armenians, Georgians and Indians.

At its zenith, during the long reign of Shah Abbas I the empire's reach comprised Iran, Iraq, Armenia, Azerbaijan, Georgia, and parts of Turkmenistan, Uzbekistan, Afghanistan, Pakistan, and Turkey.

Contacts with Europe during Abbas' reign

Abbas' tolerance towards Christians was part of his policy of establishing diplomatic links with European powers to try to enlist their help in the fight against their common enemy, the Ottoman Empire. The idea of such an anti-Ottoman alliance was not a new one — over a century before, Uzun Hassan, then ruler of part of Iran, had asked the Venetians for military aid — but none of the Safavids had made diplomatic overtures to Europe and Abbas' attitude was in marked contrast to that of his grandfather, Tahmasp I, who had expelled the English traveller Anthony Jenkinson from his court on hearing he was a Christian.[73] For his part, Abbas declared that he "preferred the dust from the shoe soles of the lowest Christian to the highest Ottoman personage."[74]

The ambassador Husain Ali Beg led the first Persian embassy to Europe (1599–1602).

Fresco in the Doge's Palace in Venice depicting Doge Mariano Grimani receiving the Persian Ambassadors, 1599

In 1599, Abbas sent his first diplomatic mission to Europe. The group crossed the Caspian Sea and spent the winter in Moscow, before proceeding through Norway, Germany (where it was received by Emperor Rudolf II) to Rome where Pope Clement VIII gave the travellers a long audience. They finally arrived at the court of Philip III of Spain in 1602. Although the expedition never managed to return to Iran, being shipwrecked on the journey around Africa, it marked an important new step in contacts between Iran and Europe and Europeans began to be fascinated by the Iranians and their culture — Shakespeare's 1601–2 *Twelfth Night*, for example, makes two references (at II.5 and III.4) to 'the Sophy', then the English term for the Shahs of Iran.[75] [76] Henceforward, the number of diplomatic missions to and fro greatly increased.[77]

The shah had set great store on an alliance with Spain, the chief opponent of the Ottomans in Europe. Abbas offered trading rights and the chance to preach Christianity in Iran in return for help against the Ottomans. But the stumbling block of Hormuz remained, a vassal kingdom which had fallen into Spanish Habsburgs hands when the King of Spain inherited the throne of Portugal in 1580. The Spanish demanded Abbas break off relations with the English East India Company before they would consider relinquishing the town. Abbas was unable to comply. Eventually Abbas became frustrated with Spain, as he did with the Holy Roman Empire, which wanted him to make his 170,000 Armenian subjects swear allegiance to the Pope but did not trouble to inform the shah when the Emperor Rudolf signed a peace treaty with the Ottomans. Contacts with the Pope, Poland and Moscow were no more fruitful.[78]

More came of Abbas' contacts with the English, although England had little interest in fighting against the Ottomans. The Sherley brothers arrived in 1598 and helped reorganise the Iranian army. The English East India Company also began to take an interest in Iran and in 1622 four of its ships helped Abbas retake Hormuz from the Portuguese in the Capture of Ormuz (1622). It was the beginning of the East India Company's long-running interest in Iran.[79]

Decline of the Safavid state

Shah Abbas the II holding a banquet for foreign dignitaries. Detail from a ceiling fresco at the Chehel Sotoun Palace in Isfahan.

In addition to fighting its perennial enemies, the Ottomans and Uzbeks, as the 17th century progressed Iran had to contend with the rise of new neighbors. Russian Muscovy in the previous century had deposed two western Asian khanates of the Golden Horde and expanded its influence into the Caucasus Mountains and Central Asia. In the east, the Mughals of India had expanded into Khorasan (now Afghanistan) at the expense of Iranian control, taking Qandahar.

More importantly, the Dutch East India company and later English/British used their superior means of maritime violence to control trade routes in the western Indian ocean. As a result, Iran was cut off from overseas links to East Africa, the Arabian peninsula, and South Asia.[80] But overland trade between Iran and South Asia grew. Many Indian merchants established a permanent presence in Iran and moved into Russia from the mid-seventeenth century.[81] Iran was also able to further develop its overland trade with North and Central Europe during the second half of the seventeenth century.[82] In the late seventeenth century, Iranian merchants established a permanent presence as far north as Narva on the Baltic sea, in what now is Estonia.[83]

The Dutch and English were still able to drain the Iranian government of much of its precious metal supplies. Except for Shah Abbas II, the Safavid rulers after Abbas I were therefore rendered ineffectual, and the Iranian government declined and finally collapsed when a serious military threat emerged on its eastern border in the early eighteenth century.[84] The end of the reign of Abbas II, 1666, thus marked the beginning of the end of the Safavid dynasty. Despite falling revenues and military threats, later shahs had lavish lifestyles. Sultan Husayn (1694–1722) in particular was known for his love of wine and disinterest in governance.[85]

The country was repeatedly raided on its frontiers — Kerman by Baloch tribes in 1698, Khorasan by the Hotakis in 1717, constantly in Mesopotamia by peninsula Arabs. Sultan Hosein tried to forcibly convert his Afghan subjects in Qandahar from Sunni to the Shi'a sect of Islam. In response, a Ghilzai Afghan chieftain named Mir Wais Hotak revolted and killed Gurgin Khan, the Safavid governor of the region, along with his army. In 1722, an Afghan army led by Mir Wais' son Mahmud advanced on the heart of the empire and defeated the government forces at the Battle of Gulnabad. He then besieged the capital of Isfahan, until Shah Sultan Husayn abdicated and acknowledged him as the new king of Persia.[86]

Map of Persia, c. 1700 by Johann Baptist Homann (1644–1724)

The tribal Afghans rode roughshod over their conquered territory for seven years but were prevented from making further gains by Nader Shah, a former slave who had risen to military leadership within the Afshar tribe in Khorasan, a vassal state of the Safavids. Nadir Shah defeated the Ghilzai Hotaki forces in the 1729 Battle of Damghan. He had removed them from power, and in 1738 conquered their last stronghold in Qandahar; in the same year he occupied Ghazni, Kabul, Lahore, and as far as Delhi in India. However, these cities were later inherited by his Abdali Afghan military commander, Ahmad Shah Durrani. Nadir had effective control under Shah Tahmasp II and then ruled as regent of the infant Abbas III until 1736 when he had himself crowned shah.

Immediately after Nadir Shah's assassination in 1747, the Safavids were re-appointed as shahs of Iran in order to lend legitimacy to the nascent Zand dynasty. However the brief puppet regime of Ismail III ended in 1760 when Karim Khan felt strong enough to take nominal power of the country as well and officially end the Safavid dynasty.

Shia Islam as the state religion

Even though Safavids were not the first Shia rulers in Iran, they played a crucial role in making Shia Islam the official religion in the whole of Iran. There were large Shia communities in some cities like Qom and Sabzevar as early as the 8th century. In the 10th and 11th centuries the Buwayhids, who were of the Zaidiyyah branch of Shia, ruled in Fars, Isfahan and Baghdad. As a result of the Mongol conquest and the relative religious tolerance of the Ilkhanids, Shia dynasties were re-established in Iran, Sarbedaran in Khorasan being the most important. The Ilkhanid ruler Öljaitü converted to Twelver Shiism in the 13th century.

Shah Suleiman I and his courtiers, Isfahan, 1670. Painter is Aliquli Jabbadar, and is kept at The St. Petersburg Institute of Oriental Studies in Russia, ever since it was acquired by Tsar Nicholas II. Note the two Georgian figures with their names at the top left.

Following his conquest of Iran, Ismail I made conversion mandatory for the largely Sunni population. The Sunni Ulema or clergy were either killed or exiled. Ismail I, brought in mainstream Ithnāʻashariyyah Shi'a religious leaders and granted them land and money in return for loyalty. Later, during the Safavid and especially Qajar period, the Shia Ulema's power increased and they were able to exercise a role, independent of or compatible with the government. Despite the Safavids' Sufi origins, most Sufi groups were prohibited, except the Nimatullahi order.

Iran became a feudal theocracy: the Shah was held to be the divinely ordained head of state and religion. In the following centuries, this religious stance cemented both Iran's internal cohesion and national feelings and provoked attacks by its Sunni neighbors.

Military and the role of Qizilbash

The Qizilbash were a wide variety of Shi'ite (*ghulāt*) and mostly Turcoman militant groups who helped found the Safavid Empire. Their military power was essential during the reign of the Shahs Ismail and Tahmasp. The Qizilbash tribes were essential to the military of Iran until the rule of Shah Abbas I- their leaders were able to exercise enormous influence and participate in court intrigues (assassinating Shah Ismail II for example).

Persian helmet

A major problem faced by Ismail I after the establishment of the Safavid state was how to bridge the gap between the two major ethnic groups in that state: the Qizilbash ("Redhead") Turcomans, the "men of sword" of classical Islamic society whose military prowess had brought him to power, and the Persian elements, the "men of the pen", who filled the ranks of the bureaucracy and the religious establishment in the Safavid state as they had done for centuries under previous rulers of Persia, be they Arabs, Mongols, or Turkmens. As Vladimir Minorsky put it, friction between these two groups was inevitable, because the Qizilbash "were no party to the national Persian tradition".

Manikin of a safavid Qezelbash soldier, exhibited in Sa'dabad

Between 1508 and 1524, the year of Ismail's death, the shah appointed five successive Persians to the office of *vakil*. When the second Persian vakil was placed in command of a Safavid army in Transoxiana, the Qizilbash, considering it a dishonor to be obliged to serve under him, deserted him on the battlefield with the result that he was slain. The fourth vakil was murdered by the Qizilbash, and the fifth was put to death by them.[49]

Reforms in the military

Shah Abbas realized that in order to retain absolute control over his empire without antagonizing the Qizilbash, he needed to create reforms that reduced the dependency that the shah had on their military support. Part of these reforms was the creation of the 3rd force within the aristocracy, but even more important in undermining the authority of the Qizilbash was the introduction of the Royal Corps into the military. This military force would serve the shah only and eventually consisted of four separate branches:[87]

- Shahsevans – these were 12 000 strong and built up from the small group of *qurchis* that Shah Abbas had inherited from his predecessor. The Shahsevans, or "Friends of the King", were Qizilbash tribesmen who had forsaken their tribal allegiance for allegiance to the shah alone.[88]
- Gulams – Tahmasp had started introducing Georgian, Armenian and Circassian slaves from the Caucasus, appointing them either in the harem or the royal household. Shah Abbas expanded this program significantly and eventually created a force of 15 000 ghulam cavalrymen.
- Musketers – realizing the advantages that the Ottomans had because of their firearms, Shah Abbas was at pains to equip both the qurchi and the ghulam soldiers with up-to-date weaponry. More importantly, for the first time in Iranian history, a substantial infantry corps of musketeers (*tofang-chis*), numbering 12 000, was created.

- Artillery Corps – with the help of Westerners, he also formed an artillery corps of 12 000 men, although this was the weakest element in his army. According to Sir Thomas Herbert, who accompanied the British embassy to Persia in 1628, the Persians relied heavily on support from the Europeans in manufacturing cannons.[89] It wasn't until a century later, when Nadir Shah became the Commander in Chief of the military that sufficient effort was put into modernizing the artillery corps and the Persians managed to excel and become self-sufficient in the manufacturing of firearms.

Despite the reforms, the Qizilbash would remain the strongest and most effective element within the military, accounting for more than half of its total strength.[89] But the creation of this large standing army, that, for the first time in Safavid history, was serving directly under the Shah, significantly reduced their influence, and perhaps any possibilities for the type of civil unrest that had caused havoc during the reign of the previous shahs.

Society

A proper term for the Safavid society is what we today can call a *meritocracy*, meaning a society in which officials were appointed on the basis of worth and merit, and not on the basis of birth. It was certainly not an oligarchy, nor was it an aristocracy. Sons of nobles were considered for the succession of their fathers as a mark of respect, but they had to prove themselves worthy of the position. This system avoided an entrenched aristocracy or a cast society.[90] There even are numerous recorded accounts of laymen that rose to high official posts, as a result of their merits.[91]

Nevertheless, the Persian society during the Safavids was that of a hierarchy, with the Shah at the apex of the hierarchical pyramid, the common people, merchants and peasants at the base, and the aristocrats in between. The term *dowlat*, which in modern Persian means "government", was then an abstract term meaning "bliss" or "felicity", and it began to be used as concrete sense of the Safavid state, reflecting the view that the people had of their ruler, as someone elevated above humanity.[92]

Also among the aristocracy, in the middle of the hierarchical pyramid, were the religious officials, who, mindful of the historic role of the religious classes as a buffer between the ruler and his subjects, usually did their best to shield the ordinary people from oppressive governments.[92]

The customs and culture of the people

Lady's clothing in the 1600s

Jean Chardin devoted a whole chapter in his book to describing the Persian character, which apparently fascinated him greatly. As he spent a large bulk of his life in Persia, he involved himself in, and took part in, their everyday rituals and habits, and eventually acquired intimate knowledge of their culture, customs and character. He admired their consideration towards foreigners, but he also stumbled upon characteristics that he found challenging. His descriptions of the public appearance, clothes and customs are corroborated by the miniatures, drawings and paintings from that time which have survived. As he describes them:[93]

> Their imagination is animated, quick and fruitful. Their memory is free and prolific. They are very favorably drawn to the sciences, the liberal and mechanical arts. Their temperament is open and leans towards sensual pleasure and self-indulgence, which makes them pay little attention to economy or business.

He then goes on:[93]

> They are very philosophical over the good and bad things in life and about expectations for the future. They are little tainted with avarice, desiring only to acquire in order to spend. They love to enjoy what is to hand and they refuse nothing which contributes to it, having no anxiety about the future which they leave to providence and fate.

But as he also experienced:[94]

> ...the Persians are dissembling, shamelessly deceitful and the greatest flatterers in the world, using great deception and insolence. They lack good faith in business dealings, in which they cheat so adeptly that one is always taken in. Hypocrisy is the usual disguise in which they proceed. They say their prayers and perform their rituals in the most devout manner. They hold the wisest and most pious conversation of which they are capable. And although they are naturally inclined to humanity, hospitality, mercy and other worldly goods, nevertheless, they do not cease feigning in order to give the semblance of being much better than they really are.

Character

It is however no question, from reading Chardin's descriptions of their manners, that he considered them to be a well educated and well behaved people, who certainly knew the strict etiquettes of social intercourse. As he describes them,[95]

Anthony Shirley and Robert Shirley (*pictured in 1622*) helped modernize the Persian Army.

> The Persians are the most civilized of the peoples of the East, and what the French are to Europe, they are to the Orient... Their bearing and countenance is the best-composed, mild, serious, impressive, genial and welcoming as far as possible. They never fail to perform at once the appropriate gestures of politeness when meeting each other... They are the most wheedling people in the world, with the most engaging manners, the most supple spirits and a language that is gentle and flattering, and devoid of unpleasant terms but rather full of circumlocutions.

Unlike Europeans, they much disliked physical activity, and were not in favor of exercise for its own sake, preferring the leisure of repose and luxuries that life could offer. Travelling was valued only for the specific purpose of getting from one place to another, not interesting them self in seeing new places and experiencing different cultures. It was perhaps this sort of attitude towards the rest of the world that accounted for the ignorance of Persians regarding other countries of the world. The exercises that they took part in were for keeping the body supple and sturdy and to acquire skills in handling of arms. Archery took first place. Second place was held by fencing, where the wrist had to be firm but flexible and movements agile. Thirdly there was horsemanship. A very strenuous form of exercise which the Persians greatly enjoyed was hunting.[96]

Entertainment

A persian miniature depicting a polo-match

Since pre-Islamic times, the sport of wrestling had been an integral part of the Iranian identity, and the professional wrestlers, who performed in Zurkhanehs, were considered important members of the society. Each town had their own troop of wrestlers, called Pahlavans. Their sport also provided the masses with entertainment and spectacle. Chardin described one such event:[97]

> The two wrestlers were covered in grease. They are present on the level ground, and a small drum is always playing during the contest for excitement. They swear to a good fight and shake hands. That done, they slap their thighs, buttocks and hips to the rhythm of the drum. That is for the women and to get themselves in good form. After that they join together in uttering a great cry and trying to overthrow each other.

As well as wrestling, what gathered the masses was fencing, tightrope dancers, puppet-players and acrobats, performing in large squares, such as the Royal square. A leisurely form of amusement was to be found in the cabarets, particularly in certain districts, like those near the mausoleum of Harun-e Velayat. People met there to drink liqueurs or coffee, to smoke tobacco or opium, and to chat or listen to poetry.[98]

Clothes and Appearances

As noted before, a key aspect of the Persian character was its love of luxury, particularly on keeping up appearances. They would adorn their clothes, wearing stones and decorate the harness of their horses. Men wore many rings on their fingers, almost as many as their wives. They also placed jewels on their arms, such as on daggers and swords. Daggers were worn at the waist. In describing the lady's clothing, he noted that Persian dress revealed more of the figure than did the European, but that women appeared differently depending on whether they were at home in the presence of friends and family, or if they were in the public. In private they usually wore a veil that only covered the hair and the back, but upon leaving the home, they would put on a large sheet, that concealed the whole of the body except from the face. They would often dye their feet and hands with henna. Their hairstyle was simple, the hair gathered back in tresses, often adorning the ends with pearls and clusters of jewels. Women with slender waists were regarded as more attractive than those with larger figures. Women from the provinces and slaves pierced their left nostrils with rings, but well-born Persian women would not do this.[99]

The most precious accessory for men was the turban. Although they lasted a long time it was necessary to have changes for different occasions like weddings and the Nowruz, while men of status never wore the same turban two days running. Clothes that became soiled in any way were changed immediately.[100]

Turks and Tajiks

Men's clothing in the 1600s

Although the Safavid rulers and citizens were of native stock and continuously reasserted their Iranian identity, the power structure of the Safavid state was mainly divided into two groups: the Turkic-speaking military/ruling elite—whose job was to maintain the territorial integrity and continuity of the Iranian empire through their leadership—and the Persian-speaking administrative/governing elite—whose job was to oversee the operation and development of the nation and its identity through their high positions. Thus came the term "Turk and Tajik", which was used by native Iranians for many generations to describe the Persianate, or Turko-Persian, nature of many dynasties which ruled over Greater Iran between the 12th and 20th centuries, in that these dynasties promoted and helped continue the dominant Persian linguistic and cultural identity of their states, although the dynasties themselves were of non-Persian (e.g. Turkic) linguistic origins. The relationship between the Turkic-speaking 'Turks' and Persian-speaking 'Tajiks' was symbiotic, yet some form of rivalry did exist between the two. As the former represented the "*people of the sword*" and the latter, "*the people of the pen*", high-level official posts would naturally be reserved for the Persians. Indeed, this had been the situation throughout Persian history, even before the Safavids, ever since the Arab conquest.[101] Shah Tahmasp introduced a change to this, when he, and the other Safavid rulers who succeeded him, sought to blur the formerly defined lines between the two linguistic groups, by taking the sons of Turkic-speaking officers into the royal household for their education in the Persian language. Consequently, they were slowly able to take on administrative jobs in areas which had hitherto been the exclusive preserve of the ethnic Persians.[102]

The third force

From 1540 and onwards, Shah Tahmasp initiated a transformation of the society by slowly constructing a new branch within the aristocracy. The campaigns that he waged against Georgia between 1540 and 1554 were primarily meant to uphold the morale and the fighting efficiency of the qizilbash military,[103] but they brought home large numbers of Georgian, Armenian and Circassian slaves. The women came to occupy prominent positions in the harems of the Safavid elite, particularly the Shah's, while the men were given special training, on completion of which they were either enrolled in one of the newly created *ghulam* regiments, or employed in the royal household.[65] Shah Abbas continued this program and greatly expanded the ghulam military corps from a few hundred to 15 000 highly trained cavalrymen.[104] He then went on to reduce the number of qizilbash provincial governorships and systematicly moved qizilbash governors to other districts, thus disrupting their ties with the local community, and reducing their power. Many were replaced by a ghulam, and within short time, Georgians, Armenians and Circassians had been appointed to many of the highest offices of state. By 1595, Allahverdi Khan, a Georgian, became one of the most powerful men in the Safavid state, when he was appointed the Governor-General of Fars, one of the richest provinces in Persia. And his power reached its peak in 1598, when he became the commander-in-chief of the armed forces.[105] Thus, this new group eventually came to constitute a powerful "third force" within the state, alongside the Tajik Persians and the Qizilbash Turks, and it only goes to prove the meritocratic society of the Safavids.

Emergence of a clerical aristocracy

An important feature of the Safavid society was the alliance that emerged between the ulama (the religious class) and the merchant community. The latter included merchants trading in the bazaars, the trade and artisan guilds (*asnāf*) and members of the quasi-religious organizations run by dervishes (*futuvva*). Because of the relative insecurity of property ownership in Persia, many private landowners secured their lands by donating them to the clergy as so called *vaqf*. They would thus retain the official ownership and secure their land from being confiscated by royal commissioners or local governors, as long as a percentage of the revenues from the land went to the ulama. Increasingly, members of the religious class, particularly the mujtahids and the seyyeds, gained full ownership of these lands, and, according to contemporary historian Iskandar Munshi, Persia started to witness the emergence of a new and significant group of landowners.[106]

Akhbaris versus Usulis

The Akhbari movement "crystalized" as a "separate movement" with the writings of Muhammad Amin al-Astarabadi (died 1627 AD). It rejected the use of reasoning in deriving verdicts and believed that only the Quran, hadith, (prophetic sayings and recorded opinions of the Imams) and consensus should be used as sources to derive verdicts (*fatāwā*). Unlike Usulis, Akhbari did and do not follow *marjas* who practice *ijtihad*.[107]

It achieved its greatest influence in the late Safavid and early post-Safavid era, when it dominated Twelver Shia Islam.[108] However, shortly thereafter Muhammad Baqir Behbahani (died 1792), along with other Usuli mujtahids, crushed the Akhbari movement.[109] It remains only a small minority in the Shia Muslim world. One result of the resolution of this conflict was the rise in importance of the concept of ijtihad and the position of the mujtahid (as opposed to other ulama) in the 18th and early 19th centuries. It was from this time that the division of the Shia world into mujtahid (those who could follow their own independent judgment) and *muqallid* (those who had to follow the rulings of a mujtahid) took place. According to author Moojan Momen, "up to the middle of the 19th century there were very few mujtahids (three or four) anywhere at any one time," but "several hundred existed by the end of the 19th century."[110]

Allamah Majlisi

Muhammad Baqir Majlisi, commonly referenced to using the title Allamah, was a highly influential scholar during the 17th century (Safavid era). Majlisi's works emphasized his desire to purge Twelver Shi`ism of the influences of mysticism and philosophy, and to propagate an ideal of strict adherence to the Islamic law (sharia).[111] Majlisi promoted specifically Shia rituals such as mourning for Hussein ibn Ali and visitation (*ziyarat*) of the tombs of the Imams and Imamzadas, stressing "the concept of the Imams as mediators and intercessors for man with God."[112]

State and government

The Safavid state was one of checks and balance, both within the government and on a local level. At the apex of this system was the Shah, with total power over the state, legitimized by his bloodline as a seyyed, or descendant of the Prophet Mohammad. So absolute was his power, that the French merchant, and later ambassador to Persia, Jean Chardin thought the Safavid Shahs ruled their land with an iron fist and often in a despotic manner.[113] To ensure transparency and avoid decisions being made that circumvented the Shah, a complex system of bureaucracy and departmental procedures had been put in place that prevented fraud. Every office had a deputy or superintendent, whose job was to keep records of all actions of the state officials and report directly to the Shah. The Shah himself exercised his own measures for keeping his ministers under control by fostering an atmosphere of rivalry and competitive surveillance. And since the Safavid society was meritocratic, and successions seldom were made on the basis of heritage, this meant that government offices constantly felt the pressure of being under surveillance and had to make sure they governed in the best interest of their leader, and not merely their own.

The Government

There probably did not exist any parliament, as we know them today. But the Portuguese ambassador to the Safavids, De Gouvea, still mentions the *Council of State*[114] in his records, which perhaps was a term for governmental gatherings of the time.

The highest level in the government was that of the Prime Minister, or Grand Vizier (*Etemad-e Dowlat*), who was always chosen from among doctors of law. He enjoyed tremendous power and control over national affairs as he was the immediate deputy of the Shah. No act of the Shah was valid without the counter seal of the Prime Minister. But even he stood accountable to a deputy (*vak'anevis*), who kept records of his decision-makings and notified the Shah. Second to the Prime Minister post were the General of the Revenues (*mostoufi-ye mamalek*), or finance minister,[115] and the *Divanbegi*, Minister of Justice. The latter was the final appeal in civil and criminal cases, and his office stood next to the main entrance to the Ali Qapu palace. In earlier times, the Shah had been closely involved in judicial proceedings, but this part of the royal duty was neglected by Shah Safi and the later kings.[116]

Next in authority were the generals: the General of the Royal Troops (the *Shahsevans*), General of the Musketeers, General of the Ghulams and The Master of Artillery. A separate official, the Commander-in-Chief, was appointed to be the head of these officials.[116]

The Royal Court

As for the royal household, the highest post was that of the *Nazir*, Court Minister. He was perhaps the closest advisor to the Shah, and, as such, functioned as his eyes and ears within the Court. His primary job was to appoint and supervise all the officials of the household and to be their contact with the Shah. But his responsibilities also included that of being the treasurer of the Shahs properties. This meant that even the Prime Minister, who held the highest office in the state, had to work in association with the Nazir when it came to managing those transactions that directly related to the Shah.[116]

Frontpage on Jean Chardin's book on his journeys to Persia, published in 1739.

The second most senior appointment was the Grand Steward (*Ichik Agasi bashi*), who would always accompany the Shah and was easily recognizable because of the great baton that he carried with him. He was responsible for introducing all guests, receiving petitions presented to the Shah and reading them if required. Next in line were the Master of the Royal Stables (*Mirakor bashi*) and the Master of the Hunt (*Mirshekar bashi*). The Shah had stables in all the principal towns, and Shah Abbas was said to have about 30 000 horses in studs around the country.[117] In addition to these, there were separate officials appointed for the caretaking of royal banquets and for entertainment.

Chardin specifically noticed the rank of doctors and astrologers and the respect that the Shahs had for them. The Shah had a dozen of each in his service and would usually be accompanied by three doctors and three astrologers, who were authorized to sit by his side on various occasions.[116] The Chief Physician (*Hakim-bashi*) was a highly considered member of the Royal court,[118] and the most revered astrologer of the court was given the title *Munajjim-bashi* (Chief Astrologer).[119]

During the first century of the dynasty, the primary court language remained Azeri,[115] although this increasingly changed after the capital was moved to Isfahan.[9]

Local governments

On a local level, the government was divided into public land and royal possessions. The public land was under the rule of local governors, or *Khans*. Since the earliest days of the Safavid dynasty, the Qizilbash generals had been appointed to most of these posts. They ruled their provinces like petty shahs and spent all their revenues on their own province, only presenting the Shah with the balance. In return, they had to keep ready a standing army at all times and provide the Shah with military assistance upon his request. It was also requested from them that they appoint a lawyer (*vakil*) to the Court who would inform them on matters pertaining to the provincial affairs.[120] Shah Abbas I intended to decrease the power of the Qizilbash by bringing some of these provinces into his direct control, creating so called *Crown Provinces* (*Khassa*). But it was Shah Safi, under influence by his Prime Minister, Saru Taqi, that initiated the program of trying to increase the royal revenues by buying land from the governors and putting in place local commissioners.[120] In time, this proved to become a burden to the people that were under the direct rule of the Shah, as these commissioners, unlike the former governors, had little knowledge about the local communities that they controlled and were primarily interested in increasing the income of the Shah. And, while it was in the governors' own interest to increase the productivity and prosperity of their provinces, the commissioners received their income directly from the royal treasury and, as such, did not care so much about investing in agriculture and local industries. Thus, the majority of the people suffered from rapacity and corruption carried out in the name of the Shah.[120]

Democratic institutions in a totalitarian society

In 16th and 17th century Iran, there existed a considerable number of local democratic institutions. Examples of such were the trade and artisan guilds, which had started to appear in Persia from the 1500s. Also, there were the quazi-religious fraternities called *futuvva*, which were run by local dervishes. Another official selected by the consensus of the local community was the *kadkhoda*, who functioned as a common law administrator.[121] The local sheriff (*kalantar*), who was not elected by the people but directly appointed by the Shah, and whose function was to protect the people against injustices on the part of the local governors, supervised the kadkhoda.[122]

Legal system

In Safavid Persia there was little distinction between theology and jurisprudence, or between divine justice and human justice, and it all went under *Islamic jurisprudence* (fiqh). The legal system was built up of two branches: civil law, which had its roots in sharia, *received wisdom*, and urf, meaning *traditional experience* and very similar to the Western form of common law. While the imams and judges of law applied civil law in their practice, urf was primarily exercised by the local commissioners, who inspected the villages on behalf of the Shah, and by the Minister of Justice (*Divanbegi*). The latter were all secular functionaries working on behalf of the Shah.[123]

The *Karkan*, a tool used for punishment of state criminals

The highest level in the legal system was the Minister of Justice, and the law officers were divided into senior appointments, such as the magistrate (*darughah*), inspector (*visir*), and recorder (*vak'anevis*). The lesser officials were the qazi, corresponding a civil lieutenant, who ranked under the local governors and functioned as judges in the provinces.

According to Chardin:[124]

> There were no particular place assigned for the administration of justice. Each magistrate executes justice in his own house in a large room opening on to a courtyard or a garden which is raised two or three feet above the ground. The Judge is seated at one end of the room having a writer and a man of law by his side.

Chardin also noted that bringing cases into court in Persia was easier than in the West. The judge (qazi) was informed of relevant points involved and would decide whether or not to take up the case. Having agreed to do so, a sergeant would investigate and summon the defendant, who was then obliged to pay the fee of the sergeant. The two parties with their witnesses pleaded their respective cases, usually without any counsel, and the judge would pass his judgment after the first or second hearing.[124]

Criminal justice was entirely separate from civil law and was judged upon *common law* administered through the Minister of Justice, local governors and the Court minister (the *Nazir*). Despite being based on *urf*, it relied upon certain sets of legal principles. Murder was punishable by death, and the penalty for bodily injuries was invariably the bastinado. Robbers had their right wrists amputated the first time, and sentenced to death on any subsequent occasion. State criminals were subjected to the *karkan*, a triangular wooden collar placed around the neck. On extraordinary occasions when the Shah took justice into his own hand, he would dress himself up in red for the importance of the event, according to ancient tradition.[123]

Economy

What fueled the growth of Safavid economy was Iran's position between the burgeoning civilizations of Europe to its west and India and Islamic Central Asia to its east and north. The Silk Road which led through northern Iran to India revived in the 16th century. Abbas I also supported direct trade with Europe, particularly England and The Netherlands which sought Persian carpet, silk and textiles. Other exports were horses, goat hair, pearls and an inedible bitter almond hadam-talka used as a spice in India. The main imports were spice, textiles (woolens from Europe, cottons from Gujarat), metals, coffee, and sugar.

Agriculture

A 19th century drawing of Isfahan

According to the historian Roger Savory, the twin bases of the domestic economy were pastoralism and agriculture. And, just as the higher levels of the social hierarchy was divided between the Turkish "men of the sword" and the Persian "men of the pen"; so were the lower level divided between the Turcoman tribes, who were cattle breeders and lived apart from the surrounding population, and the Persians, who were peasants and settled agriculturalists.[125]

The Safavid economy was to a large extent based on agriculture and taxation of agricultural products. According to the French jeweller Jean Chardin, the variety in agricultural products in Persia was unrivaled in Europe and consisted of fruits and vegetables never even heard of in Europe. Chardin was present at some feasts in Isfahan were there were more than fifty different kinds of fruit. He thought that there was nothing like it in France or Italy:[126]

> "Tobacco grew all over the country and was as strong as that grown in Brazil. Saffron was the best in the world... Melons were regarded as excellent fruit, and there were more than 50 different sorts, the finest of which came from Khorasan. And in spite of being transported for more than thirty days, they were fresh when they reached Isfahan... After melons the finest fruits were grapes and dates, and the best dates were grown in Jahrom."

Despite of this, he was disappointed when travelling the country and witnessing the abundance of land that was not irrigated, or the fertile plains that were not cultivated, something he thought was in starch contrast to Europe. He blamed this on misgovernment, the sparse population of the country, and lack of appreciation of agriculture amongst the Persians.[127]

In the period prior to Shah Abbas I, most of the land was assigned to officials (civil, military and religious). From the time of Shah Abbas onwards, more land was brought under the direct control of the shah. And since agriculture accounted to the by far largest share of tax revenue, he took measures to expand it. What remained unchanged, was the "*crop-sharing agreement*" between whom ever was the landlord, and the peasant. This agreement concisted of five elements: land, water, plough-animals, seed and labour. Each element constituted 20 per cent of the crop production, and if, for instance, the peasant provided the labour force and the animals, he would be entitled to 40 per cent of the earnings.[128] [129] According to contemporary historians, though, the landlord always had the worst of the bargain with the peasant in the crop-sharing agreements. In general, the peasants lived in comfort, and they were well paid and wore good clothes, although it was also notet that they were subject to forced labour and lived under heavy demands.[130]

Travel and Caravanserais

The Mothers Inn caravanserai in Isfahan, that was built during the reign of Shah Abass II, was a luxury resort meant for the walthiest merchants and selected guests of the shah. Today it is a luxury hotel and goes under the name of Hotel Abassi.

Horses were the most important of all the domestic animals, and the best were brought in from Arabia and Central-Asia. They were costly because of the widespread trade in them, including to Turkey and India. The next most important mount, when travelling through Persia, was the mule. Also, the camel was a good investment for the merchant, as they cost nearly nothing to feed, carried a lot weight and could travel almost anywhere.[131]

Under the governance of the strong shahs, especially during the first half of the 17th century, travelling through Persia was easy because of good roads and the caravanserais, that were strategically placed along the route. Thévenot and Tavernier commented that the Persian caravanserais were better built and cleaner than their Turkish counterparts.[132] According to Chardin, they were also more abundant than in the Mughal or Ottoman Empires, where they were less frequent but larger.[133] Caravanserais were designed especially to benefit poorer travellers, as they could stay there for as long as they wished, without payment for lodging. During the reign of Shah Abbas I, as he tried to upgrade the Silk route to improve the commersial prosperity of the Empire, an abundance of caravanserais, bridges, bazaars and roads were built, and this strategy was followed by wealthy merchants who also profitted from the increase in trade. To uphold the standard, another source of revenue was needed, and road toll, that were collected by guards (*rah-dars*), were stationed along the trading routes. They in turn provided for the safety of the travellers, and both Thevenot and Tavernier stressed the safety of travelling in 17th century Persia, and the courtesy and refinement of the policing guards.[134] The Italian traveller Pietro Della Valle was impressed by an encounter with one of these road guards:[135]

> He examined our baggage, but in the most obliging manner possible, not opening our trunks or packages, and was satisfied with a small tax, which was his due...

Foreign trade and The Silk Route

The Chehel Sotoun Palace in Isfahan was where the Shah would meet foreign dignitaries and embassies. It is famous for the frescos that cover its walls.

The Portuguese Empire and the discovery of the trading route around the Cape of Good Hope in 1487 not only hit a death blow to Venice as a trading nation, but it also hurt the trade that was going on along the Silk Route and especially the Persian Gulf. They correctly identified the three key points to control all seaborn trade between Asia and Europe; The Gulf of Aden, The Persian Gulf and the Straits of Malacca by cutting off and controlling these strategic locations with high taxation.[136] In 1602 Shah Abbas I drove the Portuguese out of Bahrain, but he needed naval assistance from the newly arrived British East India Company to finally expel them from the Strait of Hormuz and regain control of this trading route.[137] He convinced the British to assist him by allowing them to open factories in Shiraz, Isfahan and Jask.[138] [139] With the later end of the Portuguese Empire, the British, Dutch and French in particular gained easier access to Persian sea-born trade. Although they, unlike the Portuguese, did not arrive as colonisers, but as merchant adventurers. The terms of trade were not imposed on the Safavid shahs, but rather negotiated.

In the long term, however, the sea-born trade route was of less significance to the Persians than was the traditional Silk Route. Lack of investment in ship building and the navy, provided the Europeans with the opportunity to monopolize this trading route. The land-born trade would thus continue to provide the bulk of revenues to the Persian state. Much of the cash revenue came not so much from what could be sold abroad, as from the custom charges and transit dues levied on goods passing through the country.[140] Shah Abbas was determined to greatly expand this trade, but faced the problem of having to deal with the Ottomans, who controlled the two most vital routes: the route across Arabia to the Mediterranean ports, and the route through Anatolia and Istanbul. A third route was therefore devised which circumvented Ottoman territory. By travelling across the Caspian sea to the north, they would reach Russia. And with the assistans of the Muscovy Company they could cross over to Moscow, reaching Europe via Poland. This trading route proved to be of vital importance, especially during times of war with the Ottomans.[141]

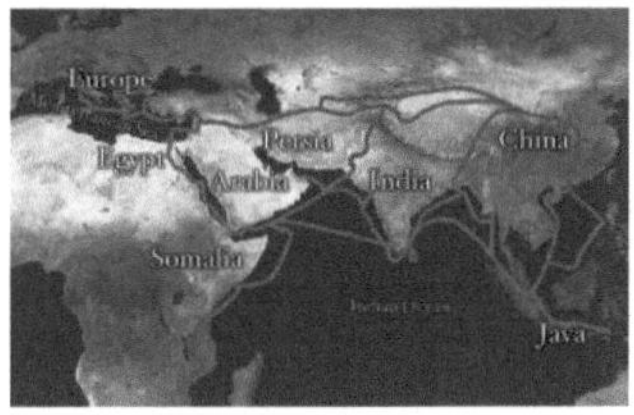

The Silk Routes

By the end of the 17th century, the Dutch had become dominant in the trade that went via The Persian Gulf, having won most trade agreements and managed to strike deals before the British or French were able to. They particularly established monopoly of the spice trade between the East Indies and Iran.[142]

The Armenian merchants and the trade of silk

The one valuable item, sought for in Europe, which Iran possessed and which could bring in silver in sufficient quantities was silk, which was produced in the northern provinces, along the Caspian coastline. The trade of this product was done by Turks and Persians to begin with, but during the 17th century the Christian Armenians became increasingly vital in the trade of this merchandise, as middlemen.[143]

The Vank Cathedral. The Armenians moved into the Jolfa district of Isfahan and were free to build their prayer houses, eventually becoming an integral part of the society.

Whereas domestic trade was largely in the hands of Persian and Jewish merchants, by late 17th century, almost all foreign trade was controlled by the Armenians.[144] They were even hired by wealthy Persian merchants to travel to Europe when they wanted to create commercial bases there, and the Armenians eventually established themselves in cities like Bursa, Aleppo, Venice, Livorno, Marseilles and Amsterdam.[143] Realizing this, Shah Abbas resettled large numbers of Armenians from the Caucasus to his capital city and provided them with loans.[143] And as the shah realized the importance of doing trade with the Europeans, he assured that the Safavid society was one with religious tolerance. The Christian Armenians thus became a commercial elite in the Safavid society and managed to survive in the tough atmosphere of business being fought over by the British, Dutch, French, Indians and Persians, by always having large capital readily available and by managing to strike harder bargains ensuring cheaper prices than what, for instance, their British rivals ever were able to.[145]

Culture

Culture within the Safavid family

Naqshe Jahan square in Isfahan is the epitome of 16th-century Iranian architecture. .

The Safavid family was a literate family from its early origin. There are extant Tati and Persian poetry from Shaykh Safi ad-din Ardabili as well as extant Persian poetry from Shaykh Sadr ad-din. Most of the extant poetry of Shah Ismail I is in Azerbaijani pen-name of Khatai.[50] Sam Mirza, the son of Shah Esmail as well as some later authors assert that Ismail composed poems both in Turkish and Persian but only a few specimens of his Persian verse have survived.[49] A collection of his poems in Azeri were published as a Divan. Shah Tahmasp who has composed poetry in Persian was also a painter, while Shah Abbas II was known as a poet, writing Azerbaijani verses.[146] Sam Mirza, the son of Ismail I was himself a poet and composed his poetry in Persian. He also compiled an anthology of contemporary poetry.[147]

Culture within the empire

Shah Abbas I recognized the commercial benefit of promoting the arts - artisan products provided much of Iran's foreign trade. In this period, handicrafts such as tile making, pottery and textiles developed and great advances were made in miniature painting, bookbinding, decoration and calligraphy. In the 16th century, carpet weaving evolved from a nomadic and peasant craft to a well-executed industry with specialization of design and manufacturing. Tabriz was the center of this industry. The carpets of Ardabil were commissioned to commemorate the Safavid dynasty. The elegantly baroque yet famously 'Polonaise' carpets were made in Iran during the 17th century.

19th century painting of the Chahar Bagh School in Isfahan

Using traditional forms and materials, Reza Abbasi (1565–1635) introduced new subjects to Persian painting — semi-nude women, youth, lovers. His painting and calligraphic style influenced Iranian artists for much of the Safavid period, which came to be known as the Isfahan school. Increased contact with distant cultures in the 17th century, especially Europe, provided a boost of inspiration to Iranian artists who adopted modeling, foreshortening, spatial recession, and the medium of oil painting (Shah Abbas II sent Zaman to study in Rome). The epic Shahnameh ("Book of Kings"), a stellar example of manuscript illumination and calligraphy, was made during Shah Tahmasp's reign. (This book was written by Ferdousi in 1000 AD for Sultan Mahmood Ghaznawi) Another manuscript is the Khamsa by Nizami executed 1539-43 by Aqa Mirak and his school in Isfahan.

Isfahan bears the most prominent samples of the Safavid architecture, all constructed in the years after Shah Abbas I permanently moved the capital there in 1598: the Imperial Mosque, Masjid-e Shah, completed in 1630, the Imam Mosque (Masjid-e Imami) the Lutfallah Mosque and the Royal Palace.

According to William Cleveland and Martin Bunton,[148] the establishment of Isfahan as the Great capital of Persia and the material splendor of the city attracted intellecutal's from all corners of the world, which contributed to the cities rich cultural life. The impressive achievements of its 400 000 residents prompted the inhabitants to coin their famous boast, "Isfahan is half the world".

The battle between Shah Ismail and Abul-khayr Khan.

Poetry stagnated under the Safavids; the great medieval ghazal form languished in over-the-top lyricism. Poetry lacked the royal patronage of other arts and was hemmed in by religious prescriptions.

The arguably most renowned historian from this time was Iskandar Beg Munshi. His *History of Shah Abbas the Great* written a few years after its subject's death, achieved a nuanced depth of history and character.

The Isfahan School – Islamic philosophy revived

Islamic philosophy[149] flourished in the Safavid era in what scholars commonly refer to the School of Isfahan. Mir Damad is considered the founder of this school. Among luminaries of this school of philosophy, the names of Iranian philosophers such as Mir Damad, Mir Fendereski, Shaykh Bahai and Mohsen Fayz Kashani standout. The school reached its apogee with that of the Iranian philosopher Mulla Sadra who is arguably the most significant Islamic philosopher after Avicenna. Mulla Sadra has become the dominant philosopher of the Islamic East, and his approach to the nature of philosophy has been exceptionally influential up to this day.[150] He wrote the *Al-Hikma al-muta'aliya fi-l-asfar al-'aqliyya al-arba'a ("The Transcendent Philosophy of the Four Journeys of the Intellect"),*[151] *a meditation on what he called 'meta philosophy' which brought to a synthesis the philosophical mysticism of Sufism, the theology of Shi'a Islam, and the Peripatetic and Illuminationist philosophies of Avicenna and Suhrawardi.*

According to the Iranologist Richard Nelson Frye:[152]

> They were the continuers of the classical tradition of Islamic thought, which after Averroes died in the Arab west. The Persians schools of thought were the true heirs of the great Islamic thinkers of the golden age of Islam, whereas in the Ottoman empire there was an intellectual stagnation, as far as the traditions of Islamic philosophy were concerned.

Medicine

The status of physicians during the Safavids stood as high as ever. Whereas neither the ancient Greeks nor the Romans accorded high social status to their doctors, Iranians had from ancient times honored their physicians, who were often appointed counselors of the Shahs. This would not change with the Arab conquest of Iran, and it was primarily the Persians that took upon them the works of philosophy, logic, medicine, mathematics, astronomy, astrology, music and alchemy.[153]

A Latin copy of *The Canon of Medicine*, dated 1484, located at the P.I. Nixon Medical Historical Library of The University of Texas Health Science Center at San Antonio, USA.

By the sixteenth century, Islamic science, which to a large extent meant Persian science, was resting on its laurels. The works of al-Razi (865-92) (known to the West as Razes) were still used in European universities as standard textbooks of alchemy, pharmacology and pediatrics. The Canon of Medicine by Avicenna (c. 980–1037) was still regarded as one of the primary textbooks in medicine throughout most of the civilized world.[154] As such, the status of medicine in the Safavid period did not change much, and relied as much on these works as ever before. Physiology was still based on the four humours of ancient and mediaeval medicine, and bleeding and purging were still the principal forms of therapy by surgeons, something even Thevenot experienced during his visit to Persia.[118]

The only field within medicine where some progress were made was pharmacology, with the compilement of the "Tibb-e Shifa'i" in 1556. This book was translated into French in 1681 by Angulus de Saint, under the name "Pharmacopoea Persica".[155]

Isfahan is Half the World

The architectural legacy of the Safavids

Painting by the French architect, Pascal Coste, visiting Persia in 1841 (from *Monuments modernes de la Perse*). In the Safavid era the Persian architecture flourished again and saw many new monuments, such as the Masjid-e Shah, part of Naghsh-i Jahan Square which is the biggest historic plaza in the world.

A new age in Iranian architecture began with the rise of the Safavid dynasty. Economically robust and politically stable, this period saw a flourishing growth of theological sciences. Traditional architecture evolved in its patterns and methods leaving its impact on the architecture of the following periods.

Indeed, one of the greatest legacies of the Safavids is the architecture. In 1598, when Shah Abbas decided to move the capital of his Persian empire from the north-western city of Qazvin to the central city of Isfahan, he initiated what would become one of the greatest programmes in Persian history; the complete remaking of the city. By choosing the central city of Isfahan, fertilized by the Zāyande roud ("The *life-giving river*"), lying as an oasis of intense cultivation in the midst of a vast area of arid landscape, he both distanced his capital from any future assaults by the Ottomans and the Uzbeks, and at the same time gained more control over the Persian Gulf, which had recently become an important trading route for the Dutch and British East India Companies.[156]

Safavid Star from ceiling of Shah Mosque, Isfahan, Iran.

The 16th-century Chehel Sotun pavilion in Qazvin, Iran. It is the last remains of the palace of the second Safavid king, Shah Tahmasp; it was heavily restored by the Qajars in the 19th century.

The Chief architect of this colossal task of urban planning was Shaykh Bahai (Baha' ad-Din al-`Amili), who focused the programme on two key features of Shah Abbas's master plan: the Chahar Bagh avenue, flanked at either side by all the prominent institutions of the city, such as the residences of all foreign dignitaries. And the Naqsh-e Jahan Square ("*Examplar of the World*").[157] Prior to the Shah's ascent to power, Persia had a decentralized power-structure, in which different institutions battled for power, including both the military (the Qizilbash) and governors of the different provinces making up the empire. Shah Abbas wanted to undermine this political structure, and the recreation of Isfahan, as a Grand capital of Persia, was an important step in centralizing the power.[158] The ingenuity of the square, or *Maidān*, was that, by building it, Shah Abbas would gather the three main components of power in Persia in his own backyard; the power of the clergy, represented by the Masjed-e Shah, the power of the merchants, represented by the Imperial Bazaar, and of course, the power of the Shah himself, residing in the Ali Qapu Palace.

Distinctive monuments like the Sheikh Lotfallah (1618), Hasht Behesht (Eight Paradise Palace) (1469) and the Chahar Bagh School(1714) appeared in Isfahan and other cities. This extensive development of architecture was rooted in Persian culture and took form in the design of schools, baths, houses, caravanserai and other urban spaces such as bazaars and squares. It continued until the end of the Qajar reign.[159]

The languages of the court, military, administration and culture

The Safavids by the time of their rise were Azerbaijani-speaking although they also used Persian as a second language. The language chiefly used by the Safavid court and military establishment was Azerbaijani.[11] [14] But the official[5] language of the empire as well as the administrative language, language of correspondence, literature and historiography was Persian.[11] The inscriptions on Safavid currency were also in Persian.[160]

Safavids also used Persian as a cultural and administrative language throughout the empire and were bilingual in Persian.[50] According to Arnold J. Toynbee,[161]

> in the heyday of the Mughal, Safawi, and Ottoman regimes New Persian was being patronized as the language of litterae humaniores by the ruling element over the whole of this huge realm, while it was also being employed as the official language of administration in those two-thirds of its realm that lay within the Safawi and the Mughal frontiers

Scene from Attar's *The Conference of the Birds*, by Habibulla Meshedi (1600).

According to John R. Perry,[162]

> In the 16th century, the Turcophone Safavid family of Ardabil in Azerbaijan, probably of Turkicized Iranian, origin, conquered Iran and established Turkic, the language of the court and the military, as a high-status vernacular and a widespread contact language, influencing spoken Persian, while written Persian, the language of high literature and civil administration, remained virtually unaffected in status and content.

According to Zabiollah Safa,[14]

> In day-to-day affairs, the language chiefly used at the Safavid court and by the great military and political officers, as well as the religious dignitaries, was Turkish, not Persian; and the last class of persons wrote their religious works mainly in Arabic. Those who wrote in Persian were either lacking in proper tuition in this tongue, or wrote outside Iran and hence at a distance from centers where Persian was the accepted vernacular, endued with that vitality and susceptibility to skill in its use which a language can have only in places where it truly belongs.

According to É. Á. Csató et al.,[34]

Prince Muhammad-Beik of Georgia by Reza Abbasi (1620)

> A specific Turkic language was attested in Safavid Persia during the 16th and 17th centuries, a language that Europeans often called Persian Turkish ("Turc Agemi", "lingua turcica agemica"), which was a favourite language at the court and in the army because of the Turkic origins of the Safavid dynasty. The original name was just turki, and so a convenient name might be Turki-yi Acemi. This variety of Persian Turkish must have been also spoken in the Caucasian and Transcaucasian regions, which during the 16th century belonged to both the Ottomans and the Safavids, and were not fully integrated into the Safavid empire until 1606. Though that language might generally be identified as Middle Azerbaijanian, it's not yet possible to define exactly the limits of this language, both in linguistic and territorial respects. It was certainly not homogenous - maybe it was an Azerbaijanian-Ottoman mixed language, as Beltadze (1967:161) states for a translation of the gospels in Georgian script from the 18th century.

According to Rula Jurdi Abisaab,[163]

> Although the Arabic language was still the medium for religious scholastic expression, it was precisely under the Safavids that hadith complications and doctrinal works of all sorts were being translated to Persian. The 'Amili (Lebanese scholars of Shi'i faith) operating through the Court-based religious posts, were forced to master the Persian language; their students translated their instructions into Persian. Persianization went hand in hand with the popularization of 'mainstream' Shi'i belief.

According to Cornelis Versteegh,[164]

> The Safavid dynasty under Shah Ismail (961/1501) adopted Persian and the Shi'ite form of Islam as the national language and religion.

Legacy

It was the Safavids who made Iran the spiritual bastion of Shi'ism against the onslaughts of Sunni Islam, and the repository of Persian cultural traditions and self-awareness of Iranianhood, acting as a bridge to modern Iran. The founder of the dynasty, Shah Isma'il, adopted the title of "Persian Emperor" *Pādišah-ī Īrān*, with its implicit notion of an Iranian state stretching from Khorasan as far as Euphrates, and from the Oxus to the southern Territories of the Persian Gulf.[165] According to Professor Roger Savory:[166] [167]

> In a number of ways the Safavids affected the development of the modern Iranian state: first, they ensured the continuance of various ancient and traditional Persian institutions, and transmitted these in a strengthened, or more 'national', form; second, by imposing Ithna 'Ashari Shi'a Islam on Iran as the official religion of the Safavid state, they enhanced the power of mujtahids. The Safavids thus set in train a struggle for power between the turban and the crown that is to say, between the proponents of secular government and the proponents of a theocratic government; third, they laid the foundation of alliance between the religious classes ('Ulama') and the bazaar which played an important role both in the Persian Constitutional Revolution of 1905–1906, and again in the Islamic Revolution of 1979; fourth the policies introduced by Shah Abbas I conduced to a more centralized administrative system.

Safavid Shahs of Iran

- Ismail I 1501–1524
- Tahmasp I 1524–1576
- Ismail II 1576–1578
- Mohammed Khodabanda 1578–1587
- Abbas I 1587–1629
- Safi 1629–1642
- Abbas II 1642–1666
- Suleiman I 1666–1694
- Sultan Hoseyn I 1694–1722
- Tahmasp II 1722–1732
- Abbas III 1732–1736

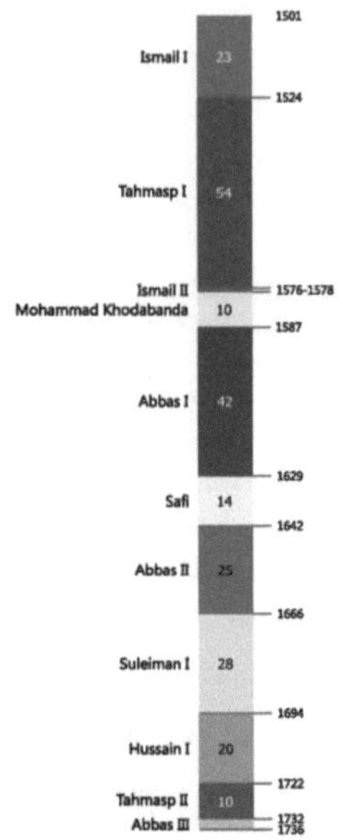

Safavid dynasty timeline

See also

- Persianate states
- List of Shi'a Muslim dynasties
- Safavid conversion of Iran from Sunnism to Shiism

References and notes

[1] *Safavid dynasty* (http://www.britannica.com/EBchecked/topic/516019/Safavid-dynasty). Britannica. .

[2] *Safavid Persia* (http://books.google.com/books?q="Safavid+Persia"). Books. Google. .

[3] *...the Order of the Lion and the Sun, a device which, since the 17 century at least, appeared on the national flag of the Safavids the lion representing 'Ali and the sun the glory of the Shi'i faith*, Mikhail Borisovich Piotrovskiĭ, J. M. Rogers, Hermitage Rooms at Somerset House, Courtauld Institute of Art, *Heaven on earth: Art from Islamic Lands : Works from the State Hermitage Museum and the Khalili Collection*, Prestel, 2004, p. 178.

[4] Ingvild Flaskerud (26 November 2010). *Visualizing Belief and Piety in Iranian Shiism* (http://books.google.com/books?id=fH1pvM0AdNIC&pg=PA182). Continuum International Publishing Group. pp. 182–183. ISBN 978-1-4411-4907-7. . Retrieved 24 July 2011.

[5] Roemer, H. R. (1986). "The Safavid Period". *The Cambridge History of Iran*, Vol. 6: The Timurid and Safavid Periods. Cambridge: Cambridge University Press, pp. 189–350. ISBN 0-521-20094-6, p. 331: "Depressing though the condition in the country may have been at the time of the fall of Safavids, they cannot be allowed to overshadow the achievements of the dynasty, which was in many respects to prove essential factors in the development of Persia in modern times. These include the maintenance of Persian as the official language and of the present-day boundaries of the country, adherence to the Twelever Shi'i, the monarchical system, the planning and architectural features of the urban centers, the centralised administration of the state, the alliance of the Shi'i Ulama with the merchant bazaars, and the symbiosis of the Persian-speaking population with important non-Persian, especially Turkish speaking minorities".

[6] Rudi Matthee, " Safavids (http://www.iranicaonline.org/articles/safavids)" in *Encyclopædia Iranica*, accessed on April 4, 2010. "The Persian focus is also reflected in the fact that theological works also began to be composed in the Persian language and in that Persian verses replaced Arabic on the coins." "The political system that emerged under them had overlapping political and religious boundaries and a core language, Persian, which served as the literary tongue, and even began to replace Arabic as the vehicle for theological discourse".

[7] Ronald W Ferrier, *The Arts of Persia*. Yale University Press. 1989, p. 9.

[8] John R Perry, "Turkic-Iranian contacts", *Encyclopædia Iranica*, January 24, 2006: "...written Persian, the language of high literature and civil administration, remained virtually unaffected in status and content"

[9] Cyril Glassé (ed.), *The New Encyclopedia of Islam*, Lanham, Maryland: Rowman & Littlefield Publishers, revised ed., 2003, ISBN 0-7591-0190-6, p. 392: "Shah Abbas moved his capital from Qazvin to Isfahan. His reigned marked the peak of Safavid dynasty's achievement in art, diplomacy, and commerce. It was probably around this time that the court, which originally spoke a Turkic language, began to use Persian"

[10] Arnold J. Toynbee, *A Study of History*, V, pp. 514-15. excerpt: "in the heyday of the Mughal, Safawi, and Ottoman regimes New Persian was being patronized as the language of *literae humaniores* by the ruling element over the whole of this huge realm, while it was also being employed as the official language of administration in those two-thirds of its realm that lay within the Safawi and the Mughal frontiers"

[11] Mazzaoui, Michel B; Canfield, Robert (2002). "Islamic Culture and Literature in Iran and Central Asia in the early modern period" (http://books.google.com/books?id=qwwoozMU0LMC&pg=PA86#PPA87,M1). *Turko-Persia in Historical Perspective*. Cambridge University Press. pp. 86–7. ISBN 0-521-52291-9, ISBN 978-0-521-52291-5. . "Safavid power with its distinctive Persian-Shi'i culture, however, remained a middle ground between its two mighty Turkish neighbors. The Safavid state, which lasted at least until 1722, was essentially a "Turkish" dynasty, with Azeri Turkish (Azerbaijan being the family's home base) as the language of the rulers and the court as well as the Qizilbash military establishment. Shah Ismail wrote poetry in Turkish. The administration nevertheless was Persian, and the Persian language was the vehicle of diplomatic correspondence (insha'), of belles-lettres (adab), and of history (tarikh)."

[12] Ruda Jurdi Abisaab. "Iran and Pre-Independence Lebanon" in Houchang Esfandiar Chehabi, *Distant Relations: Iran and Lebanon in the Last 500 Years*, IB Tauris 2006, p. 76: "Although the Arabic language was still the medium for religious scholastic expression, it was precisely under the Safavids that hadith complications and doctrinal works of all sorts were being translated to Persian. The 'Amili (Lebanese scholars of Shi'i faith) operating through the Court-based religious posts, were forced to master the Persian language; their students translated their instructions into Persian. Persianization went hand in hand with the popularization of 'mainstream' Shi'i belief."

[13] Savory, Roger (2007). *Iran Under the Safavids* (http://books.google.com/books?id=v4Yr4foWFFgC&pg=PA213). Cambridge University Press. p. 213. ISBN 0-521-04251-8, ISBN 978-0-521-04251-2. . "*qizilbash normally spoke Azari brand of Turkish at court, as did the Safavid shahs themselves; lack of familiarity with the Persian language may have contributed to the decline from the pure classical standards of former times*"

[14] Zabiollah Safa (1986), "Persian Literature in the Safavid Period", *The Cambridge History of Iran*, vol. 6: The Timurid and Safavid Periods. Cambridge: Cambridge University Press, ISBN 0-521-20094-6, pp. 948–65. P. 950: "In day-to-day affairs, the language chiefly used at the Safavid court and by the great military and political officers, as well as the religious dignitaries, was Turkish, not Persian; and the last class of persons wrote their religious works mainly in Arabic. Those who wrote in Persian were either lacking in proper tuition in this tongue, or wrote outside Iran and hence at a distance from centers where Persian was the accepted vernacular, endued with that vitality and susceptibility to skill in its use which a language can have only in places where it truly belongs."

[15] Price, Massoume (2005). *Iran's Diverse Peoples: A Reference Sourcebook* (http://books.google.com/books?id=gzpdq679oJwC&pg=PA66). ABC-CLIO. p. 66. ISBN 1-57607-993-7, ISBN 978-1-57607-993-5. . "*The Shah was a native Turkic speaker and wrote poetry in the Azerbaijani language.*"

[16] *The New Encyclopedia of Islam*, Ed. Cyril Glassé, (Rowman & Littlefield Publishers, 2008), 449.

[17] Ferrier, RW, *A Journey to Persia: Jean Chardin's Portrait of a Seventeenth-century Empire*, p. ix.

[18] Helen Chapin Metz. *Iran, a Country study*. 1989. University of Michigan, p. 313.

[19] Emory C. Bogle. *Islam: Origin and Belief*. University of Texas Press. 1989, p. 145.

[20] Stanford Jay Shaw. History of the Ottoman Empire. Cambridge University Press. 1977, p. 77.

[21] Andrew J. Newman, Safavid Iran: *Rebirth of a Persian Empire*, IB Tauris (March 30, 2006).

[22] RM Savory, *Safavids, Encyclopedia of Islam*, 2nd ed.

[23] "Peoples of Iran" (http://www.iranicaonline.org/articles/iran-v1-peoples-survey) *Encyclopædia Iranica*. RN Frye.

[24] RM Savory. Ebn Bazzaz. (http://www.iranicaonline.org/articles/ebn-bazzaz) *Encyclopædia Iranica*

[25] Peter B. Golden: An Introduction to the History of the Turkic Peoples; In: Osman Karatay, Ankara 2002, p.321

[26] Aptin Khanbaghi (2006) *The Fire, the Star and the Cross: Minority Religions in Medieval and Early*. London & New York. IB Tauris. ISBN 1-84511-056-0, pp. 130-1

[27] Anthony Bryer. "Greeks and Türkmens: The Pontic Exception", *Dumbarton Oaks Papers, Vol. 29* (1975), Appendix II - Genealogy of the Muslim Marriages of the Princesses of Trebizond

[28] *Why is there such confusion about the origins of this important dynasty, which reasserted Iranian identity and established an independent Iranian state after eight and a half centuries of rule by foreign dynasties?* RM Savory, *Iran under the Safavids* (Cambridge University Press, Cambridge, 1980), p. 3.

[29] Alireza Shapur Shahbazi (2005), "The History of the Idea of Iran", in Vesta Curtis ed., Birth of the Persian Empire, IB Tauris, London, p. 108: "Similarly the collapse of Sassanian Eranshahr in AD 650 did not end Iranians' national idea. The name "Iran" disappeared from official records of the Saffarids, Samanids, Buyids, Saljuqs and their successor. But one unofficially used the name Iran, Eranshahr, and similar national designations, particularly Mamalek-e Iran or "Iranian lands", which exactly translated the old Avestan term Ariyanam Daihunam. On the other hand, when the Safavids (not Reza Shah, as is popularly assumed) revived a national state officially known as Iran, bureaucratic usage in the Ottoman empire and even Iran itself could still refer to it by other descriptive and traditional appellations".

[30] In the pre-Safavid written work Safvat as-Safa (oldest manuscripts from 1485 and 1491), the origin of the Safavids is tracted to Piruz Shah Zarin Kolah who is called a Kurd from Sanjan, while in the post-Safavid manuscripts, this portion has been excised and Piruz Shah Zarin Kollah is made a descendant of the Imams. R Savory, "Ebn Bazzaz" in *Encyclopædia Iranica*). In the Silsilat an-nasab-i Safawiya (composed during the reign of Shah Suleiman, 1667–94), by Hussayn ibn Abdal Zahedi, the ancestry of the Safavid was purported to be tracing back to Hijaz and the first Shi'i Imam as follows: Shaykh Safi al-din Abul Fatah Eshaq ibn (son of) Shaykh Amin al-Din Jabrail ibn Qutb al-din ibn Salih ibn Muhammad al-Hafez ibn Awad ibn Firuz Shah Zarin Kulah ibn Majd ibn Sharafshah ibn Muhammad ibn Hasan ibn Seyyed Muhammad ibn Ibrahim ibn Seyyed Ja'afar ibn Seyyed Muhammad ibn Seyyed Isma'il ibn Seyyed Muhammad ibn Seyyed Ahmad 'Arabi ibn Seyyed Qasim ibn Seyyed Abul Qasim Hamzah ibn Musa al-Kazim ibn Ja'far As-Sadiq ibn Muhammad al-Baqir ibn Imam Zayn ul-'Abedin ibn Hussein ibn Ali ibn Abi Taleb Alayha as-Salam. There are differences between this and the oldest manuscript of Safwat as-Safa. Seyyeds have been added from Piruz Shah Zarin Kulah up to the first Shi'i Imam and the nisba "Al-Kurdi" has been excised. The title/name "Abu Bakr" (also the name of the first Caliph and highly regarded by Sunnis) is deleted from Qutb ad-Din's name. Ŝource: Husayn ibn Abdāl

Zāhedī, 17th cent. Silsilat al-nasab-i Safavīyah, nasabnāmah-'i pādishāhān bā ‘uzmat-i Safavī, ta'līf-i Shaykh Husayn pisar-i Shaykh Abdāl Pīrzādah Zāhedī dar 'ahd-i Shāh-i Sulaymnān-i Safavī. Berlīn, Chāpkhānah-'i Īrānshahr, 1343 (1924), 116 pp. Original Persian: یفص خیش نبا ضوع نبا ظفاحلا دمحم نبا حلاص نبا نیدلا بطق نب لیئاربج نیدلا نیما خیش نبا قحسا حتفلا وبا نیدلا دمحم دیس نب رفعج دیس نبا میهاربا نبا دمحم دیس نبا نسح نبا دمحم نبا هاشفرش نبا دمحم نبا هال نیرز هاشزوریف رفعج نبا مظا ل ا یسوم نب هزمح مساقلا وبا دیس نب مساق دیس نب یبارعا دمحا دیس نب دمحم دیس نب لیعمسا دیس نبا مالسلا. هیلع بلاطیبا نبا یلع نبا نیسح نب نیدباعلا نیز ماما نبا رقابلا دمحم نبا قداصلا

[31] R.M. Savory, "Safavid Persia" in: Ann Katherine Swynford Lambton, Peter Malcolm Holt, Bernard Lewis, *The Cambridge History of Islam*, Cambridge University Press, 1977. p. 394: "They (Safavids after the establishment of the Safavid state) fabricated evidence to prove that the Safavids were Sayyids."

[32] F. Daftary, "Intellectual Traditions in Islam", I.B.Tauris, 2001. p. 147: "But the origins of the family of Shaykh Safi al-Din go back not to Hijaz but to Kurdistan, from where, seven generations before him, Firuz Shah Zarin-kulah had migrated to Adharbayjan"

[33] Tamara Sonn. *A Brief History of Islam*, Blackwell Publishing, 2004, p. 83, ISBN 1-4051-0900-9

[34] É. Á. Csató, B. Isaksson, C Jahani. *Linguistic Convergence and Areal Diffusion: Case Studies from Iranian, Semitic and Turkic*, Routledge, 2004, p. 228, ISBN 0-415-30804-6.

[35] Minorsky, V (2009). "Adgharbaydjan (Azarbaydjan)" (http://www.encislam.brill.nl/). In Berman, P; Bianquis, Th; Bosworth, CE et al.. *Encyclopedia of Islam* (2nd ed.). NL: Brill. . "After 907/1502, Adharbayjan became the chielf bulwark and rallying ground of the Safawids, themselves natives of Ardabil and originally speaking the local Iranian dialect"

[36] Roger M. Savory. "Safavids" in Peter Burke, Irfan Habib, Halil İnalcık: *History of Humanity-Scientific and Cultural Development: From the Sixteenth to the Eighteenth Century*, Taylor & Francis. 1999, p. 259: "From the evidence available at the present time, it is certain that the Safavid family was of indigenous Iranian stock, and not of Turkish ancestry as it is sometimes claimed. It is probable that the family originated in Persian Kurdistan, and later moved to Azerbaijan, where they adopted the Azari form of Turkish spoken there, and eventually settled in the small town of Ardabil sometimes during the eleventh century."

[37] E. Yarshater, "Iran", . *Encyclopædia Iranica*. "The origins of the Safavids are clouded in obscurity. They may have been of Kurdish origin (see R. Savory, Iran Under the Safavids, 1980, p. 2; R. Matthee, "Safavid Dynasty" at iranica.com), but for all practical purposes they were Turkish-speaking and Turkified. "

[38] John L. Esposito, *The Oxford History of Islam*, Oxford University Press US, 1999. pp 364: "To support their legitimacy, the Safavid dynasty of Iran (1501-1732) devoted a cultural policy to estbalish their regime as the reconstruction of the historic Iranian monarchy. To the end, they commisioned [*sic?*] elaborate copies of the Shahnameh, the Iranian national epic, such as this one made for Tahmasp in the 1520s."

[39] Ira Marvin Lapidus, *A history of Islamic Societies*, Cambridge University Press, 2002, 2nd edition. pg 445: To bolster the prestige of the state, the Safavid dynasty sponsered an Iran-Islamic style of culture concentrating on court poetry, painting, and monumental architecture that symbolized not only the Islamic credentials of the state but also the glory of the ancient Persian traditions."

[40] Colin P. Mitchell, "ṬAHMĀSP I" in *Encyclopædia Iranica*. "Shah Ṭahmāsp's own brother, Sām Mirzā, wrote the Taḏkera-yetoḥfa-ye sāmi, in which he mentioned 700 poets during the reigns of the first two Safavid rulers. Sām Mirzā himself was an ardent poet, writing 8,000 verses and a Šāh-nāma dedicated to his brother, Ṭahmāsp (see Sām Mirzā, ed. Homāyun-Farroḵ, 1969)."

[41] See: Willem Floor, Hasan Javadi(2009), *The Heavenly Rose-Garden: A History of Shirvan & Daghestan* by Abbas Qoli Aqa Bakikhanov, Mage Publishers, 2009.(see Sections on Safavids quoting poems of Shah Tahmasp I)

[42] Kathryn Babayan, *Mystics, Monarchs and Messiahs: Cultural Landscapes of Early Modern Iran*, Cambridge, Massachusetts ; London : Harvard University Press, 2002. p. 143: "It is true that during their revolutionary phase (1447-1501), Safavi guides had played on their descent from the family of the Prophet. The hagiography of the founder of the Safavi order, Shaykh Safi al-Din Safvat al-Safa written by Ibn Bazzaz in 1350-was tampered with during this very phase. An initial stage of revisions saw the transformation of Safavi identity as Sunni Kurds into Arab blood descendants of Muhammad."

[43] "The Iranian Language of Azerbaijan" (http://www.iranicaonline.org/articles/azerbaijan-vii) E. Yarshater, *Encyclopædia Iranica*

[44] Ehsan Yarshater, *Encyclopædia Iranica*, Book 1, p. 240 (http://www.iranicaonline.org/articles/azerbaijan-vii).

[45] Peter Charanis. "Review of Emile Janssens' *Trébizonde en Colchide*", *Speculum*, Vol. 45, No. 3 (July 1970), p. 476.

[46] Anthony Bryer, *open citation*, p. 136.

[47] Virani, Shafique N. *The Ismailis in the Middle Ages: A History of Survival, A Search for Salvation* (New York: Oxford University Press), 2007, p.113.

[48] The writer Ṛūmlu documented the most important of them in his history.

[49] "Ismail Safavi" *Encyclopædia Iranica*

[50] V. Minorsky, "The Poetry of Shāh Ismā'īl I", *Bulletin of the School of Oriental and African Studies*, University of London 10/4 (1942): 1006–53.

[51] Richard Tapper. "Shahsevan in Safavid Persia", *Bulletin of the School of Oriental and African Studies*, University of London, Vol. 37, No. 3, 1974, p. 324.

[52] Lawrence Davidson, Arthur Goldschmid, *A Concise History of the Middle East*, Westview Press, 2006, p. 153.

[53] *Britannica Concise* (http://concise.britannica.com/ebc/article-9377424/Safavid-dynasty). "Safavid Dynasty", Online Edition 2007.

[54] George Lenczowski, "Iran under the Pahlavis", Hoover Institution Press, 1978, p. 79: "Ismail Safavi, descendant of the pious Shaykh Ishaq Safi al-Din (d. 1334), seized Tabriz assuming the title of Shahanshah-e-Iran".

[55] Stefan Sperl, C. Shackle, Nicholas Awde, "Qasida poetry in Islamic Asia and Africa", Brill Academic Pub; Set Only edition (February 1996), p. 193: "Like Shah Ni'mat Allah-i Vali he hosted distinguished visitors among them Ismail Safavi, who had proclaimed himself

Shahanshah of Iran in 1501 after having taken Tabriz, the symbolic and political capital of Iran".
[56] Heinz Halm, Janet Watson, Marian Hill, *Shi'ism*, translated by Janet Watson, Marian Hill, Edition: 2, illustrated, published by Columbia University Press, 2004, p. 80: "...he was able to make his triumphal entry into Alvand's capital Tabriz. Here he assumed the ancient Iranian title of King of Kings (Shahanshah) and setup up Shi'i as the ruling faith"
[57] Virani, Shafique N. *The Ismailis in the Middle Ages: A History of Survival, A Search for Salvation* (New York: Oxford University Press), 2007, 113.
[58] H.R. Roemer, The Safavid Period, in Cambridge History of Iran, Vol. VI, Cambridge University Press 1986, p. 339: "Further evidence of a desire to follow in the line of Turkmen rulers is Ismail's assumption of the title "Padishah-i-Iran"
[59] "Iranian identity iii. Medieval Islamic period" in Encyclopedia Iranica (http://www.iranicaonline.org/articles/iranian-identity-iii-medieval-islamic-period): "The Safavid kings called themselves, among other appellations, the "heart of the shrine of ʿAli" (*kalb-e āstān-e ʿAli*), while assuming the title of *Šāhanšāh* (the king of kings) of Persia/Iran". Quote 2: "Even Ottoman sultans, when addressing the Āq Quyunlu and Safavid kings, used such titles as the "king of Iranian lands" or the "sultan of the lands of Iran" or "the king of kings of Iran, the lord of the Persians" or the "holders of the glory of Jamšid and the vision of Faridun and the wisdom of Dārā." They addressed Shah Esmaʿil as: "the king of Persian lands and the heir to Jamšid and Kay-ḵosrow" (Navāʾi, pp. 578, 700–2, 707). During Shah ʿAbbās's reign (q.v.) the transformation is complete and Shiʿite Iran comes to face the two adjacent Sunni powers: the Ottoman Empire to the west and the Kingdom of Uzbeks to the east."
[60] Savory, R, Iran under the Safavids, pp. 60–4.
[61] Richards, JF, The Mughal Empire, p. 11.
[62] Riazul Islam; *Indo-Persian relations* (Tehran 1970), pp. 22–47.
[63] John F Richards, Gordon Johnson (1996), Cambridge University Press, The Mughal Empire.
[64] Savory, R, Iran under the Safavids, p. 66.
[65] Blow, D; Shah Abbas: The ruthless king who became an Iranian legend, p. 9.
[66] Savory, p. 57.
[67] Savory, pp. 129-31.
[68] DM Lang. "Georgia and the Fall of the Safavi Dynasty", *Bulletin of the School of Oriental and African Studies, University of London, Vol. 14, No. 3, Studies Presented to Vladimir Minorsky by His Colleagues and Friends* (1952), pp. 523–39.
[69] (http://www.iranica.com/articles/v7/v7f4/v7f446.html)
[70] Dzhalilov, O Dzh (1967) (in Russian). *Kurdski geroicheski epos Zlatoruki Khan (The Kurdish heroic epic Gold-hand Khan)*. Moscow. ISBN 0-89158-296-7.
[71] (JPEG) *Islamic Groups* (http://www.lib.utexas.edu/maps/commonwealth/commonwealth_islamic_groups.jpg). University of Texas. .
[72] *Encyclopædia Iranica*, "Abbas I the Great", p. 75.
[73] Laurence Lockhart in *The Legacy of Persia* ed. A. J. Arberry (Oxford University Press, 1953), p. 347.
[74] Nahavandi and Bomati p. 114.
[75] Shakespeare, William (2001-04). *Twelfth Night* (http://books.google.com/?id=DNKKsveMfhQC&pg=PA177&lpg=PA177&dq=the+sophy+twelfth+night#v=onepage&q=the sophy twelfth night&f=false). Books. Eng., UK: Google. ISBN 978-0-7426-5294-1. .
[76] Richard Wilson, "When Golden Time Convents" (http://www.informaworld.com/smpp/content~content=a923903386~db=all~jumptype=rss): Twelfth Night and Shakespeare's Eastern Promise, *Shakespeare*, Volume 6, Issue 2 June 2010, pp. 209–26.
[77] Nahavandi, Bomati pp. 128–30.
[78] Nahavandi, Bomati, pp. 130–7.
[79] Nahavandi, Bomati, pp. 161–2.
[80] Subrahmanyam, Sanjay (1988). "Persians, pilgrims, and Portuguese: The travails of Masulipatnam shipping in the western Indian ocean, 1590-1665". *Modern Asian Studies* **22** (3): 503–530. doi:10.1017/S0026749X00009653.
[81] Dale, Stephen Frederic (1994). *Indian merchants and Eurasian trade, 1600-1750*. Cambridge. pp. 80–100, 122–123.
[82] Kotilaine, Jarmo T. (2005). *Russia's foreign trade and economic expansion in the seventeenth century: Windows on the world*. Leiden. pp. 330–360, 450–485.
[83] Utz, Axel (2011). *Cultural exchange, imperialist violence, and pious missions: Local perspectives from Tanjavur and Lenape country, 1720-1760* (http://search.proquest.com/docview/902171220) (Ph.D. thesis). Pennsylvania State University. pp. 84–85, 93-94. .
[84] Floor, Willem; Clawson, Patrick (2000). "Safavid Iran's search for silver and gold". *International Journal of Middle East Studies* **32** (3): 345–368. doi:10.1017/S0020743800021139.
[85] Mottahedeh, Roy, *The Mantle of the Prophet: Religion and Politics in Iran*, One World, Oxford, 1985, 2000, p. 204.
[86] Axworthy pp.39-55
[87] Blow, David; Shah Abbas: The ruthless king who became an Iranian legend, pp. 37–8.
[88] "Shahsavan Tribes" (http://www.events.ir/no002/002d.htm), Dr P. Shahsavand, Professor of Sociology at Islamic Azad University — *Events* Magazine, Cultural, Economical and General Events of Iran (retrieved 4 Sep 2007).
[89] Blow, p. 38.
[90] Savory, Roger, Iran under the Safavids, p. 183.
[91] Sir E. Denison Ross, Sir Anthony Sherley and his Persian Adventure, pp. 219–20.
[92] Savory, R, Iran under the Safavids, p. 177.

[93] Ferrier, RW, *A journey to Persia: Jean Chardin's portrait of a seventeenth-century empire*, p. 110.
[94] Ferrier, p. 111.
[95] Ferrier; p. 111-113.
[96] Ferrier; p. 114-115.
[97] Ferrier; p. 116.
[98] Ferrier; p. 117- 118.
[99] Ferrier; pp. 120- 124.
[100] Ferrier; p. 124.
[101] Axworthy, Michael; *History of Iran* (2010).
[102] Savory; 184-5.
[103] Savory; Iran under the Safavids; p. 65
[104] Blow; p. 37.
[105] Savory; p. 82.
[106] Savory, pp. 185–6.
[107] Nasr, Vali (2006). *The Shia revival : how conflicts within Islam will shape the future.* Norton. p. 69. ISBN 978-0-393-06211-3
[108] Momen, Moojan (1985). *An introduction to Shi'i Islam : the history and doctrines of Twelver Shi'ism.* G. Ronald. p. 127. ISBN 0-85398-201-5
[109] Momen, Moojan (1985). *An introduction to Shi'i Islam : the history and doctrines of Twelver Shi'ism.* G. Ronald. p. 222. ISBN 0-85398-201-5
[110] Momen, Moojan (1985). *An introduction to Shi'i Islam : the history and doctrines of Twelver Shi'ism.* G. Ronald. p. 204. ISBN 0-85398-201-5
[111] Momen, Moojan (1985). *An introduction to Shi'i Islam : the history and doctrines of Twelver Shi'ism.* G. Ronald. p. 115. ISBN 0-85398-201-5
[112] Momen, Moojan (1985). *An introduction to Shi'i Islam : the history and doctrines of Twelver Shi'ism.* G. Ronald. p. 116. ISBN 0-85398-201-5
[113] Ferrier, R. W.; A Journey to Persia: Jean Chardin's Portrait of a Seventeenth-century Empire; pp 71-71.
[114] Blow, p. 173.
[115] Blow, David. Shah Abbas: the ruthless king who became an Iranian legend, p. 165.
[116] Ferrier, pp. 80–2.
[117] Blow, p. 170.
[118] Savory, Roger, Iran under the Safavids, p. 221.
[119] Blow, p. 175.
[120] Ferrier; pp. 85-89.
[121] Malcolm; vol II, p. 456.
[122] Savory; p. 182.
[123] Ferrier, RW, A journey to Persia: Jean Chardin's Portrait of a Seventeenth-century Empire, pp. 90–4.
[124] Ferrier p. 91.
[125] Savory, R.; Iran under the Safavids; pp. 186-7.
[126] Ferrier, R. W.; *A journey to Persia: Jean Chardin's portrait of a seventeenth-century Empire*; p. 24.
[127] Ferrier; p. 23.
[128] Savory; p.187.
[129] Blow, D.; Shah Abbas: The ruthless king who became an Iranian legend; p. 211.
[130] Lambton, A. K. S.; *Landlord and Peasant in Persia* (Oxford 1953); p 127-8.
[131] Ferrier; pp. 25-6.
[132] Savory; p.190.
[133] Ferrier; p. 31.
[134] Savory; p. 191.
[135] Blow; p. 210.
[136] Savory, R; Iran under the Safavids; pp. 193-5.
[137] Blow, D; Shah Abbas: The ruthless king who became an Iranian legend; pp. 113-131.
[138] Blow; chapter: "English adventurers at the servise of Shah Abbas."
[139] Savory; p. 195.
[140] Blow; p. 212.
[141] Savory; p. 196.
[142] Savory; pp. 199-200.
[143] Blow; p. 213.
[144] Savory; p. 195-8
[145] Savory; p. 202.

[146] E. Yarshater, Language of Azerbaijan (http://www.iranica.com/newsite/search/searchpdf.isc?ReqStrPDFPath=/home1/iranica/articles/v3_articles/azerbaijan/language_azerbaijan&OptStrLogFile=/home/iranica/public_html/logs/pdfdownload.html), vii., Persian language of Azerbaijan", *Encyclopædia Iranica*, v, pp. 238–45, Online edition.
[147] Emeri "van" Donzel, *Islamic Desk Reference*, Brill Academic Publishers, 1994, p. 393.
[148] William L. Cleveland and Martin P. Bunton, *A History of the Modern Middle East* (Westview Press, 2000), 2nd ed., pp. 56-57.
[149] Dabashi, H. (1996) 'Mir Damad and the Founding of the School of Isfahan', in SH Nasr and O. Leaman (eds) History of Islamic Philosophy, London: Routledge, ch. 34, 597–634.
[150] Rizvi, Sajjad (Summer 2009). "Mulla Sadra" (http://plato.stanford.edu/archives/sum2009/entries/mulla-sadra/). In Zalta, Edward N. *The Stanford Encyclopedia of Philosophy*. .
[151] Nasr, Seyyed Hossein, **Sadr al-Din Shirazi and his Transcendent Theosophy, Background, Life and Works**, 2nd ed., Tehran: Institute for Humanities and Cultural Studies.
[152] RN Frye, *The Golden Age of Persia*, Phoenix Press, 2000, p. 234
[153] Savory, Roger: Iran under the Safavids, pp. 220–5.
[154] Savory, pp. 220.
[155] Savory, p. 222.
[156] Savory, Roger; *Iran under the Safavids*, p. 155.
[157] Sir Roger Stevens; *The Land of the Great Sophy*, p. 172.
[158] Savory; chpt: *The Safavid empire at the height of its power under Shāh Abbas the Great (1588-1629)*
[159] Jodidio, Philip, *Iran: Architecture For Changing Societies*:Umberto Allemandi (August 2, 2006).
[160] Ronald W. Ferrier, *The Arts of Persia*, Yale University Press, 1989, p. 199.
[161] Arnold J. Toynbee, *A Study of History*, V, pp. 514-15.
[162] John R. Perry, "Turkic-Iranian contacts", *Encyclopædia Iranica*, January 24, 2006.
[163] Ruda Jurdi Abisaab. "Iran and Pre-Independence Lebanon" in Houchang Esfandiar Chehabi, *Distant Relations: Iran and Lebanon in the Last 500 Years*, I.B. Tauris (2006), p. 76.
[164] Cornelis Henricus Maria Versteegh, *The Arabic Language*, Columbia University Press, 1997, p. 71.
[165] Hillenbrand R., *Islamic Art and Architecture*, London (1999), ISBN 0-500-20305-9, p. 228.
[166] Savory, RM. "Rise of a Shi'i State in Iran and New Orientation in Islamic Thought and Culture" (http://www.unesco.org/culture/humanity/html_eng/volume5.htm). *'UNESCO: History of Humanity*. **5: From the Sixteenth to the Eighteenth Century**. London, New York: Routledge. p. 263. .
[167] Mujtahid: A *mujtahid* in Arabic means a person who qualified to engange in *ijtihad*, or interpretation of religious texts. *Ithna 'ashari* is the number twelve in Arabic, signifying Twelver Imami Shi'i Islam. Ulama: Arabic for religious scholars.

Literature

- M.I. Marcinkowski (tr.),*Persian Historiography and Geography: Bertold Spuler on Major Works Produced in Iran, the Caucasus, Central Asia, India and Early Ottoman Turkey*, M. Ismail Marcinkowski, Singapore: Pustaka Nasional, 2003, ISBN 9971-77-488-7.
- M.I. Marcinkowski (tr., ed.),*Mirza Rafi'a's Dastur al-Muluk: A Manual of Later Safavid Administration. Annotated English Translation, Comments on the Offices and Services, and Facsimile of the Unique Persian Manuscript*, M. Ismail Marcinkowski, Kuala Lumpur, ISTAC, 2002, ISBN 983-9379-26-7.
- M.I. Marcinkowski,*From Isfahan to Ayutthaya: Contacts between Iran and Siam in the 17th Century*, M. Ismail Marcinkowski, Singapore, Pustaka Nasional, 2005, ISBN 9971-77-491-7.
- Hakim Syed Zillur Rahman, "Safavi Ahad Main Ilm Tashreeh Ka Mutala (a book in Urdu on Studies of History of anatomy during Safavid dynasty), Tibbi Academy, Aligarh, India, 1983, 96 pp.
- "The Voyages and Travels of the Ambassadors" (http://depts.washington.edu/uwch/silkroad/texts/olearius/travels.html), Adam Olearius, translated by John Davies (1662),

External links

- History of the Safavids on Iran Chamber (http://www.iranchamber.com/history/safavids/safavids.php)
- "Safavid dynasty" (http://www.iranicaonline.org/articles/safavids), *Encyclopædia Iranica* by Rudi Matthee
- The History Files: Rulers of Persia (http://www.historyfiles.co.uk/KingListsMiddEast/EasternPersia.htm)
- BBC History of Religion (http://www.bbc.co.uk/religion/religions/islam/history/safavidempire_1.shtml)
- Iranian culture and history site (http://www.iranian.ws/7000-6.htm)
- Artistic and cultural history of the Safavids from the Metropolitan Museum of Art (http://www.metmuseum.org/toah/hd/safa/hd_safa.htm)
- History of Safavid art (http://www.art-arena.com/safavidart.htm)
- A Study of the Migration of Shi'i Works from Arab Regions to Iran at the Early Safavid Era. (http://www.imamreza.net/eng/imamreza.php?print=589)
- Why is Safavid history important? (http://www.iranchamber.com/history/articles/reflections_safavid_history_historiography1.php) (Iran Chamber Society)
- Historiography During the Safawid Era (http://www.imamreza.net/eng/imamreza.php?id=3021)
- NEWMAN, Andrew J. *Safavid Iran - Rebirth of a Persian Empire* (http://gunturkbetikevi.files.wordpress.com/2011/01/safavid_iran-rebirth_of_a_persian_empire.pdf). New York: I.B. Tauris, 2009.
- "IRAN ix. RELIGIONS IN IRAN (2) Islam in Iran (2.3) Shiʿism in Iran Since the Safavids: Safavid Period" (http://www.iranicaonline.org/articles/iran-ix23-shiism-in-iran-since-the-safavids), *Encyclopædia Iranica* by Hamid Algar

Sunni Islam

The Al-Azhar University in Cairo, Egypt, is a major centre of Sunni Islamic learning in the world.

Sunni Islam (pron.: /ˈsuːni/ or /ˈsʊni/) is the largest branch of Islam; its adherents are referred to in Arabic as *ʾAhlu-s-Sunnati wa-l-Jamāʿah* (Arabic: أهل السنة والجماعة), "people of the tradition of Muhammad and the consensus of the Ummah" or *ʾAhlu-s-Sunnah* (Arabic: أهل السنة). For short, in English, they are known as **Sunni Muslims**, **Sunnis**, **Sunnites** or simply **Muslims**.

Sunni Islam is sometimes referred to as the orthodox version of the religion.[1] [2] The word "Sunni" comes from the term *Sunnah* (Arabic: سنة), which refers to the sayings and actions of Muhammad that are recorded in hadiths (collections of narrations regarding prophet Muhammad).[3]

The primary hadiths Al-Kutub Al-Sittah (the six books), in conjunction with the Quran, form the basis of all jurisprudence methodologies within Sunni Islam. Laws are derived from the text of the Quran and the hadith, in addition to using methods of juristic reasoning (like qiyas) and consensus (ijma). There are a multitude of scholarly opinions in each field; however, these can be summarised as either derived from the four major schools of thought (Madh'hab) or from an expert scholar who exercises independent derivation of Islamic Law (ijtihad). Both are considered valid as differences of opinion (which were present at the time of the early Muslims, the Salaf).

Etymology

Sunnī (Classical Arabic: سُنِّي /ˈsunniː/) is a broad term derived from *sunnah* (سُنَّة /ˈsunna/, pl. سُنَن *sunan* /ˈsunæn/), means "habit" or "usual practice".[4] The Muslim usage of this term refers to the sayings and living habits of Muhammad. In its full form, this branch of Islam is referred to as "Ahl al-Sunnah wa Jama'ah" (literally, "People of the *Sunah* and the Community"). People claiming to follow the Sunnah who can demonstrate that they have no action or belief against the prophetic Sunnah can consider themselves to be Sunni Muslims.

History

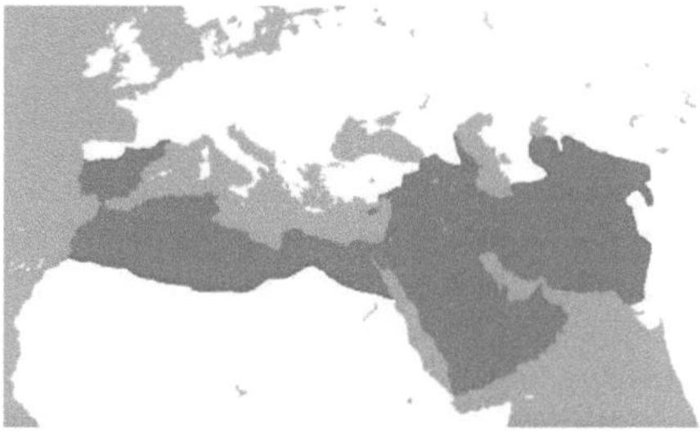

Umayyad Caliphate at its greatest extent

After the death of Muhammad, Muslims who accepted Abu Bakr as the first Caliph became known as the Ahlus Sunnah wa`l Jamah in order to differentiate them from the sect of Shiaism who were the group of Sahabah that said Ali should have been Caliph.

The first four caliphs are known among Sunnis as the *Four Rightly Guided Caliphs*. The first was Abu Bakr Siddique, the second was Umar ibn al-Khattāb, the third was Uthman ibn Affan, and the fourth was Ali ibn Abi Talib.[5]

After the first four caliphs, the Caliphate was upheld as a political system by dynasties such as the Umayyads, the Abbasids, and the Ottomans. It was also upheld for relatively short periods of time by other competing dynasties in al-Andalus, North Africa and Egypt.

Mustafa Kemal Atatürk abolished the system of the Ottoman Caliphate after Abdülmecid II was officially deposed and expelled from Turkey, whereby the Republic of Turkey was founded in 1923 upon secular principles. To this day, the Ottoman Empire is regarded as the last major Islamic Caliphate.

Adherents

Sunnis believe that the companions of the prophet Muhammad were the best of the Muslims. This belief is based on Prophetic traditions such as one narrated by Abdullah, son of Masud, in which Muhammad said: "The best of the people are my generation, then those who come after them, then those who come after them." Support for this view is also found in Qur'an,[6] according to Sunnis. Sunnis also believe that the companions were true believers since it was the companions who were given the task of compiling the Quran. Furthermore, narrations that were narrated by the companions are considered by Sunnis to be a primary source of the Muslim faith.

Estimates of the world Sunni population has been estimated by some analysts to be "over 75 percent," though the majority of statistical counts put the portion of Sunni adherent between 85 and 90 percent.[7]

Schools of law

There are several intellectual traditions within the field of Islamic law. These varied traditions reflect differing viewpoints on some laws and obligations within Islamic law. While one school of thought may see a certain act as a religious obligation, another may see the same act as optional. These schools of thought aren't regarded as sects; rather, they represent differing viewpoints on issues that are not considered the core of Islamic belief.

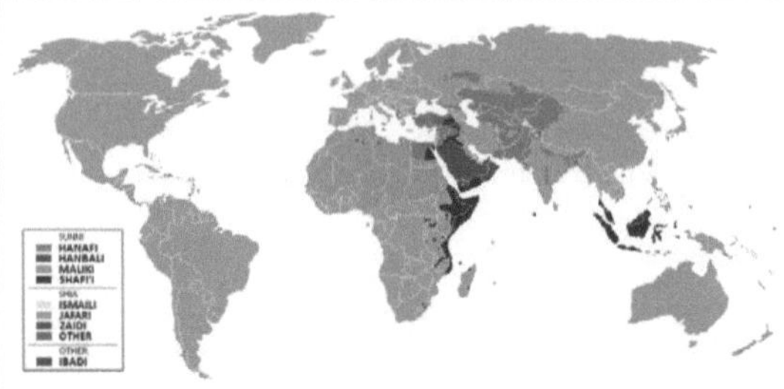

Distribution of Sunni, Shia, and Ibadi branches of Islam

Historians have differed regarding the exact delineation of the schools based on the underlying principles they follow. Many traditional scholars saw Sunni Islam in two groups: Ahl al-Ra'i, or "people of reason," due to their emphasis on scholarly judgment and discourse; and Ahl al-Hadith, or people of traditions, due to their emphasis on restricting juristic thought to only what is found in scripture.[8] Ibn Khaldun defined the Sunni schools as three: the Hanafi school representing reason, the Zahiri school representing tradition, and a broader, middle school encompassing the Shafi'i, Maliki and Hanbali schools.[9]

Hanafi

Abu Hanifah (699 — 767 CE / 80 — 148 AH) was the founder of the Hanafi school. He was born in Kufa, Iraq[10] [11] in a family of Persian ancestry.[12] [13] Sunni Muslims of Iraq, Syria, Lebanon, Bangladesh, Mauritius, Pakistan, India, Afghanistan, Central Asia, the Muslim areas of Southern Russia, the Caucasus, most of the Muslim areas of the Balkans and Turkey and parts of Egypt all follow this school of jurisprudence. The Barelwi and Deobandi movements, the two largest Islamic movements in South Asia, are both Hanafi, as is the Tablighi Jamaat.

Maliki

Malik ibn Anas (c. 711 – 795CE / 93 – 179 AH) developed his Islamic jurisprudence in the holy city of Medina. His school has been adopted by most North African and West African countries such as upper Egypt, Libya, Tunisia, Algeria, Morocco, Senegal, Mali, Nigeria, Mauritania and Sudan along with parts of the Persian Gulf. The Maliki school of jurisprudence forms the official state legal codes of Kuwait, Bahrain and the United Arab Emirates. The Murabitun World Movement also follows the Maliki school.

In the past, the school also predominated in Sicily and parts of Spain. One of greatest historical centers of Maliki teaching, especially during the ninth, tenth and eleventh centuries, is the Mosque of Uqba.[14] [15]

The Mosque of Uqba also known as the Great Mosque of Kairouan was, in particular during the 9th, 10th and 11th centuries, an important center of Islamic learning with an emphasis on the Maliki Madh'hab.[14] The Mosque of Uqba is located in the city of Kairouan in Tunisia.

Shafi'i

Muhammad ibn Idris ash-Shafi`i (767 — 820 CE / 150 — 204 AH) was a student of Malik. He taught in Iraq and then in Egypt. The Shafi'i madhab today is the dominant school of jurisprudence in Yemen, Lower Egypt, Djibouti, Eritrea, Somalia, Sudan, Syria, the Palestinian Territories, Jordan, the North Caucasus, Kurdistan, Maldives, Malaysia, Brunei and Indonesia.

It is also practised by minority communities in Kuwait, 'Asir Province in Saudi Arabia, the United Arab Emirates, Palestine, the Swahili Coast, Mauritius, Singapore, South Africa, Thailand, Vietnam, Cambodia, the Philippines, Sri Lanka, Ethiopia, Kazakhstan (by Chechens) and the Indian states of Kerala (most of the Mappilas), Karnataka (Bhatkal, Mangalore and Coorg districts), Maharashtra (by Konkani Muslims), and Tamil Nadu.

Hanbali

Ahmad bin Hanbal (780 — 855 CE / 164 — 241 AH) was the namesake of the Hanbali school of thought in fiqh. He was born in Baghdad and learned extensively from Shafi'i. The Hanbali school of jurisprudence is followed predominantly in Saudi Arabia and Qatar as well as minority communities in Syria and Iraq. The majority of the Salafist movement, though not all adherents, tend to follow the Hanbali school.

Zahiri

Dawud al-Zahiri (815 - 883 CE / 200 - 270 AH) was the founder of the Zahiri school of thought. He lived in Baghdad, though his family was attributed to Isfahan. Today, the school does not comprise a majority in any country, instead being followed by minority communities in Morocco and Pakistan. In the past, the school was followed by the majority in Mesopotamia, Portugal, the Balearic Islands, North Africa and parts of Spain.

Differences in the Schools

Interpreting Islamic law by deriving specific rulings - such as how to pray - is known as fiqh, commonly termed jurisprudence. A madh'hab is a particular tradition of interpreting this jurisprudence. These schools possess different focuses, such as specific evidence (Shafi'i and Hanbali) or general principles (Hanafi and Maliki) derived from specific evidences. As these schools represent clearly spelled out methodologies for interpreting Islamic law, there has been little change in the methodology with regard to each school. All four madh'habs are recognised viable legal schools and are not seen as in error in contrast to one another, each school has their evidences and differences are respected as being *Ikhtilaf*; this is because all the schools are united upon the usul (*base*) issues but generally differ on furu' (*branches*).

As the social and economic environment changes, new rulings are derived - this is known as Ijtihad - this can be from a Mujtahid who specialises in one or more Madh'habs. For example, when tobacco appeared, it was considered disliked because of its smell. When medical information showed that smoking was dangerous, most jurists took the view that it is forbidden. Current issues include social topics such as downloading pirated software and scientific issues such as cloning.

The Six pillars of Iman

Sunni Islam has six articles of faith known as the six pillars of *iman* that all Sunni Muslims are united upon in belief, along with the 105 key points of creed mentioned in "Aṭ-Ṭaḥāwī's Islamic Theology".[16]

- Reality of one God (see Tawhid)
- Existence of angels of God
- Authority of the books of God
- Following the prophets of God
- Preparation for and belief in the Day of Judgment
- Supremacy of God's will, i.e. belief in predestination good or bad is from God alone

Sunni theological traditions

Some Islamic scholars faced questions that they felt were not explicitly answered in the *Qur'an* and *Sunnah*, especially questions with regard to philosophical conundra such as the nature of God, the existence of human free will, or the eternal existence of the *Qur'an*. Various schools of theology and philosophy developed to answer these questions, each claiming to be true to the *Qur'an* and the Muslim tradition (*sunnah*). Among Sunni Muslims, various schools of thought in theology began to be born out of the sciences of kalam in opposition to the textualists who stood by affirming texts without delving into philosophical speculation as they saw it as an innovation in Islam. The following were the three dominant schools of theology that grew. All three of these are accepted by Muslims around the globe, and are considered within "Islamic Orthodoxy". The key beliefs of Sunni Islam are all agreed upon (being the six pillars of Iman) and codified in the treatise on Aqeedah by Imam Ahmad ibn Muhammad al-Tahawi in his Aqeedat Tahawiyyah.

Maturidi

Founded by Abu Mansur al-Maturidi (died 944). Maturidiyyah was a minority tradition until it was accepted by the Turkish tribes of Central Asia (previously they had been Ash'ari and followers of the Shafi'i school, it was only later on migration into Anatolia that they became Hanafi and followers of the Maturidi creed). One of the tribes, the Seljuk Turks, migrated to Turkey, where later the Ottoman Empire was established.[17] Their preferred school of law achieved a new prominence throughout their whole empire although it continued to be followed almost exclusively by followers of the Hanafi school while followers of the Shafi and Maliki schools within the empire followed the Ash'ari and Athari schools of thought. Thus, wherever can be found Hanafi followers, there can be found the Maturidi creed. Maturidis argue that the proof of God's existence can be derived through pure reason.

Ash'ari

Founded by Abu al-Hasan al-Ash'ari (873–935). This theological school of Aqeedah was embraced by plenty of Muslim scholars and developed in parts of the Islamic world throughout history, Imam al-Ghazali wrote on the creed discussing it and agreeing upon some of its principles.[18]

Ash'ari theology stresses divine revelation over human reason. Contrary to the Mu'tazilites, they say that ethics cannot be derived from human reason, but that God's commands, as revealed in the *Qur'an* and the *Sunnah* (the practices of Muhammad and his companions as recorded in the traditions, or *hadith*), are the sole source of all morality and ethics.

Regarding the nature of God and the divine attributes, the Ash'ari rejected the Mu'tazili position that all Qur'anic references to God as having real attributes were metaphorical. The Ash'aris insisted that these attributes were as they "best befit his majesty". The Arabic language is a wide language in which one word can have 15 different meanings, so the Ash'aris endeavor to find the meaning that best befits Allah and is not contradicted by the Qur'an. Therefore when Allah states in the Holy Qur'an, "He who does not resemble any of this creation," this clearly means Allah can't be attributed with body parts because he created body parts. This is one way which differentiates these Muslims from most Christians and Jews. Ash'aris tend to stress divine omnipotence over human free will and they believe that the Qur'an is eternal and uncreated.

Athari

Athari (Arabic: أثري), or "textualism", is derived from the Arabic word *athar*, literally meaning "remnant", and also referring to "narrations". Their disciples are called the Atharis or *al-Atharia*. The Atharis are considered to be one of three Sunni schools of Aqidah.

The Athari methodology of textual interpretation is to avoid delving into any extensive theological speculation. They believe in Allah and his attributes in the exact fashion that they were mentioned in the Quran, the Sunnah, and by the Sahabah. They do not attempt to further interpret the aforementioned texts by giving a literal meaning like in Ẓāhirīya (literalism) or the Tashbih (simile or likening), nor through tahrif (distortion), nor ta`weel (allegory or metaphor), nor ta'teel (denial). They avoid entering into deep rational philosophical discussions of matters relating to Islamic beliefs that are not supported by the Quran, the Sunnah or the understanding of the Sahabah with specific wording; rather, their discussion and presentation of beliefs revolves entirely around textual evidences found in these three main sources, while remaining cautious to avoid taking the path of the Ẓāhirīs (literalists) either. The Atharis believe this to be the methodology adhered to by the first three generations of Muslims (i.e. the Salaf), therefore making it the school of Sunni Aqeedah that they believe is adhering to the truth and keeping to the balanced middle path of Islam.

The codifier of the Athari Aqeedah was the great Islamic Classical Scholar Ahmad ibn Hanbal[19] who today is perhaps better known for his School of Jurisprudence than his school of Aqeedah. Prominent proponents of Classical Atharism today include Dr. Yusuf Al-Qaradawi and Mauritanian scholar Sheikh Muhammad Al-Hassan Ad-Dedew, amongst others.[20] [21]

Sunni view of hadith

The Qur'an as it exists today in book form was compiled by Muhammad's companions (Sahaba) within a handful of months of his death, and is accepted by all Muslim denominations. However, there were many matters of belief and daily life that were not directly prescribed in the Qur'an, but were actions that were observed by Muhammad and the early Muslim community. Later generations sought out oral traditions regarding the early history of Islam, and the practices of Muhammad and his first followers, and wrote them down so that they might be preserved. These recorded oral traditions are called *hadith*. Muslim scholars have through the ages sifted through the *hadith* and evaluated the chain of narrations of each tradition, scrutinizing the trustworthiness of the narrators and judging the strength of each *hadith* accordingly.

Al-Kutub Al-Sittah

Al-Kutub al-Sittah translates as "the Six Books". Most Sunni Muslims accept the *hadith* collections of Bukhari and Muslim as the most authentic (*sahih,* or correct), and while accepting all hadiths verified as authentic, grant a slightly lesser status to the collections of other recorders. There are, however, four other collections of *hadith* that are also held in particular reverence by Sunni Muslims, making a total of six:

- Sahih al-Bukhari of Muhammad al-Bukhari
- Sahih Muslim of Muslim ibn al-Hajjaj
- Sunan al-Sughra of Al-Nasa'i
- Sunan Abu Dawud of Abu Dawood
- Sunan al-Tirmidhi of Al-Tirmidhi
- Sunan Ibn Majah of Ibn Majah

There are also other collections of *hadith* which also contain many authentic *hadith* and are frequently used by scholars and specialists. Examples of these collections include:

- Musannaf of Abd al-Razzaq of ʻAbd ar-Razzaq as-Sanʻani
- Musnad of Ahmad ibn Hanbal
- Mustadrak of Al Haakim
- Muwatta of Imam Malik
- Sahih Ibn Hibbaan
- Sahih Ibn Khuzaymah of Ibn Khuzaymah
- Sunan al-Darimi of Al-Darimi

Notes

[1] *Gale Encyclopedia of the Mideast & N. Africa* (http://www.answers.com/topic/sunni). . "The largest branch in Islam, sometimes referred to as "orthodox Islam"; its full name is ahl al-Sunna wa aljamaʿa (the people of Sunna and consensus), and it represents about 90 percent of the world Muslim population."
[2] "Sunni and Shia Islam" (http://lcweb2.loc.gov/cgi-bin/query/r?frd/cstdy:@field(DOCID+af0060)). Library of Congress Country Studies. . Retrieved December 17, 2011.
[3] "Sunna" (http://www.merriam-webster.com/dictionary/sunna). Merriam-Webster. . Retrieved 2010-12-17. "*the body of Islamic custom and practice based on Muhammad's words and deeds*"
[4] Sunnah (http://www.usc.edu/schools/college/crcc/engagement/resources/texts/muslim/reference/glossary/term.SUNNAH.html), Center for Muslim-Jewish Engagement
[5] Tore Kjeilen. "Lexic Orient.com" (http://lexicorient.com/e.o/ali.htm). Lexic Orient.com. . Retrieved 2011-06-05.
[6] Qur'an, 9:100
[7] See:

- Eastern Europe Russia and Central Asia (http://books.google.co.uk/books?id=NI1G_9j1AhcC&pg=PA51&dq=&hl=en&ei=yQDiTt_qE9C28QOKn9yMBA&sa=X&oi=book_result&ct=result&resnum=3&ved=0CEgQ6AEwAjgK#v=onepage&q=world muslim 80% sunni&f=false) "some 80% of the worlds Muslims are Sunni"
- "Religions" (https://www.cia.gov/library/publications/the-world-factbook/fields/2122.html). *The World Factbook*. Central Intelligence Agency (CIA). . Retrieved 8 December 2011. "Sunni Islam accounts for over 75% of the world's Muslim population"
- Sue Hellett;U.S. should focus on sanctions against Iran (http://www.galesburg.com/newsnow/x917961022/Sue-Hulett-U-S-should-focus-on-sanctions-against-Iran) "Sunnis make up over 75 percent of the world's Muslim population"
- Iran, Israel and the United States (http://books.google.co.uk/books?id=nVcmKRyID2EC&pg=PA15&dq=&hl=en&ei=e4XoTt-TDIaL8gOG692iCg&sa=X&oi=book_result&ct=result&resnum=6&ved=0CFEQ6AEwBTgU#v=onepage&q=sunni 75%&f=false) "Sunni, accounts for over 75% of the Islamic population"
- A dictionary of modern politics (http://books.google.co.uk/books?id=qHXbGOUuF9YC&pg=PA252&dq=&hl=en&ei=5wHiTpzMGMrT8gPIl52FBA&sa=X&oi=book_result&ct=result&resnum=2&ved=0CEMQ6AEwATgU#v=onepage&q=world muslim 80% sunni&f=false) "probably 80% of the worlds Muslims are Sunni"
- "Mapping the Global Muslim Population: A Report on the Size and Distribution of the World's Muslim Population" (http://pewforum.org/Muslim/Mapping-the-Global-Muslim-Population.aspx). *Pew Research Center*. October 7, 2009. . Retrieved 2010-08-24. "Of the total Muslim population, 10-13% are Shia Muslims and 87-90% are Sunni Muslims."
- "Quick guide: Sunnis and Shias" (http://news.bbc.co.uk/2/hi/middle_east/6213248.stm). BBC News. 2011-12-06. . Retrieved December 18, 2011. "The great majority of Muslims are Sunnis - estimates suggest the figure is somewhere between 85% and 90%."
- "Tension between Sunnis, Shiites emerging in USA" (http://www.usatoday.com/news/religion/2007-09-24-muslim-tension_N.htm). USA Today. 2007-09-24. . Retrieved December 18, 2011. "Among the world's estimated 1.4 billion Muslims, about 85% are Sunni and about 15% are Shiite."
- Sunni Islam: Oxford Bibliographies Online Research Guide (http://books.google.co.uk/books?id=D5_N97bAiJ0C&lpg=PP1&pg=PA3#v=onepage&q&f=false) "Sunni Islam is the dominant division of the global Muslim community, and throughout history it has made up a substantial majority (85 to 90 percent) of that community."

[8] Murtada Mutahhari, The Role of Ijtihad in Legislation (http://www.al-islam.org/al-tawhid/ijtihad-legislation.htm), Al-Tawhid (http://www.al-islam.org/al-tawhid/) volume IV, No.2, Publisher: Islamic Thought Foundation (http://www.itf.org.ir/)
[9] Meinhaj Hussain, A New Medina, The Legal System (http://www.grandestrategy.com/2012/01/6655434312-chapter-nine-new-medina.html), Grande Strategy, January 5th, 2012
[10] Josef W. Meri, *Medieval Islamic Civilization: An Encyclopedia*, 1 edition, (Routledge: 2005), p.5
[11] Hisham M. Ramadan, *Understanding Islamic Law: From Classical to Contemporary*, (AltaMira Press: 2006), p.26
[12] S. H. Nasr(1975), "The religious sciences", in R.N. Frye, the Cambridge History of Iran, Volume 4, Cambridge University Press. pg 474: "Abū Ḥanīfah, who is often called the "grand imam"(al-Imam al-'Azam) was Persian

[13] Cyril Glasse, "The New Encyclopedia of Islam", Published by Rowman & Littlefield, 2008. pg 23: "Abu Hanifah, a Persian, was one of the great jurists of Islam and one of the historic Sunni Mujtahids"
[14] Wilfrid Scawen Blunt and Riad Nourallah, *The future of Islam*, Routledge, 2002, page 199
[15] Ira Marvin Lapidus, *A history of Islamic societies*, Cambridge University Press, 2002, page 308
[16] "Sunni Islam Afterlife and Salvation" (http://www.patheos.com/Library/Sunni-Islam/Beliefs/Afterlife-and-Salvation.html). .
[17] "Maturidiyyah" (http://philtar.ucsm.ac.uk/encyclopedia/islam/sunni/matur.html). *Philtar*. . Retrieved 2006-04-01.
[18] J. B. Schlubach. "Fethullah Gülen and Al-Ghazzali on Tolerance" (http://www.fethullahgulenforum.org/articles/13/fethullah-gulen-al-ghazzali-on-tolerance). . Retrieved 2010-01-07.
[19] http://www.tutorgigpedia.com/ed/Islamic_schools_and_branches
[20] http://www.suhaibwebb.com/islam-studies/suhaib-where-do-you-stand-what-kind-of-muslim-are-you/
[21] http://www.suhaibwebb.com/multimedia/unity-in-action-shaykh-al-qaradawi-defends-the-creed-of-al-azhar-university/

Further reading

- Branon Wheeler, Applying the Canon in Islam: The Authorization and Maintenance of Interpretive Reasoning in Ḥanafī Scholarship (http://books.google.com.pk/books?id=slLpouSlzPcC&printsec=frontcover&source=gbs_atb), SUNY Press, 1996

External links

- Books Related of Ahle Sunnat Wa Jamat Aqaid (http://www.faizaneattar.net)
- Books relating to belief of ahl as-Sunnat (http://www.hakikatkitabevi.com.tr/english/english.htm)
- Ahl as-sunnat belief (http://www.ourreligionislam.com/detail.asp?Aid=4279)
- Translation and Detailed Commentary on Quran (http://www.classicalislamgroup.co.uk/index.php?view=tafseer)
- SunniPath - Study Islam Online (http://www.sunnipath.com/)

Hotaki dynasty

Hotaki Empire

← 

← 

1709–1738  →

Flag

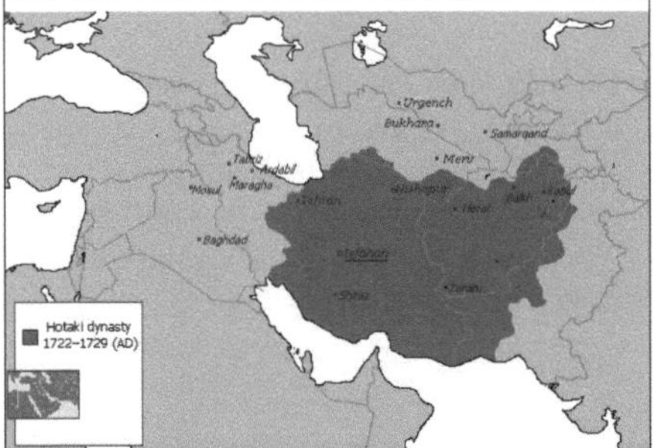

Hotaki Empire at its peak (1722-1729)

Capital		Kandahar Isfahan
Languages		Pashto Persian
Religion		Sunni Islam
Government		Absolute Monarchy
Emir		
-	1709–1715	Mir Wais Hotak
-	1715–1717	Abdul Aziz Hotak
-	1717–1725	Mahmud Hotaki
-	1725–1730	Ashraf Hotaki
-	1725–1738	Hussain Hotaki
History		
-	Revolt by Mir Wais Hotak	April 1709
-	Siege of Kandahar	24 March 1738

Today part of	Afghanistan Iran Pakistan Turkmenistan

The **Hotaki dynasty** was established in April 1709 by Mir Wais Hotak after leading a successful revolution against the Persian Safavids in Kandahar.[1] It was an Afghan monarchy that lasted until 1738 when Nader Shah of Khorasan defeated Hussain Hotaki during the long siege of Kandahar.[2] At its peak, the Hotaki dynasty ruled over a wide area which is now Iran, Afghanistan, and northwestern Pakistan.

In 1715, Mir Wais died of a natural cause and his brother Abdul Aziz succeeded the monarchy. He was quickly followed by Mahmud who conquered Persia in 1722 and ruled it until his death in 1725. Ashraf Hotaki took over but was defeated at the Battle of Damghan in 1729 by the Afsharids. Hussain Hotaki became the last ruler until he was also defeated in 1738.

Rise to power

A miniature from Padshahnama depicting Safavid and Mughal soldiers in 1638. Similar scenes were often witnessed at Kandahar between the Persians and the Mughals since the mid 16th century.

Kandahar province was ruled by the Shi'a Safavids during the early 18th century and the native Afghan tribes living in the area were Sunni Muslims. Immediately to the east began the Sunni Mughul Empire, who occasionally fought wars with the Safavids over the territory of southern Afghanistan.[3]

In 1704, the Safavid Shah Husayn appointed George XI (*Gurgīn Khān*), who is believed to have converted to Islam.[4] Gurgin began imprisoning and executing Afghans, especially those suspected of organizing rebellions. One of those arrested and imprisoned was Mir Wais who belonged to an influential Hotak family in Kandahar. Mir Wais was sent as a prisoner to the Persian court in Isfahan but the charges against him were dismissed by Shah Husayn, so he was sent back to his native land as a free man.[5]

In April 1709, Mir Wais along with his followers revolted against the Safavid rule at Kandahar. The uprising began when Gurgīn Khān and his escort were killed during a feast that was organized by Mir Wais at his farmhouse outside the city. It is reported that drinking of wine was involved. Next, Mirwais ordered the killings of the remaining Persian military officials in the region. The Afghans then defeated a twice as large Persian army that had been dispatched from Isfahan (capital of the Safavids), one which included Qizilbash and Georgian troops.[6]

> *Several half-hearted attempts to subdue the rebellious city having failed, the Persian Government despatched Khusraw Khán, nephew of the late Gurgín Khán, with an army of 30,000 men to effect its subjugation, but in spite of an initial success, which led the Afgháns to offer to surrender on terms, his uncompromising attitude impelled them to make a fresh desperate effort, resulting in the complete defeat of the Persian army (of whom only some 700 escaped) and the death of their general. Two years later, in A.D. 1713, another Persian army commanded by Rustam Khán was also defeated by the rebels, who thus secured possession of the whole province of Qandahár.*[6]
>
> —Edward G. Browne, 1924

Kandahar (*Candahar*) during the Afsharid and Mughal period.

Refusing the title of king, Mirwais was called "Prince of Qandahár and General of the national troops" by his Afghan countrymen. He died peacefully in November 1715 from natural causes and was succeeded by his brother Abdul Aziz; the latter was murdered later by Mirwais' son Mahmud. In 1720, Mahmud's Afghan forces crossed the deserts of Sistan and captured Kerman.[6] His plan was to conquer the Persian capital, Isfahan.[7] After defeating the Persian army at the Battle of Gulnabad on March 8, 1722, he proceeded to and sacked the city of Isfahan.[8] On October 23, 1722, Sultan Husayn abdicated and acknowledged Mahmud as the new Shah of Persia.[9]

Majority of the Persian people, however, rejected the Afghan regime as usurping. For the next seven years until 1729, the Hotakis became the de facto rulers of Persia, but the southern and eastern areas of Afghanistan still remained under their control until 1738.

The Hotaki dynasty was a troubled and violent one as internecine conflict made it difficult to establish permanent control. The dynasty lived under great turmoil due to bloody succession feuds that made their hold on power tenuous, and after the massacre of thousands of civilians in Isfahan – including more than three thousand religious scholars, nobles, and members of the Safavid family – the Hotaki dynasty was eventually removed from power in Persia.[10] On the other hand, the Afghans had also been suppressed by the Safavid government under Gurgin Khan before their uprising in 1709.[5]

Decline

Ashraf Hotaki, who took over the monarchy following Shah Mahmud's death in 1725, and his soldiers were defeated in the October 1729 Battle of Damghan by Nader Shah, a soldier of fortune from the Sunni Turkmen background and the founder of the Afsharid dynasty that replace the Safavids in Persia. Nader Shah had driven out the remaining Ghilzai forces from Persia and began enlisting the Abdali Afghans of Farah and Kandahar in his military. Nader Shah's forces (among them were Ahmad Shah Abdali and his 4,000 Abdali troops) conquered Kandahar in 1738. They besieged and destroyed the last Hotaki seat of power, which was held by Hussain Hotaki (or Shah Hussain).[7][11] Nader Shah then built a new town nearby, named after himself, "Naderabad". The Abdalis were also restored to the general area of Kandahar, with the Ghilzai's being pushed back to their former stronghold of Kalat-i Ghilzai--an arrangement that lasts to the present day.

List of rulers

Name	Picture	Reign started	Reign ended
Mirwais Hotak *Woles Mashar*		1709	1715
Abdul Aziz Hotak *Emir*		1715	1717

Mahmud Hotaki *Shah*		1717	1725
Ashraf Hotaki *Shah*		1725	1729
Hussain Hotaki *Emir*		1729	1738

See also

- Durrani dynasty
- Safavid dynasty
- Delhi Sultanate

References

[1] Malleson, George Bruce (1878). *History of Afghanistan, from the Earliest Period to the Outbreak of the War of 1878* (http://books.google.com/books?id=pqNGBEmHUd4C&lpg=PP1&pg=PA227#v=onepage&q&f=false). London: Elibron.com. p. 227. ISBN 1-4021-7278-8, 9781402172786. . Retrieved 2010-09-27.

[2] "AN OUTLINE OF THE HISTORY OF PERSIA DURING THE LAST TWO CENTURIES (A.D. 1722-1922)" (http://persian.packhum.org/persian/pf?file=90001014&ct=33). *Edward Granville Browne*. London: Packard Humanities Institute. p. 33. . Retrieved 2010-09-24.

[3] Romano, Amy (2003). *A Historical Atlas of Afghanistan* (http://books.google.com/books?id=TKUxyVCrYn0C&lpg=PP1&pg=PA28#v=onepage&q&f=false). The Rosen Publishing Group. p. 28. ISBN 0-8239-3863-8, 9780823938636. . Retrieved 2010-10-17.

[4] *Nadir Shah and the Afsharid Legacy*, **The Cambridge history of Iran: From Nadir Shah to the Islamic Republic**, Ed. Peter Avery, William Bayne Fisher, Gavin Hambly and Charles Melville, (Cambridge University Press, 1991), p. 11.

[5] Otfinoski, Steven Bruce (2004). *Afghanistan* (http://books.google.com/books?id=fbXmk-EauHIC&lpg=PP1&pg=PA8#v=onepage&q&f=false). Infobase Publishing. p. 8. ISBN 0-8160-5056-2, 9780816050567. . Retrieved 2010-09-27.

[6] "AN OUTLINE OF THE HISTORY OF PERSIA DURING THE LAST TWO CENTURIES (A.D. 1722-1922)" (http://persian.packhum.org/persian/pf?file=90001014&ct=29). *Edward Granville Browne*. London: Packard Humanities Institute. p. 29. . Retrieved 2010-09-24.

[7] "Last Afghan empire" (http://www.britannica.com/EBchecked/topic/7798/Afghanistan/21392/Last-Afghan-empire). *Louis Dupree, Nancy Hatch Dupree and others*. Encyclopædia Britannica. . Retrieved 2010-09-24.

[8] "Account of British Trade across the Caspian Sea" (http://www.conflicts.rem33.com/images/Georgia/Allen_14.htm). *Jonas Hanway*. Centre for Military and Strategic Studies. . Retrieved 2010-09-27.

[9] Axworthy pp.39-55

[10] "AN OUTLINE OF THE HISTORY OF PERSIA DURING THE LAST TWO CENTURIES (A.D. 1722-1922)" (http://persian.packhum.org/persian/pf?file=90001014&ct=31). *Edward Granville Browne*. London: Packard Humanities Institute. p. 31. . Retrieved 2010-09-24.

[11] "AFGHANISTAN x. Political History" (http://www.iranicaonline.org/articles/afghanistan-x-political-history). *D. Balland*. Encyclopaedia Iranica. . Retrieved 2010-09-24.

External links

- Encyclopædia Britannica Online - Last Afghan empire (http://www.britannica.com/EBchecked/topic/7798/Afghanistan/21392/Last-Afghan-empire)

Afsharid dynasty

The Afsharid kingdom

1736–1796

Flag

Coat of arms

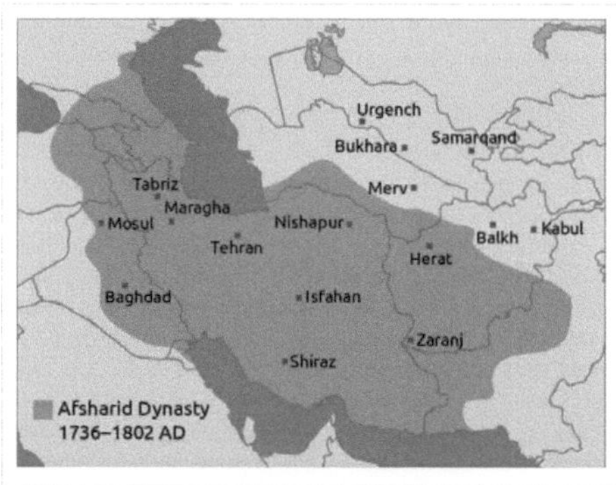

Afsharid Kingdom at its height

Capital	Mashhad
Languages	Persian (official and court language)[1] Turkic (court, Nader Shah's native language)[2] [3]
Government	Monarchy
Shah	Nader Shah (first)
	Shahrokh (last)
History	
- Afsharid dynasty begins	1736

-	Afsharid dynasty ends	1796

The **Afsharids** (Persian: افشاریان) were members of an Iranian dynasty of Turkic[4] origin, from Khorasan, who ruled Persia in the 18th century. The dynasty was founded in 1736 by the military commander Nader Shah, who deposed the last member of the Safavid dynasty and proclaimed himself King of Iran. During Nader's reign, Iran reached its greatest extent since the Sassanid Empire. After his death, most of his empire was divided between the Zands and the Durranis, and Afsharid rule was confined to a small local state in Khorasan. Finally, the Afsharid dynasty was overthrown by Mohammad Khan Qajar in 1796.

The dynasty was named after the Turkic Afshar tribe to which Nader belonged. The Afshars had migrated from Turkestan to Azerbaijan in the 13th century. In the early 17th century, the Persian Shah Abbas the Great moved many Afshars from Azerbaijan to Khorasan to defend the north-eastern borders of his state against the Uzbeks. Nader belonged to the Qereqlu branch of the Afshars.[5]

Foundation of the dynasty

Nader Shah was born (as Nader Qoli) into a humble semi-nomadic family of Khorasan, where he became a local warlord.[6] His path to power began when the Ghilzai Shah Mahmud overthrew the weak Safavid shah Sultan Husayn in 1722. At the same time, Ottoman and Russian forces seized Persian land. Nader joined forces with Sultan Husayn's son Tahmasp II and led the resistance against the Ghilzai Afghans, driving their leader Ashraf Khan out of the capital in 1729 and establishing Tahmasp on the throne. Nader fought to regain the lands lost to the Ottomans and restore Persian control of the Afghans. While he was away in the east fighting the Ghilzais, Tahmasp allowed the Ottomans to retake territory in the west. Nader, disgusted, had Tahmasp deposed in favour of his baby son Abbas III in 1732. Four years later, after he had recaptured most of the lost Persian lands, Nader was confident enough to have himself proclaimed shah in his own right at a ceremony on the Moghan Plain.[7]

Nader's conquests and the succession problem

Nader initiated a new religious policy aimed at reconciling Shia with Sunni Islam. The Safavid dynasty had relied heavily on the support of Shi'ites, but many soldiers in Nader's army were Sunnis. Nader also wanted to set himself up as a rival of the Ottoman sultan for supremacy within the Muslim world, which would have been impossible had he remained an orthodox Shi'ite.[8]

Soon afterwards Nader waged a war against the Afghans and captured Kandahar. In 1738, he invaded Mughal India, massacred 30,000 of the inhabitants of Delhi and in a single campaign captured an incredible amount of wealth, including the legendary Peacock Throne and the Koh-i-Nor diamond.[9]

Sindh silver rupee found in south east Pakistan, under the Afsharid Shahs of Iran 18th century.

After his return from India, Nader fell out with his eldest son Reza Qoli Mirza, who had ruled Persia during his father's absence. Having heard a rumour that Nader was dead, he had prepared to seize the throne by having the Safavid royal captives, Tahmasp and his son Abbas, executed. Nader was not pleased with the young man's behaviour and humiliated him by removing him from the post of viceroy. Nader became increasingly despotic, taxing his subjects heavily to pay for his military campaigns, and his health decayed. When there was an assassination attempt on him during an expedition to Daghestan, Nader blamed Reza and in 1742 had him blinded so he could not succeed to the throne [10] Nader's despotism and excessive demands for tax provoked many revolts. In 1747 while on his way to crush one of them, he was assassinated by two of his own officers. Iran was soon to descend into civil war.[11]

Civil war and downfall of the Afsharids

After Nader's death, his nephew Ali Qoli (who may have been involved in the assassination plot) seized the throne and proclaimed himself Adil Shah ("The Just Shah"). He ordered the execution of all Nader's sons and grandsons, with the exception of the 13-year old Shahrokh, the son of Reza Qoli.[12] Meanwhile, Nadir's former treasurer, Ahmad Shah Abdali, had declared his independence by founding the Durrani Empire in Khorasan. In the process, the eastern territories were lost and in the following decades became part of Afghanistan, the successor-state to the Durrani Empire.

Adil made the mistake of sending his brother Ebrahim to secure the capital Isfahan. Ebrahim decided to set himself up as a rival, defeated Adil in battle, blinded him and took the throne. Adil had reigned for less than a year. Meanwhile a group of army officers freed Shahrokh from prison in Mashhad and proclaimed him shah in October 1748. Ebrahim was defeated and died in captivity in 1750 and Adil was also put to death at the request of Nader Shah's widow. Shahrokh was briefly deposed in favour of another puppet ruler Soleyman II but, although blinded, Shahrokh was restored to the throne by his supporters. He reigned in Mashhad and from the 1750s his territory was mostly confined to Khorasan. In 1796 Mohammad Khan Qajar, the founder of the Qajar dynasty, seized Mashhad and tortured Shahrokh to force him to reveal the whereabouts of Nader Shah's treasures. Shahrokh died of his injuries soon after and with him the Afsharid dynasty came to an end.[13] [14] Shahrokh's descendants continue into the 21st century under the Afshar Naderi surname.

List of Afsharid Monarchs

- Nader Shah (1736–1747)
- Adil Shah (1747–1748)
- Ebrahim (1748)
- Shahrukh (1748–1796)

Family tree

Imam Qoli (*d.*1704)

Ebrahim Khan (*k.*1738)

Nader Shah 1736-1747

Adil Shah 1747-1748

Ebrahim 1748

Reza Qoli Mirza (*b.*1719-*k.*1747

Shahrukh 1748-1796

Nader Mirza (*k.*1803)

See also

- List of kings of Persia
- List of Shi'a Muslims dynasties

References

[1] Homa Katouzian, "Iranian history and politics", Published by Routledge, 2003. pg 128: "*Indeed, since the formation of the Ghaznavids state in the tenth century until the fall of Qajars at the beginning of the twentieth century, most parts of the Iranian cultural regions were ruled by Turkic-speaking dynasties most of the time. At the same time, the official language was Persian, the court literature was in Persian, and most of the chancellors, ministers, and mandarins were Persian speakers of the highest learning and ability*"

[2] Axworthy *Iran; Empire of the Mind* (Penguin) p. 159: "he [Nader] and his Safavid predecessors were of Turkic origin and spoke a Turkic language at court..."

[3] Michael Axworthy *The Sword of Persia* (I.B. Tauris, 2006), p. 19

[4] Jamie Stokes, Anthony Gorman, Encyclopedia of the Peoples of Africa and the Middle East, Infobase Publishing, 2010, p.11, Online Edition (http://books.google.com/books?id=stl97FdyRswC&printsec=frontcover&hl=en#v=onepage&q&f=false)

[5] *Cambridge History of Iran* Volume 7, pp.2-4

[6] Encyclopedia Iranica (http://www.iranica.com/newsite/index.isc?Article=http://www.iranica.com/newsite/articles/ot_grp10/ot_nadershah_20060329.html)

[7] Michael Axworthy *Iran: Empire of the Mind* (Penguin, 2008) pp.153-156

[8] Axworthy *Iran* p.157

[9] Axworthy *Iran* pp.158-159

[10] Axworthy *Iran* pp.160-161

[11] Axworthy *Iran* p.165

[12] *Cambridge History* p.59

[13] Axworthy p.168

[14] *Cambridge History* pp.60-62

External links

- "*Afsharids*", Encyclopedia Iranica (http://www.iranica.com/newsite/articles/v1f6/v1f6a023.html) (mostly about Asharids after Nader Shah)

Shia Islam

The Imam Hussein Shrine in Karbala, Iraq is a holy site for Shia Muslims

Shia Islam (Arabic: شيعة, *Shī'ah*) is the second largest denomination of Islam. Adherents of Shia Islam are called **Shi'ites** or **Shias**. "Shia" is the short form of the historic phrase *Shī'atu 'Alī* (شيعة علي), meaning "followers", "faction", or "party" of Muhammad's son-in-law and cousin Ali, whom the Shia believe to be Muhammad's successor. Like other branches of Islam, Shia Islam is based on the teachings of the Quran and the message of the Islamic prophet Muhammad.[1] [2] In contrast to other types, the Shia believe that only God has the right to choose a representative to safeguard Islam, the Quran and *sharia*. Thus the Shias look to Ali, Muhammad's son-in-law, whom they consider divinely appointed, as the rightful successor to Muhammad, and the first imam. The Shia extend this belief to Muhammad's family, the *Ahl al-Bayt* ("the People of the House"), and certain individuals among his descendants, known as *imams*, who have special spiritual and political authority over the community.

Although there were many Shia branches throughout history, modern Shia Islam is divided into three main branches.[3] The largest Shia sect in the early 21st century is the *Ithna ashariyya*,[4] commonly referred to in English as the Twelvers, while smaller branches include the Ismaili and Zaidi ("Fivers").[5] Alevis, Alawis and Druze are also considered sects of Shia Islam.

Etymology

The word *Shia* (Classical Arabic: شيعة *shīʻah* /ˈʃiːʕa/) means follower[6] and is the short form of the historic phrase *shīʻatu ʻAlī* (شيعة علي /ˈʃiːʕatu ˈʕaliː/), meaning "followers of Ali", "faction of Ali", or "party of Ali".[7] *Shi'a* and *Shiism* are forms used in English, while *Shi'ite* or *Shiite*, as well as *Shia*, refer to its adherents.

Beliefs

Imamate

Succession of Ali

The Investiture of Ali at Ghadir Khumm (MS Arab 161, fol. 162r, AD 1309/8 Ilkhanid manuscript illustration)

Shia Muslims believe that just as a prophet is appointed by God alone, only God has the prerogative to appoint the successor to his prophet. They believe God chose Ali to be Muhammad's successor, infallible, the first caliph (*khalifa*, head of state) of Islam. Muhammad, before his death, designated Ali as his successor.

Ali was Muhammad's first cousin and closest living male relative as well as his son-in-law, having married Muhammad's daughter Fatimah.[8] [9] Ali would eventually become the fourth Muslim caliph.[10]

After the last pilgrimage, Muhammad ordered the gathering of Muslims at the pond of Khumm and it was there Muhammad nominated Ali to be his successor. The hadith of the pond of Khumm was narrated on 18th of Dhu al-Hijjah of 10 AH in the Islamic calendar (March 10, 632 AD) at a place called Ghadir Khumm, located near the city of al-Juhfah, Saudi Arabia.[11] Muhammad there stated:

> Oh people! Reflect on the Quran and comprehend its verses. Look into its clear verses and do not follow its ambiguous parts, for by Allah, none shall be able to explain to you its warnings and its mysteries, nor shall anyone clarify its interpretation, other than the one that I have grasped his hand, brought up beside myself, [and lifted his arm,] the one about whom I inform you that whomever I am his master (Mawla), this Ali is his master (Mawla); and he is Ali Ibn Abi Talib, my brother, the executor of my will (Wasiyyi), whose appointment as your guardian and leader has been sent down to me from Allah, the mighty and the majestic.
>
> — Muhammad, **'The Farewell Sermon**[12]

Shia Muslims believe this to be Muhammad's appointment of Ali as his successor.

Ali's caliphate

When Muhammad died in 632 CE, Ali and Muhammad's closest relatives made the funeral arrangements. While they were preparing his body, Abu Bakr, Umar, and Abu Ubaidah ibn al Jarrah met with the leaders of Medina and elected Abu Bakr as caliph. Ali and his family were dismayed, but accepted the appointment for the sake of unity in the early Muslim community.[8] It was not until the murder of the third caliph, Uthman, in 657 CE that the Muslims in Medina invited Ali to become the fourth caliph,[8] and he established his capital in Kufah, in present-day Iraq.[7]

Ali's rule over the early Muslim community was often contested, and wars were waged against him. As a result, he had to struggle to maintain his power against the groups who betrayed him after giving allegiance to his succession, or those who wished to take his position. This dispute eventually led to the First Fitna, which was the first major civil war within the Islamic Caliphate. The Fitna began as a series of revolts fought against the first Imam, Ali ibn Abi Talib, caused by the assassination of his political predecessor, Uthman ibn Affan. While the rebels who accused Uthman of nepotism affirmed Ali's *khilafa* (caliph-hood), they later turned against him and fought him.[8] Ali ruled from 656 CE to 661 CE,[8] when he was assassinated[9] while prostrating in prayer (*sujud*). Ali's main rival Muawiyah then claimed the caliphate.[13]

Hasan

Upon the death of Ali, his elder son Hasan became leader of the Muslims of Kufa, and after a series of skirmishes between the Kufa Muslims and the army of Muawiyah, Hasan agreed to cede the caliphate to Muawiyah and maintain peace among Muslims. Hasan then retired to Medina, where in 50 AH he died by poisoning.

Hussein

Imam Ali Mosque in Najaf, Iraq, where Ali is buried.

Hussein, Ali's younger son and brother to Hasan, initially resisted calls to lead the Muslims against Muawiyah and reclaim the caliphate. In 680 CE Muawiyah died and passed the caliphate to his son Yazid; Yazid asked Hussein to allegiance (baya) him which Hussein rejected. Ali's faction saw this as a betrayal of the peace treaty, having expected the caliphate to return to Ali's line upon Muawiyah's death. There was a groundswell of support in Kufa for Hussein to return there and take his position as caliph and imam, so Hussein collected his family and followers in Medina and set off for Kufa. En route to Kufa, he was blocked by an army of Yazid's men near Karbala (modern Iraq), and Hussein and approximately 72 of his family and followers were killed in the Battle of Karbala.

The Shia regard Hussein as martyr (*shahid*), and count him as an imam from the Ahl al-Bayt. They view Hussein as the defender of Islam from annihilation at the hands of Yazid I. Hussein is the last imam following Ali whom all Shia sub-branches mutually recognise.[14] The Battle of Karbala is often cited as the definitive break between the Shia and Sunni sects of Islam, is commemorated each year by Shia Muslims on the Day of Ashura.

Imamate of the Ahl al-Bayt

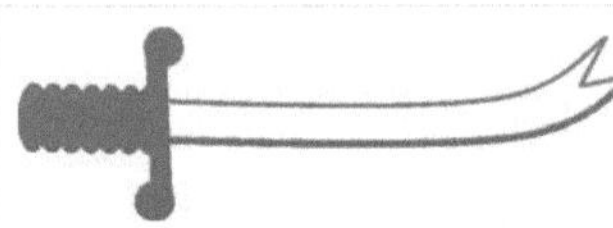
A fictional representation of the Sword of Ali, the Zulfiqar, two swords were captured from the temple of the pagan polytheist god Manāt during the Raid of Sa'd ibn Zaid al-Ashhali. Muhammad gave them to Ali, saying that one of them was Al-Dhulfiqar, which became the famous sword of Ali and a symbol of the Shia Islam[15]

Most of the early Shia as well as Zaydis differed only marginally from mainstream Sunnis in their views on political leadership, but it is possible in this sect to see a refinement of Shia doctrine. Early Sunnis traditionally held that the political leader must come from the tribe of Muhammad—namely, the Quraysh. The Zaydis narrowed the political claims of the Ali's supporters, claiming that not just any descendant of Ali would be eligible to lead the Muslim community (*ummah*) but only those males directly descended from Muhammad through the union of Ali and Fatimah. But during the Abbasid revolts, other Shia, who came to be known as Imamiyyah (followers of the imams), followed the theological school of Ja'far al-Sadiq. They asserted a more exalted religious role for imams and insisted that, at any given time, whether in power or not, a single male descendant of Ali and Fatimah was the divinely appointed imam and the sole authority, in his time, on all matters of faith and law. To those Shia, love of the imams and of their persecuted cause became as important as belief in God's oneness and the mission of Muhammad.[4]

Later most of Shia, including Twelver and Ismaili, became Imamis. Imami Shia believe that Imams are the spiritual and political successors to Muhammad.[4] Imams are human individuals who not only rule over the community with justice, but also are able to keep and interpret the divine law and its esoteric meaning. The words and deeds of Muhammad and the imams are a guide and model for the community to follow; as a result, they must be free from error and sin, and must be chosen by divine decree, or *nass*, through Muhammad.[16] [17]

According to this view, there is always an Imam of the Age, who is the divinely appointed authority on all matters of faith and law in the Muslim community. Ali was the first imam of this line, the rightful successor to Muhammad, followed by male descendants of Muhammad through his daughter Fatimah.[4]

This difference between following either the Ahl al-Bayt (Muhammad's family and descendants) or Caliph Abu Bakr has shaped Shia and non-Shia views on some of the Quranic verses, the hadith (narrations from Muhammad) and other areas of Islam. For instance, the collection of hadith venerated by Shia Muslims is centered on narrations by members of the Ahl al-Bayt and their supporters, while some hadith by narrators not belonging to or supporting the Ahl al-Bayt are not included (those of Abu Hurairah, for example who is a lier, Ibn Asakir in his Ta'rikh Kabir and Muttaqi in his Kanzu'l-Umma report that Caliph Umar lashed him, rebuked him, and forbade him to narrate hadith from the Holy Prophet. Umar said: "Because you narrate hadith in large numbers from the Holy Prophet, you are fit only for attributing lies to him. (That is, one expects a wicked man like you to utter only lies about the Holy Prophet.) So you must stop narrating hadith from the Prophet; otherwise, I will send you to the land of Dus." (A clan in Yemen, to which Abu Huraira belonged.)). According to Sunnis, Ali was the fourth successor to Abu Bakr, while the Shia maintain that Ali was the first divinely sanctioned "Imam," or successor of Muhammad. The seminal event in Shia history is the martyrdom in 680 CE at the Battle of Karbala of Ali's son Hussein ibn Ali, who led a non-allegiance movement against the defiant caliph (71 of Hussein's followers were killed as well). Hussein came to symbolize resistance to tyranny.

It is believed in Twelver and Ismaili Shia Islam that 'aql, divine wisdom, was the source of the souls of the prophets and imams and gave them esoteric knowledge called *ḥikmah* and that their sufferings were a means of divine grace to their devotees.[4] [18] [19] Although the Imam was not the recipient of a divine revelation, he had a close relationship with God, through which God guides him, and the imam in turn guides the people. Imamate, or belief in the divine guide, is a fundamental belief in the Twelver and Ismaili Shia branches and is based on the concept that God would not leave humanity without access to divine guidance.[20]

Theology

The Shia Islamic faith is vast and inclusive of many different groups.[7] Shia theological beliefs, and religious practises such as prayers slightly differ from the Sunnis. While all Muslims pray five times daily, Shias have the option of always combining *Dhuhr* with *Asr* and *Maghrib* with *Isha'*, as there are three distinct times mentioned in the Quran. The Sunnis tend to combine only under certain circumstances.[21] [22] Shia Islam embodies a completely independent system of religious interpretation and political authority in the Muslim world.[23] [24] The Shia identity emerged during the lifetime of Muhammad,[25] and Shia theology was formulated in the 2nd century AH, or after Hijra (8th century CE).[26] The first Shia governments and societies were established by the end of the 3rd century AH/9th century CE. The 4th century AH /10th century CE has been referred to by Louis Massignon as 'the Shiite Ismaili century in the history of Islam'.[27]

Hadith

The Shia believe that the status of Ali is supported by numerous hadith, including the Hadith of the pond of Khumm, Hadith of the two weighty things, Hadith of the pen and paper, Hadith of the invitation of the close families, and Hadith of the Twelve Successors. In particular, the Hadith of the Cloak is often quoted to illustrate Muhammad's feeling towards Ali and his family by both Sunni and Shia scholars. Shias prefer hadith attributed to the Ahl al-Bayt and close associates, and have their own separate collection of hadiths.[28] [29]

Profession of faith

The Shia version of the Shahada, the Islamic profession of faith, differs from that of the Sunni. The Sunni Shahada states *There is no god except Allah, Muhammad is the messenger of the God*, but to this the Shia append *Ali is his (Muhammad's)* wali *(caretaker)*. This phrase embodies the Shia emphasis on the inheritance of authority though Muhammad's lineage. The three clauses of the Shia Shahada thus address *tawhid* (the unity of God), *nubuwwah* (the prophethood of Muhammad), and *imamah* (imamate, the leadership of the faith).

Kalema at Qibla of the Mosque of Ibn Tulun in Cairo, Egypt with phrase "Ali-un-Waliullah"

Muhammad with Ali in Fatimid Shia heritage

pierced medallion just over the main entrance gate, Aqmar Mosque, Cairo

There are historical mosque of Fatimid period located in Cairo. The Aqmar Mosque is notable for its façade, which is elaborately decorated with inscriptions and geometric carving. There is inscription seen in pierced medallion just over the main entrance gate with quran verse 33;33 encircling names of Muhammad and Ali. It is 33:33: "O people of the house (Ahle-bayt), god only desires to put away from you abomination and with cleansing to cleanse you." Which Shia regards as evidence of claim for Imamat of Ali and his decedents who lost caliphate to win the invisible ones? [30]

In addition to above combination of Mohammad and Ali are further prominent in the decoration of the in the left half of façade of the mosque, here name of Ali is in center and encircled by five linked Muhammad.

Inscription of 5 'Muhammad'chain and Ali in middle at left wing of Aqmar Mosque, Cairo

chamfered Aqmar corner with three niches

There is chamfered corner appear three niches having verse 16:128: "verily god is with (top), those who are god fearing (bottom right), and with those who are good doers (bottom left)". On both sides of top niches there are roundel one having name of Muhammad and another of Ali. Hence top portion can read as having message: "Verily God is with Muhammad and Ali".[31]

Infallibility

Ismah is the concept of infallibility or "divinely bestowed freedom from error and sin" in Islam.[32] Muslims believe that Muhammad and other prophets in Islam possessed *ismah*. Twelver and Ismaili Shia Muslims also attribute the quality to Imams as well as to Fatimah, daughter of Muhammad, in contrast to the Zaidi, who do not attribute 'ismah to the Imams.

After Muhammad, Ali is credited as the first male to accept Islam

According to Shia theologians, infallibility is considered a rational necessary precondition for spiritual and religious guidance. They argue that since God has commanded absolute obedience from these figures they must only order that which is right. The state of infallibility is based on the Shia interpretation of the verse of purification.[33] [34] Thus, they are the most pure ones, the only immaculate ones preserved from, and immune to, all uncleanness.[35] It does not mean that supernatural powers prevent them from committing a sin, but due to the fact that they have an absolute belief in God, that they find themselves in the presence of God.[36]

They also have a complete knowledge of God's will. They are in possession of all knowledge brought by the angels to the prophets (*nabi*) and the messengers (*rasul*). Their knowledge encompasses the totality of all times. They thus act without fault in religious matters.[37] Shias regard Ali as the successor of Muhammad not only ruling over the community in justice, but also interpreting Islamic practices and its esoteric meaning. Hence he was regarded as being free from error and sin (infallible), and appointed by God by divine decree (*nass*) to be the first Imam.[38] Ali is known as "perfect man" (*al-insan al-kamil*) similar to Muhammad, according to Shia viewpoint.[39]

Occultation

The Occultation is a belief in some forms of Shia Islam that a messianic figure, a disappeared Imam known as the Mahdi, will one day return alongside Jesus and fill the world with justice. According to the Twelver Shia, the main goal of the Mahdi will be to establish an Islamic state and to apply Islamic laws that were revealed to Muhammad.[40]

Some Shia, such as the Zaidi and Nizari Ismaili, do not believe in the idea of the Occultation. The groups which do believe in it differ as to which lineage of the Imamate is valid, and therefore which individual has gone into occultation. They believe there are many signs that will indicate the time of his return.

Twelver Shia Muslims believe that Imam Mahdi (the twelfth imam, Muhammad al-Mahdi) is already on Earth, is in occultation and will return at the end of time. Fatimid/ Bohra/ Dawoodi Bohra believe the same but for their 21st Imam Tayyib. Whereas Sunnis believe the future Mahdi has not yet arrived on Earth.[41]

History

Historians dispute the origin of Shia Islam, with many Western scholars positing that Shiism began as a political faction rather than a truly religious movement.[4] [42] However, other scholars disagree, considering this concept or religious-political separation to be an anachronistic application of a Western concept.[43] Following the Battle of Karbala, as various Shi'a-affiliated groups diffused in the emerging Islamic world, several nations arose based around a Shi'a leadership or population.

- Idrisids (788 to 985 CE): a Zaydi dynasty in what is now Morocco
- Fatimids (909–1171 CE): Controlled much of North Africa, the Levant, parts of Arabia and Mecca and Medina.
- Uqaylids (990 to 1096 CE): a Shi'a Arab dynasty with several lines that ruled in various parts of Al-Jazira, northern Syria and Iraq.
- Buyids (934–1055 CE): at its peak consisted of large portions of modern Iraq and Iran.
- Ilkhanate (1256–1335): a Mongol khanate established in Persia in the 13th century, considered a part of the Mongol Empire. The Ilkhanate was based, originally, on Genghis Khan's campaigns in the Khwarezmid Empire in 1219–1224, and founded by Genghis's grandson, Hulagu, in territories which today comprise most of Iran, Iraq, Afghanistan, Turkmenistan, Armenia, Azerbaijan, Georgia, Turkey, and Pakistan. The Ilkhanate initially embraced many religions, but was particularly sympathetic to Buddhism and Christianity. Later Ilkhanate rulers, beginning with Ghazan in 1295, embraced Shia Islam.
- Bahmanis (1347–1527 CE): a Shia Muslim state of the Deccan in southern India and one of the great medieval Indian kingdoms.[44] Bahmanid Sultanate was the first independent Islamic Kingdom in South India.[45]

Safavids

A major turning point in Shia history was the Safavid dynasty (1501–1736) in Persia.

- The ending of the relative mutual tolerance between Sunnis and Shias that existed from the time of the Mongol conquests onwards and the resurgence of antagonism between the two groups.
- The beginning of the emergence of an independent body of *ulama* capable of taking a political stand different from the policies of the state.
- The growth in importance of Iranian centers of religious learning and change from Twelver Shiaism being a predominantly Arab phenomenon.[46]
- The growth of the Akhbari School which preached that only the Quran, hadith are to be bases for verdicts, rejecting the use of reasoning.

With the fall of the Safavids, the state in Persia – including the state system of courts with government-appointed judges (qadis) – became much weaker. This gave the Sharia courts of *mujtahids* an opportunity to fill in the slack and enabled "the ulama to assert their judicial authority." The Usuli School also increased in strength at this time.[47]

Community

Demographics

It is variously estimated that 10–20%[48] [49] [50] [51] of the world's Muslims are Shia. They may number up to 200 million as of 2009.[49] The Shia majority countries are Iran, Iraq, Azerbaijan and Bahrain.[52] They also constitute 36.3% of entire local population and 38.6% of the local Muslim population of the Middle East.[53]

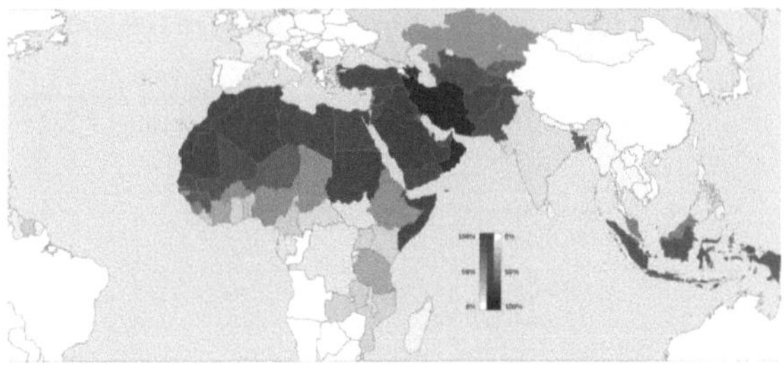
The Shia majority countries are Iran, Iraq, Azerbaijan, and Bahrain; all are coloured in red.

Shia Muslims constitute 27% of the population in Lebanon,[54] over 45% of the population in Yemen,[55] 20-40% of the population in Kuwait[49] [56] , over 20% in Turkey,[49] [57] 10–20% of the population in Pakistan,[49] and 10-19% of Afghanistan's population.[58] [59]

Distribution of Sunni and Shia branches of Islam

Saudi Arabia hosts a number of distinct Shia communities, including the Twelver Baharna in the Eastern Province and Nakhawila of Medina, and the Ismaili Sulaymani and Zaidiyyah of Najran. Estimations put the number of Shiite citizens at 2-4 million, accounting for roughly 15% of the local population.[60]

Significant Shia communities exist in the coastal regions of West Sumatra and Aceh in Indonesia (see Tabuik). The Shia presence is negligible elsewhere in Southeast Asia, where Muslims are predominantly Shafi'i Sunnis.

A significant Shia minority is present in Nigeria, made up of modern-era converts to a syncretic Shia movement, centered around Kano State. East Africa holds several populations of Ismaili Shia, primarily descendants of immigrants from South Asia during the colonial period, such as the Khoja.

According to Shia Muslims, one of the lingering problems in estimating Shia population is that unless Shia form a significant minority in a Muslim country, the entire population is often listed as Sunni. The reverse, however, has not held true, which may contribute to imprecise estimates of the size of each sect. For example, the 1926 rise of the House of Saud in Arabia brought official discrimination against Shia.[61]

List of nations with Shia population

Figures indicated in the first three columns below are based on the October 2009 demographic study by the Pew Research Center report, *Mapping the Global Muslim Population.*[49] [50]

Nations with over 100,000 Shia[49] [50]

Country	Shia population[49] [50]	Percent of Muslim population that is Shia[49] [50]	Percent of global Shia population[49] [50]	Minimum estimate/claim	Maximum estimate/claim
Iran	66,000,000 – 70,000,000	90–95	37–40		
Iraq	19,000,000 – 22,000,000	65–70	11–12		
Pakistan	17,000,000 – 26,000,000	10–15	10–15		43,250,000[62] – 57,666,666[63] [64]
India	16,000,000 – 24,000,000	10–15	9–14		40,000,000[65] – 50,000,000.[66]
Yemen	8,000,000 – 10,000,000	35–40	5		
Turkey	7,000,000 – 11,000,000	10–15	4–6		
Azerbaijan	5,000,000 – 7,000,000	65–75	3–4		85% of total population[67]
Afghanistan	3,000,000 – 4,000,000	10–15	<2		15–19% of total population[58]
Syria	3,000,000 – 4,000,000	15–20	<2		
Nigeria	<4,000,000	<5	<2		5-10 million[68]
Saudi Arabia	3,000,000 – 4,000,000	15–22	<1		
Lebanon	1,000,000 – 1,600,000[69]	30-35[70] [71] [72]	<1		Estimated, no official census.[73]
Tanzania	<2,000,000	<10	<1		
Kuwait	500,000 – 700,000	30–35	<1		35–40% of total population[56]
Germany	400,000 – 600,000	10–15	<1		
Bahrain	375,000 – 400,000	66–70	<1	375,000 (66%[74] of citizen population)	400,000 (70%[75] of citizen population)
Tajikistan	400,000	7	<1		
United Arab Emirates	300,000 – 400,000	10	<1		
United States	200,000 – 400,000	10–15	<1		
Oman	100,000 – 300,000	5–10	<1		948,750[76]

United Kingdom	100,000 – 300,000	10–15	<1		
Bulgaria	100,000	10–15	<1		
Qatar	100,000	10	<1		

Persecution

The history of Sunni-Shia relations has often involved violence, dating back to the earliest development of the two competing sects. Militarily established and holding control over the Umayyad government, many Sunni rulers perceived the Shia as a threat – both to their political and religious authority.[77]

The Sunni rulers under the Umayyads sought to marginalize the Shia minority, and later the Abbasids turned on their Shia allies and imprisoned, persecuted, and killed them. The persecution of the Shia throughout history by Sunni co-religionists has often been characterized by brutal and genocidal acts. Comprising only about 10–15% of the entire Muslim population, the Shia remain a marginalized community to this day in many Sunni Arab dominant countries without the rights to practice their religion and organize.[78]

At various times Shia groups have faced persecution.[79] [80] [81] [82] [83] [84] In 1514 the Ottoman sultan, Selim I, ordered the massacre of 40,000 Anatolian Shia.[85] According to Jalal Al-e-Ahmad, "Sultan Selim I carried things so far that he announced that the killing of one Shiite had as much otherworldly reward as killing 70 Christians."[86] In 1801 the Al Saud-Wahhabi armies attacked and sacked Karbala, the Shia shrine in eastern Iraq that commemorates the death of Husayn.[87]

In March 2011, the Malaysian government declared the Shia a 'deviant' sect and banned them from promoting their faith to other Muslims, but left them free to practise it themselves.[88]

Holidays

Both Sunni and Shia, celebrate the following annual holidays:

- Eid ul-Fitr, which marks the end of fasting during the month of Ramadan
- Eid al-Adha, which marks the end of the Hajj or pilgrimage to Mecca

Shia Muslims in Bahrain hitting their chests during the time of Muharram in remembrance of the hardships Hussein ibn Ali went through.

The following days are some of the most important holidays observed by Shia Muslims:

- Eid al-Ghadeer, which is the anniversary of the Ghadir Khum, the occasion when Muhammad announced Ali's Imamate before a multitude of Muslims.[89] Eid al-Ghadeer is held on the 18th of Dhu al-Hijjah.
- The Mourning of Muharram and the Day of Ashura for Shia commemorates Hussein ibn Ali's martyrdom. Imam Husayn was a grandson of Muhammad who was killed by Yazid ibn Muawiyah. Ashurah is a day of deep mourning which occurs on the 10th of Muharram.
- Arba'een commemorates the suffering of the women and children of Hussein ibn Ali's household. After Hussein was killed, they were marched over the desert, from Karbala (central Iraq) to Shaam (Damascus, Syria). Many children (some of whom were direct descendants of Muhammad) died of thirst and exposure along the route. Arba'een occurs on the 20th of Safar, 40 days after Ashurah.
- Mawlid, Muhammad's birth date. Unlike Sunni Muslims, who celebrate 12th of Rabi' al-awwal as Muhammad's birthday or deathday (because they said that birth & death both is in this week), Shia Muslims celebrate the 17th of the month, which also coincides with the birth date of the sixth imam, Ja'far al-Sadiq.[90] Note that, not all

Sunni Muslims celebrate Muhammad's birthday, stating it as a bid'ah.

- Fatimah's birthday on 20th of Jumada al-Thani. It's also considered as the "Women and Mothers' day".
- Ali's birthday on 13th of Rajab. It's also considered as the "Men and Fathers' day".
- Mid-Sha'ban is the birth date of the 12th and final Imam of Twelvers, Muhammad al-Mahdi. It is celebrated by Shia Muslims on the 15th of Sha'aban.
- Laylat al-Qadr, anniversary of the night of the revelation of the Quran.
- Eid al-Mubahila celebrates a meeting between the Ahl al-Bayt (household of Muhammad) and a Christian deputation from Najran. Al-Mubahila is held on the 24th of Dhu al-Hijjah.

Holy sites

The holiest sites common to all Muslims are Mecca, Medina and Jerusalem. For Shias, the Imam Husayn Shrine, Al Abbas Mosque in Karbala, and Imam Ali Mosque in Najaf are also highly revered.

Imam Husayn Shrine in Karbala. Two tall minarets of the Al Abbas Mosque are also seen in the picture.

Other venerated sites include Wadi-us-Salaam cemetery in Najaf, Al-Baqi' cemetery in Medina, Imam Reza shrine in Mashhad, Kadhimiya Mosque in Kadhimiya, Al-Askari Mosque in Samarra, Sahla Mosque and Great Mosque of Kufa in Kufa and several other sites in the cities of Qom, Susa and Damascus.

Most of the holy Islamic sites in Saudi Arabia have been destroyed by Wahhabis and the Saudi royal family, the most notable being the shrines and tombs in the Al-Baqi' cemetery in 1925.[91] In 2006 a bomb destroyed the shrine of Al-Askari Mosque.[92]

Branches

The Shia faith throughout its history split over the issue of the Imamate. The largest branch are the Twelvers, followed by the Zaidi and Ismaili. All three groups follow a different line of Imamate.

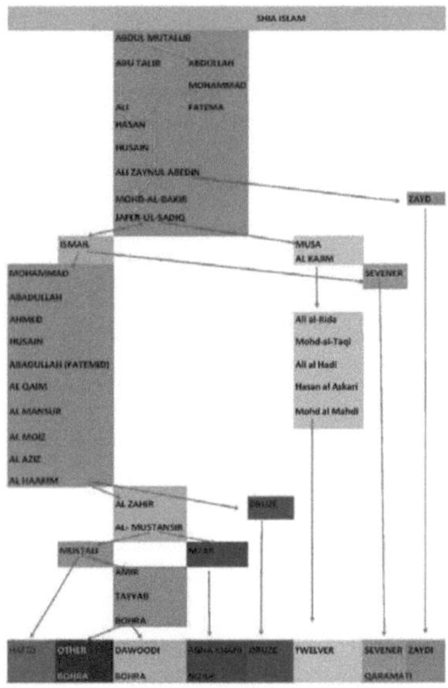

Branching of Shi'a Islam at a glance

Twelver

Twelver Shia or the **Ithnā'ashariyyah**' is the largest branch of Shia Islam, and the term *Shia Muslim* often refers to the Twelvers by default. The term *Twelver* is derived from the doctrine of believing in twelve divinely ordained leaders, known as The Twelve Imams. Twelver Shia are also known as *Imami* or *Ja'fari*, originated from the name of the 6th Imam, Ja'far al-Sadiq, who elaborated the twelver jurisprudence.[93]

Twelvers constitute the majority of the population in Iran (90%),[94] Azerbaijan (85%),[7] [67] Bahrain (70%), Iraq (65%) and Lebanon (65% of Muslims).[95] [96] [97]

Doctrine

Twelver doctrine is based on five principles.[98] These five principles known as *Usul ad-Din* are as follow:[99] [100]

1. **Monotheism**, God is one and unique.
2. **Justice**, the concept of moral rightness based on ethics, fairness, and equity, along with the punishment of the breach of said ethics.
3. **Prophethood**, the institution by which God sends emissaries, or prophets, to guide mankind.
4. **Leadership**, a divine institution which succeeded the institution of Prophethood. Its appointees (imams) are divinely appointed.
5. **Last Judgment**, God's final assessment of humanity.

More specifically, these principles are known as Usul al-Madhhab (principles of the Shia sect) according to Twelver Shias which differ from Daruriyat al-Din (Necessities of Religion) which are principles in order for one to be a Muslim. The Necessities of Religion do not include Leadership (Imamah) as it is not a requirement in order for one to be recognized as a Muslim. However, this category, according to Twelver scholars like Ayatollah al-Khoei, does include belief in God, Prophethood, the Day of Resurrection and other "necessities" (like belief in angels). In this regard, Twelver Shias draw a distinction in terms of believing in the main principles of Islam on the one hand, and specifically Shia doctrines like Imamah on the other.

The Twelve Imams

The *Twelve Imams* are the spiritual and political successors to Muhammad for the Twelvers.[4] According to the theology of Twelvers, the successor of Muhammad is an infallible human individual who not only rules over the community with justice but also is able to keep and interpret the divine law and its esoteric meaning. The words and deeds of Muhammad and the imams are a guide and model for the community to follow; as a result, they must be free from error and sin, and Imams must be chosen by divine decree, or *nass*, through Muhammad.[16] [17] Each Imam was the son of the previous Imam, with the exception of Hussein ibn Ali, who was the brother of Hasan ibn Ali.[4] The twelfth and final Imam is Muhammad al-Mahdi, who is believed by the Twelvers to be currently alive and in occultation.[20]

Jurisprudence

The Twelver jurisprudence is called *Ja'fari jurisprudence*. In this jurisprudence *Sunnah* is considered to be the oral traditions of Muhammad and their implementation and interpretation by the twelve Imams. There are three schools of Ja'fari jurisprudence: Usuli, Akhbari, and Shaykhi. The Usuli school is by far the largest of the three. Twelver groups that do not follow Ja'fari jurisprudence include the Alawi, Alevi, Bektashi, and Qizilbash.

In Ja'fari jurisprudence, there are ten ancillary pillars, known as *Furu' ad-Din*, which are as follows:[101]

1. *Prayer*
2. *Fasting*
3. *Pilgrimage*
4. *Alms giving*
5. *Struggle*
6. *Directing others towards good*
7. *Directing others away from evil*
8. *Alms giving* (One Fifth) *(20% tax on yearly earnings after deduction of household and commercial expenses.)*
9. *Love those who are in God's path*
10. *Disassociation with those who oppose God*

According to Twelvers, defining and interpretation of Islamic jurisprudence is the responsibility of Muhammad and the twelve Imams. As the 12th Imam is in occultation, it is the duty of clerics to refer to the Islamic literature such as the Quran and hadith and identify legal decisions within the confines of Islamic law to provide means to deal with current issues from an Islamic perspective. In other words, Twelver clerics provide Guardianship of the Islamic Jurisprudence, which was defined by Muhammad and his twelve successors. This process is known as *Ijtihad* and the clerics are known as *Marja'*, meaning reference. The labels *Allamah* and *Ayatollah* are in use for Twelver clerics.

Zaidi

Zaidiyya, **Zaidism** or **Zaydi** is the second largest branch of Shia islam. It is a Shia school named after Zayd ibn Ali. Followers of the Zaidi fiqh are called Zaidis (or occasionally **Fivers**). However, there is also a group called *Zaidi Wasītīs* who are Twelvers (see below). Zaidis constitute roughly 40–45% of the population of Yemen.[102]

Doctrine

The Zaydis, Twelvers and Ismailis recognize the same first four Imams; however, the Zaidis recognise Zayd ibn Ali as the fifth. After the time of Zayd ibn Ali, the Zaidis recognized that any descendant of Hasan ibn Ali or Hussein ibn Ali could be Imam after fulfilling certain conditions.[103] Other well-known Zaidi Imams in history were Yahya ibn Zayd, Muhammad al-Nafs az-Zakiyah and Ibrahim ibn Abdullah. In matters of Islamic jurisprudence, the Zaydis follow Zayd ibn Ali's teachings which are documented in his book *Majmu'l Fiqh* (in Arabic: مجموع الفِقه). Al-Hadi ila'l-Haqq Yahya, founder of the Zaydi state in Yemen, instituted elements of the jurisprudential tradition of the Sunni Muslim jurist Abū Ḥanīfa, and as a result, Zaydi jurisprudence today continues somewhat parallel to that of the Hanafis.

The Zaidi doctrine of Imamah does not presuppose the infallibility of the Imam nor that the Imams receive divine guidance. Zaidis also do not believe that the Imamate must pass from father to son but believe it can be held by any Sayyid descended from either Hasan ibn Ali or Hussein ibn Ali (as was the case after the death of Hasan ibn Ali). Historically, Zaidis held that Zayd was the rightful successor of the 4th Imam since he led a rebellion against the Umayyads in protest of their tyranny and corruption. Muhammad al-Baqir did not engage in political action, and the followers of Zayd believed that a true Imam must fight against corrupt rulers.

Timeline

The Idrisids (Arabic: ةسرادألا) were Arab[104] Zaydi Shia[105] [106] [107] [108] [109] [110] dynasty in the western Maghreb ruling from 788 to 985 C.E., named after its first sultan, Idris I.

A Zaydi state was established in Gilan, Deylaman and Tabaristan (northern Iran) in 864 C.E. by the Alavids;[111] it lasted until the death of its leader at the hand of the Samanids in 928 C.E. Roughly forty years later the state was revived in Gilan and survived under Hasanid leaders until 1126 C.E. Afterwards, from the 12th to 13th centuries, the Zaydis of Deylaman, Gilan and Tabaristan then acknowledged the Zaydi Imams of Yemen or rival Zaydi Imams within Iran.[112]

The Buyids were initially Zaidi[113] as well as the Banu Ukhaidhir rulers of al-Yamama in the 9th and 10th centuries.[114] The leader of the Zaydi community took the title of Caliph. As such, the ruler of Yemen was known as the Caliph, al-Hadi Yahya bin al-Hussain bin al-Qasim ar-Rassi Rassids (a descendant of Hasan ibn Ali the son of Ali) who, at Sa'dah, in 893-7 CE, founded the Zaydi Imamate, and this system continued until the middle of the 20th century, when the revolution of 1962 CE deposed the Zaydi Imam. The founding Zaidism of Yemen was of the Jarudiyya group; however, with increasing interaction with Hanafi and Shafi'i rites of Sunni Islam, there was a shift from the Jarudiyya group to the Sulaimaniyya, Tabiriyya, Butriyya or Salihiyya groups.[115] Zaidis form the second dominant religious group in Yemen. Currently, they constitute about 40–45% of the population in Yemen. Ja'faris and Isma'ilis are 2–5%.[116] In Saudi Arabia, it is estimated that there are over 1 million Zaydis (primarily in the western provinces).

Currently the most prominent Zaydi movement is Houthis movement, known by the name of *Shabab Al Mu'mineen* (Believing Youth). They have been the subject of an ongoing campaign against them by the Yemeni Government in which the army has lost 743 men, and thousands of innocent civilians have been killed or displaced by government forces causing a grave humanitarian crisis in north Yemen.[117]

Ismaili

Ismailis get their name from their acceptance of Isma'il ibn Jafar as the divinely appointed spiritual successor (Imam) to Ja'far al-Sadiq, wherein they differ from the Twelvers, who accept Musa al-Kadhim, younger brother of Isma'il, as the true Imam.

After the death or Occultation of Muhammad ibn Ismaill in the 8th century, the teachings of Ismailism further transformed into the belief system as it is known today, with an explicit concentration on the deeper, esoteric meaning (*bāṭin*) of the faith. With the eventual development of Twelverism into the more literalistic *(zahir)* oriented Akhbari and later Usuli schools of thought, Shiaism developed in two separate directions: the metaphorical Ismailli group focusing on the mystical path and nature of God and the divine manifestation in the personage of the "Imam of the Time" as the "Face of God", with the more literalistic Twelver group focusing on divine law (*sharī'ah*) and the deeds and sayings (*sunnah*) of Muhammad and his successors (the *Ahlu l-Bayt*), who as A'immah were guides and a light to God.[118]

Though there are several sub-groupings within the Ismailis, the term in today's vernacular generally refers to the Nizari community, who are followers of the Aga Khan and the largest group among the Ismailiyyah. Another community which falls under the Isma'il's are the Dawoodi Bohras, lead by a *Da'i al-Mutlaq* as representative of a hidden imam. While there are many other branches with extremely differing exterior practices, much of the spiritual theology has remained the same since the days of the faith's early Imams. In recent centuries Ismailis have largely been an Indo-Iranian community,[119] but they are found in India, Pakistan, Syria, Palestine, Saudi Arabia,[120] Yemen, China,[121] Jordan, Uzbekistan, Tajikistan, Afghanistan, East Africa and South Africa, and have in recent years emigrated to Europe, Australia, New Zealand, and North America.[122]

Ismaili Imams

After the death of Isma'il ibn Jafar, many Ismailis believed the line of Imamate ended and that one day the messianic Mahdi, whom they believed to be Muhammad ibn Ismail, would return and establish an age of justice. One group included the violent Qarmatians, who had a stronghold in Bahrain. In contrast, some Ismailis believed the Imamate *did* continue, and that the Imams were in occultation and still communicated and taught their followers through a network of dawah "Missionaries".

In 909, Ubayd Allah al-Mahdi Billah, a claimant to the Ismaili Imamate, established the Fatimid Caliphate. During this period, three lineages of imams formed. The first branch, known today as the Druze, began with Al-Hakim bi-Amr Allah. Born in 386 AH (985), he ascended as ruler at the age of eleven. The typical religiously tolerant Fatimid Empire saw much persecution under his reign. When in 411 AH (1021) his mule returned without him, soaked in blood, a religious group that was forming in his lifetime broke off from mainstream Ismailism and did not acknowledge his successor. Later to be known as the Druze, they believe al-Hakim to be the incarnation of God and the prophesied Mahdi who would one day return and bring justice to the world.[123] The faith further split from Ismailism as it developed very unusual doctrines which often class it separately from both Ismailiyyah and Islam.

The second split occurred following the death of Ma'ad al-Mustansir Billah in 487 AH (1094). His rule was the longest of any caliph in any Islamic empire. Upon his passing away, his sons, Nizar the older, and Al-Musta'li, the younger, fought for political and spiritual control of the dynasty. Nizar was defeated and jailed, but according to Nizari tradition, his son escaped to Alamut, where the Iranian Ismaili had accepted his claim.[124] From here on, the Nizari Ismaili community has continued with a present, living Imam.

The Mustaali line split again between the Taiyabi (Dawoodi Bohra is its main branch) and the Hafizi. The former claim that At-Tayyib Abi l-Qasim (son of Al-Amir bi-Ahkami l-Lah) and the imams following him went into a period of anonymity (*Dawr-e-Satr*) and appointed a Da'i al-Mutlaq to guide the community, in a similar manner as the Ismaili had lived after the death of Muhammad ibn Ismail. The latter (Hafizi) claimed that the ruling Fatimid Caliph was the Imam, and they died out with the fall of the Fatimid Empire.

Pillars

Ismailis have categorized their practices which are known as *seven pillars*:

- Walayah (Guardianship)
- Taharah (Purity)
- Salah (Prayer)
- Zakāt (Charity)
- Sawm (Fasting)
- Hajj (Pilgrimage)
- Jihad (Struggle)

The Shahada (profession of faith) of the Shia differs from that of Sunnis due to mention of Ali[125]

Contemporary leadership

The Nizaris place importance on a scholarly institution because of the existence of a present Imam. The Imam of the Age defines the jurisprudence, and his guidance may differ with Imams previous to him because of different times and circumstances. For Nizari Ismailis, the Imam is Karim al-Husayni Aga Khan IV. The Nizari line of Imams has continued to this day as an unending line.

Divine leadership has continued in the Bohra branch through the institution of the "Unrestricted Missionary" Dai. According to Bohra tradition, before the last Imam, At-Tayyib Abi l-Qasim, went into seclusion, his father, the 20th Imam Al-Amir bi-Ahkami l-Lah, had instructed Al-Hurra Al-Malika the Malika (Queen consort) in Yemen to appoint a vicegerent after the seclusion – the *Unrestricted Missionary*, who as the Imam's vicegerent has full authority to govern the community in all matters both spiritual and temporal while the lineage of Mustaali-Tayyibi Imams remains in seclusion (Dawr-e-Sitr). The three branches of the Mustaali, the Alavi Bohra, Sulaimani Bohra and Dawoodi Bohra, differ on who the current Unrestricted Missionary is.

See also

- List of extinct Shi'a sects
- List of Shia books
- List of Shi'a Muslim scholars of Islam
- List of Shi'a Muslims
- Shia Crescent
- Nikah mut'ah
- Bada'
- Rafida
- Hosay
- Hosay massacre
- al-Qaeda

Notes

[1] "Esposito, John. "What Everyone Needs to Know about Islam." Oxford University Press, 2002 I ISBN 978-0-19-515713-0. p. 40
[2] "From the article on Shii Islam in Oxford Islamic Studies Online" (http://www.oxfordislamicstudies.com/article/opr/t125/e2189?_hi=26&_pos=238). Oxfordislamicstudies.com. . Retrieved 2011-05-04.
[3] "Esposito, John. "What Everyone Needs to Know about Islam" Oxford University Press, 2002. ISBN 978-0-19-515713-0. p.40
[4] "Shīʿite" (http://www.britannica.com/EBchecked/topic/540503/Shiite). Encyclopædia Britannica Online. 2010. . Retrieved 2010-08-25.
[5] Tabataba'i (1979), p. 76
[6] Duncan S. Ferguson, (2010), *Exploring the Spirituality of the World Religions* (http://books.google.com/books?id=BPwHem3bV9sC&pg=PA192&dq=shia+means+follower+Duncan+S.+Ferguson&hl=en&ei=-00XTvfeGYao8QP91tUU&sa=X&oi=book_result&ct=result&resnum=1&ved=0CCgQ6AEwAA#v=onepage&q&f=false), p.192
[7] The New Encyclopædia Britannica, Jacob E. Safra, Chairman of the Board, 15th Edition, Encyclopædia Britannica, Inc., 1998, ISBN 0-85229-663-0, Vol 10, p. 738
[8] Merriam-Webster's Encyclopedia of World Religions, Wendy Doniger, Consulting Editor, Merriam-Webster, Incorporated, Springfield, MA 1999, ISBN 0-87779-044-2, LoC: BL31.M47 1999, p. 525
[9] "Esposito, John. "What Everyone Needs to Know about Islam" Oxford University Press, 2002. ISBN 978-0-19-515713-0. p. 46
[10] The New Encyclopædia Britannica, Jacob E. Safra, Chairman of the Board, 15th Edition, Encyclopædia Britannica, Inc., 1998, ISBN 0-85229-663-0, Vol 22, p. 17.
[11] "Event of Ghadir Khumm" (http://www.al-islam.org/ghadir/incident.htm). Al-islam.org. . Retrieved 2011-05-04.
[12] The Last Sermon of Muhammad by Shia Accounts
[13] The New Encyclopædia Britannica, Jacob E. Safra, Chairman of the Board, 15th Edition, Encyclopædia Britannica, Inc., 1998, ISBN 0-85229-663-0, Vol 10, p. tid738
[14] Discovering Islam: making sense of Muslim history and society (2002) Akbar S. Ahmed
[15] Religious trends in pre-Islamic Arabic poetry, By Ghulam Mustafa (Hafiz.), Pg 11 (http://books.google.co.uk/books?id=tMVkAAAAMAAJ), Author writes: Similarly, swords were also placed on the Idols, as it is related that Harith b. Abi Shamir, the Ghassanid king, had presented his two swords, called Mikhdham and Rasub, to the image of the goddess, Manat....to note that the famous

sword of Ali, the fourth caliph, called Dhu-al-Fiqar, was one of these two swords
[16] Nasr (1979), p.10
[17] Momen (1985), p. 174
[18] Corbin 1993, pp. 45-51
[19] Nasr (1979), p. 15
[20] Gleave, Robert. "Imamate". *Encyclopaedia of Islam and the Muslim world; vol.1*. MacMillan. ISBN 0-02-865604-0.
[21] "Learn to do Shia Prayer - Islamic Prayer - Shia Salat" (http://www.revertmuslims.com/sala.html). Revertmuslims.com. . Retrieved 2011-05-04.
[22] "Joining Prayers and Other related Issues" (http://www.al-islam.org/encyclopedia/chapter7/5.html). Al-islam.org. . Retrieved 2011-05-04.
[23] "Druze and Islam" (http://americandruze.com/Druze And Islam.html). American Druze.com. . Retrieved 2010-08-12.
[24] "Ijtihad in Islam" (http://alqazwini.org/qazwini_org/articles/by_articles/ijtihad.htm). AlQazwini.org. . Retrieved 2010-08-12.
[25] "Shi'ite Islam," by Allamah Sayyid Muhammad Husayn Tabataba'i, translated by Sayyid Husayn Nasr, State University of New York Press, 1975, p. 24
[26] Dakake (2008), pp. 1 and 2
[27] In his "Mutanabbi devant le siècle ismaëlien de l'Islam", in Mém. de l'Inst Français de Damas, 1935, p.
[28] "The Complete Idiot's Guide to World Religions," Brandon Toropov, Father Luke Buckles, Alpha; 3rd edition, 2004, ISBN 978-1-59257-222-9, p. 135
[29] "Shi'ite Islam" by Allamah Sayyid Muhammad Husayn Tabataba'i (1979), pp. 41–44
[30] Williams, Caroline. 1983. *The Cult of 'Alid Saints in the Fatimid Monuments of Cairo, Part I: The Mosque of al-Aqmar*. Muqarnas 1:37-52 and EI, first ed.,s.v. "shia'.
[31] Williams, Caroline. 1983. *The Cult of 'Alid Saints in the Fatimid Monuments of Cairo, Part I: The Mosque of al-Aqmar*. Muqarnas 1:37-52
[32] Dabashi, *Theology of Discontent*, p.463
[33] Qur'an 33:33
[34] Momen (1985), p. 155
[35] Corbin (1993), pp. 48 and 49
[36] Dabashi (2006), p. 463 (http://books.google.com/books?id=sTFdNNQP4ewC&pg=PA463&vq=Infallible&dq=Dabashi+Theology+of+Discontent&source=gbs_search_r&cad=1_1&sig=ACfU3U1qqg-6MZbV5y-TTuwAxtj26CIQhg)
[37] Corbin (1993), p. 48
[38] Motahhari, Perfect man, Chapter 1 (http://www.al-islam.org/perfectman/1.htm)
[39] How do Sunnis and Shi'as differ theologically? (http://www.bbc.co.uk/religion/religions/islam/subdivisions/sunnishia_1.shtml) Last updated 2009-08-19, BBC religions
[40] Nasr, Sayyed Hossein. "Expectation of the Millennium : Shiìsm in History,", State University of New York Press, 1989, p. 19, ISBN 978-0-88706-843-0
[41] "Comparison of Shias and Sunnis" (http://www.religionfacts.com/islam/comparison_charts/islamic_sects.htm). Religionfacts.com. . Retrieved 2011-05-04.
[42] See: Lapidus p. 47, Holt p. 72
[43] Jafri, S.H Mohammad. "The Origin and Early Development of Shi'a Islam,", Oxford University Press, 2002, p. 6, ISBN 978-0-19-579387-1
[44] "The Five Kingdoms of the Bahmani Sultanate" (http://orbat.com/site/cimh/kings_master/kings/ibrahimII_adil_shahi/5_provinces.html). orbat.com. . Retrieved 2007-01-05.
[45] Ansari, N.H. "Bahmanid Dynasty" (http://www.iranica.com/newsite/home/index.isc) *Encyclopaedia Iranica*
[46] Momen, Moojan, *An Introduction to Shi'i Islam*, Yale University Press, 1985, p.123
[47] Momen, Moojan, *An Introduction to Shi'i Islam*, Yale University Press, 1985, p.191, 130
[48] "Shī'ite" (http://www.britannica.com/EBchecked/topic/540503/Shiite). Encyclopædia Britannica Online. 2010. . Retrieved 2010-08-25.
[49] "Mapping the Global Muslim Population: A Report on the Size and Distribution of the World's Muslim Population" (http://pewforum.org/Muslim/Mapping-the-Global-Muslim-Population(6).aspx). *Pew Research Center*. October 7, 2009. . Retrieved 2010-08-25.
[50] Miller, Tracy, ed. (10 2009) (PDF). *Mapping the Global Muslim Population: A Report on the Size and Distribution of the World's Muslim Population* (http://pewforum.org/newassets/images/reports/Muslimpopulation/Muslimpopulation.pdf). Pew Research Center. . Retrieved 2009-10-08.
[51] "Religions" (https://www.cia.gov/library/publications/the-world-factbook/fields/2122.html). *CIA*. The World Factbook. 2010. . Retrieved 2010-08-25.
[52] "Quick guide: Sunnis and Shias" (http://news.bbc.co.uk/1/hi/world/middle_east/6213248.stm). *BBC News*. 2006-12-11. .
[53] *Atlas of the Middle East* (Second ed.). U.S.A: National Geographic. 2008 (published 15 April). pp. 80–81. ISBN 978-1-4262-0221-6
[54] "International Religious Freedom Report 2010" (http://www.state.gov/j/drl/rls/irf/2010/148830.htm). U.S. Government Department of State. . Retrieved 2010-11-17.
[55] "How many Shia?" (http://islamicweb.com/beliefs/cults/shia_population.htm). Islamicweb.com. . Retrieved 2011-05-04.
[56] "The actual percentage of Kuwaiti Shiites is 40 percent" (http://www.arabtimesonline.com/NewsDetails/tabid/96/smid/414/ArticleID/144073/reftab/69/Default.aspx). ArabTimesOnline. . Retrieved 2010-07-19.

[57] Shankland, David (2003). *The Alevis in Turkey: The Emergence of a Secular Islamic Tradition* (http://books.google.com/?id=lFFRzTqLp6AC&pg=PP1&lpg=PP1&dq=Religion+in+Turkey). Routledge (UK). ISBN 0-7007-1606-8. .

[58] "Shia women too can initiate divorce" (http://lcweb2.loc.gov/frd/cs/profiles/Afghanistan.pdf). Library of Congress Country Studies on Afghanistan. August 2008. . Retrieved 2010-08-27. "*Religion: Virtually the entire population is Muslim. Between 80 and 85 percent of Muslims are Sunni and 15 to 19 percent, Shia.*"

[59] "Afghanistan" (https://www.cia.gov/library/publications/the-world-factbook/fields/2122.html?countryName=Afghanistan&countryCode=af&regionCode=sas&#af). *Central Intelligence Agency (CIA)*. The World Factbook on Afghanistan. . Retrieved 2010-08-27. "Religions: Sunni Muslim 80%, Shia Muslim 19%, other 1%"

[60] al-Qudaihi, Anees (2009-03-24). "Saudi Arabia's Shia press for rights" (http://news.bbc.co.uk/2/hi/7959531.stm). BBC Arabic Service. . Retrieved 24 March 2009.

[61] "Discrimination towards Shia in Saudi Arabia" (http://www.wsws.org/articles/2001/oct2001/saud-o08.shtml). Wsws.org. 2001-10-08. . Retrieved 2011-05-04.

[62] "CIA - The World Factbook" (https://www.cia.gov/library/publications/the-world-factbook/geos/pk.html). Cia.gov. . Retrieved 2011-05-04.

[63] "Violence Against Pakistani Shias Continues Unnoticed | International News" (http://www.islamicinsights.com/news/international-news/violence-against-pakistani-shias-continues-unnoticed.html). Islamic Insights. . Retrieved 2011-05-04.

[64] Taliban kills Shia school children in Pakistan (http://www.presstv.ir/detail.aspx?id=86937&sectionid=351020401)

[65] "Shia women too can initiate divorce" (http://timesofindia.indiatimes.com/city/lucknow/Shia-women-too-can-initiate-divorce/articleshow/334804.cms). The Times of India. November 6, 2006. . Retrieved 2010-06-21.

[66] "Talaq rights proposed for Shia women" (http://www.dnaindia.com/india/report_talaq-rights-proposed-for-shia-women_1062327). Daily News and Analysis, www.dnaindia.com. 5 November 2006. . Retrieved 2010-06-21.

[67] Administrative Department of the President of the Republic of Azerbaijan - Presidential Library - Religion (http://files.preslib.az/projects/remz/pdf_en/atr_din.pdf)

[68] Nigeria: 'No Settlement With Iran Yet' (http://www.getnewsinfo.com/), Paul Ohia, allAfrica - This Day, 16 November 2010

[69] Hazran, Yusri. *The Shiite Community in Lebanon:* From Marginalization to Ascendancy, *Brandeis University*

[70] Hassan, Farzana. *Prophecy and the Fundamentalist Quest*, page 158

[71] Corstange, Daniel M. *Institutions and Ethnic politics in Lebanon and Yemen*, page 53

[72] Dagher, Carole H. *Bring Down the Walls: Lebanon's Post-War Challenge*, page 70

[73] Growth of the world's urban and rural population:n1920-2000, Page 81. United Nations. Dept. of Economic and Social Affairs

[74] "UK FCO" (http://www.fco.gov.uk/en/travel-and-living-abroad/travel-advice-by-country/country-profile/middle-east-north-africa/bahrain/). UK FCO. . Retrieved 3 March 2012.

[75] "Why Bahrain blew up" (http://www.nypost.com/p/news/opinion/opedcolumnists/why_bahrain_blew_up_NkYx4h4E1m80WGe2tXfsrI). New York Post. 2011-02-17. . Retrieved 2011-02-22.

[76] Top 15 Countries with Highest Proportion of Shiites in the Population (http://www.adherents.com/largecom/com_shiite.html), *7 July 1999*

[77] "The Origins of the Sunni/Shia split in Islam" (http://www.islamfortoday.com/shia.htm). Islamfortoday.com. . Retrieved 2011-05-04.

[78] Nasr,Vali (2006). *The Shia Revival: How Conflicts Within Islam Will Shape the Future.* W.W. Norton & Company Inc. ISBN 978-0-393-06211-3 p. 52-53

[79] *(Ya'qubi; vol.lll, pp. 91–96*, and *Tarikh Abul Fida', vol. I, p. 212.)*

[80] *The Psychologies in Religion, E. Thomas Dowd and Stevan Lars Nielsen, chapter 14* (http://books.google.com/?id=PcKBtc8bymoC&pg=PA237&dq=shia+persecution). Books.google.com. 2006-02-22. ISBN 978-0-8261-2856-0. . Retrieved 2011-05-04.

[81] "Basra handover completed" (http://www.inthenews.co.uk/news/autocodes/countries/iraq/basra-handover-completed-$1179488.htm). Inthenews.co.uk. . Retrieved 2011-05-04.

[82] Maddox, Bronwen (2006-12-30). "Hanging will bring only more bloodshed" (http://www.timesonline.co.uk/article/0,,30809-2523714,00.html). *The Times* (London). . Retrieved 2010-05-23.

[83] "Al-Ahram Weekly | Region | Shi'ism or schism" (http://weekly.ahram.org.eg/2004/681/re2.htm). Weekly.ahram.org.eg. 2004-03-17. . Retrieved 2011-05-04.

[84] The Shia, Ted Thornton, NMH, Northfield Mount Hermon (http://www.nmhschool.org/tthornton/mehistorydatabase/shia.php)

[85] George C. Kohn (2007.) *Dictionary of Wars*. Infobase Publishing. p.385. ISBN 0-8160-6577-2

[86] Al-e Ahmad, Jalal. *Plagued by the West* (*Gharbzadegi*), translated by Paul Sprachman. Delmor, NY: Center for Iranian Studies, Columbia University, 1982.

[87] " Saudi Arabia – The Saud Family and Wahhabi Islam (http://countrystudies.us/saudi-arabia/7.htm)". Library of Congress Country Studies.

[88] Malaysia bans Shias for promoting their faith (http://www.dailytimes.com.pk/default.asp?page=2011\03\10\story_10-3-2011_pg4_2)

[89] Paula Sanders (1994), *Ritual, politics, and the city in Fatimid Cairo* (http://books.google.com/books?id=9fnBFANHMn4C&pg=PA121&dq=ghadir+khumm&hl=en&ei=T3YYTurVBZHKswbu6cnqBA&sa=X&oi=book_result&ct=result&resnum=2&ved=0CC4Q6AEwAQ#v=onepage&q=ghadir khumm&f=false), p.121

[90] Bernard Trawicky, Ruth Wilhelme Gregory, (2002), *Anniversaries and holidays* (http://books.google.com/books?id=gDbKexa1jfcC&pg=PA233&dq=muhammad+birth+17+rabi+awal&hl=en&ei=7XEYTrGNEY30sgby7NiuDw&sa=X&oi=book_result&ct=result&

resnum=1&ved=0CCsQ6AEwAA#v=onepage&q=muhammad birth 17 rabi awal&f=false), p.233

[91] Laurence Louër (2008), *Transnational Shia politics: religious and political networks in the Gulf* (http://books.google.com/books?id=1vcfDuatockC&pg=PA22&dq=shia+holy+baqi+destroyed+wahhabi&hl=en&ei=P2UYTo3YA8r2sgbzhbXMDw&sa=X&oi=book_result&ct=result&resnum=1&ved=0CCgQ6AEwAA#v=onepage&q=shia holy baqi destroyed wahhabi&f=false), p.22

[92] Karen Dabrowska, Geoff Hann, (2008), *Iraq Then and Now: A Guide to the Country and Its People* (http://books.google.com/books?id=DhJ3lRnXyXcC&pg=PA239&dq=askari+mosque+bomb+2006&hl=en&ei=UGgYTvueE8zHsgagrOiZDw&sa=X&oi=book_result&ct=result&resnum=2&ved=0CCsQ6AEwAQ#v=onepage&q=askari mosque bomb 2006&f=false), p.239

[93] Vincent J. Cornell (2007), *Voices of Islam: Voices of tradition* (http://books.google.com/books?id=g5LNUS0ciAAC&pg=PA237&dq=jafar+sadiq+jurisprudence+shia&hl=en&ei=f-QYTsvBJoTysgb9nYyXBA&sa=X&oi=book_result&ct=result&resnum=4&ved=0CDYQ6AEwAw#v=onepage&q&f=false), p.237

[94] "Esposito, John. "What Everyone Needs to Know about Islam" Oxford University Press, 2002. ISBN 978-0-19-515713-0. p. 45.

[95] Esposito, John. "What Everyone Needs to Know about Islam" Oxford University Press, 2002. ISBN 978-0-19-515713-0. p. 45

[96] Religion in Bahrain (http://www.globalsecurity.org/military/world/gulf/bahrain-religion.htm)

[97] Challenges For Saudi Arabia Amidst Protests In The Gulf – Analysis (http://www.eurasiareview.com/25032011-challenges-for-saudi-arabia-amidst-protests-in-the-gulf-analysis/)

[98] Shiite doctrine Encyclopedia Iranica Retrieved 2011-07-08 (http://www.iranicaonline.org/articles/shiite-doctrine)

[99] Joanne Richter, (2006), *Iran the Culture* (http://books.google.com/books?id=X9fxXEdeIP8C&pg=PA7&dq=usul+din+shiite+five&hl=en&ei=glYXTrXZBoW28QPvweAW&sa=X&oi=book_result&ct=result&resnum=5&ved=0CDwQ6AEwBA#v=onepage&q=usul al-din&f=false), p.7]

[100] Mulla Bashir Rahim, *An Introduction to Islam* (http://www.al-islam.org/begin/intro/rahim.html), by Ahlul Bayt Digital Islamic Library Project

[101] *Iran the Culture* (http://books.google.com/books?id=X9fxXEdeIP8C&pg=PA7&dq=usul+din+shiite+five&hl=en&ei=glYXTrXZBoW28QPvweAW&sa=X&oi=book_result&ct=result&resnum=5&ved=0CDwQ6AEwBA#v=onepage&q=usul al-din&f=false) Joanne Richter (2007), p.7

[102] http://www.yemenincanada.ca/map.php , http://www.library.uu.nl/wesp/populstat/Asia/yemeng.htm

[103] Sunni-Shi'i Schism: Less There Than Meets the Eye (http://www.wrmea.com/backissues/0591/9105024.htm) 1991 Page 24

[104] Hodgson, Marshall (1961). *Venture of Islam*. Chicago: University of Chicago Press. pp. 262

[105] Ibn Abī Zar' al-Fāsī, 'Alī ibn 'Abd Allāh (1340). *Rawḍ al-Qirṭās: Anīs al-Muṭrib bi-Rawd al-Qirṭās fī Akhbār Mulūk al-Maghrib wa-Tārīkh Madīnat Fās*. ar-Rabāṭ: Dār al-Manṣūr (published 1972). pp. 38

[106] "" (http://hespress.com/?browser=view&EgyxpID=5116).

[107] Introduction to Islamic theology and law, By Ignác Goldziher, Bernard Lewis, pg.218 (http://books.google.com.au/books?id=6zeStDQZOSgC&pg=PA218&dq=Idrisid+dynasty+zaydi&as_brr=3&cd=1#v=onepage&q=Idrisid dynasty zaydi&f=false)

[108] Encyclopedia of Religion and Ethics, Part 24, By James Hastings, pg.844 (http://books.google.com.au/books?id=XBwOF6jXBdIC&pg=PA844&dq=Idrisid+dynasty+zaidi&as_brr=3&cd=1#v=onepage&q=&f=false)

[109] The Idrisids (http://www.iis.ac.uk/view_article.asp?ContentID=101310)

[110] Shi'ah tenets concerning the question of the imamate (http://www.muslimphilosophy.com/ik/Muqaddimah/Chapter3/Ch_3_25.htm)

[111] Article by Sayyid 'Ali ibn 'Ali Al-Zaidi, A short History of the Yemenite Shi'ites (2005) Referencing: Iranian Influence on Moslem Literature

[112] Article by Sayyid 'Ali ibn 'Ali Al-Zaidi, A short History of the Yemenite Shi'ites (2005) Referencing: Encyclopedia Iranica

[113] Walker, Paul Ernest (1999). *Hamid Al-Din Al-Kirmani: Ismaili Thought in the Age of Al-Hakim*. Ismaili Heritage Series. **3**. London ; New York: I.B. Tauris in association with the Institute of Ismaili Studies.. pp. 13. ISBN 1-86064-321-3

[114] Madelung, W. "al-Ukhayḍir." (http://www.brillonline.nl/subscriber/entry?entry=islam_SIM-7693) Encyclopaedia of Islam. Edited by: P. Bearman, Th. Bianquis, C.E. Bosworth, E. van Donzel and W.P. Heinrichs. Brill, 2007. Brill Online. 07 December 2007 (registration required)

[115] Article by Sayyid 'Ali ibn 'Ali Al-Zaidi, A short History of the Yemenite Shi'ites (2005)

[116] "Universiteit Utrecht Universiteitsbibliotheek" (http://www.library.uu.nl/wesp/populstat/Asia/yemeng.htm). Library.uu.nl. . Retrieved 2011-05-04.

[117] Shia Population of the Middle East (http://gulf2000.columbia.edu/images/maps/MidEastReligionCore_lg.jpg)

[118] "Shaykh Ahmad al-Ahsa'i" (http://www-personal.umich.edu/~jrcole/ahsai1.htm). . Retrieved 2007-04-25.

[119] Nasr, Vali, *The Shia Revival*, Norton, (2006), p. 76

[120] "Congressional Human Rights Caucus Testimony – NAJRAN, The Untold Story" (http://lantos.house.gov/HoR/CA12/Human+Rights+Caucus/Briefing+Testimonies/107/TESTIMONY+OF+ALI+H.+ALYAMI.htm). . Retrieved 2007-01-08.

[121] "News Summary: China; Latvia" (http://jmm.aaa.net.au/articles/11253.htm). . Retrieved 2007-06-01.

[122] Daftary, Farhad (1998). *A Short History of the Ismailis*. Edinburgh, UK: Edinburgh University Press. pp. 1–4. ISBN 0-7486-0687-4.

[123] "al-Hakim bi Amr Allah: Fatimid Caliph of Egypt" (http://baheyeldin.com/history/al-hakim-bi-amr-allah-fatimid-caliph-of-egypt.html). . Retrieved 2007-04-24.

[124] Daftary, Farhad (1998). *A Short History of the Ismailis*. Edinburgh, UK: Edinburgh University Press. pp. 106–108. ISBN 0-7486-0687-4.

[125] "Encyclopedia of the Middle East" (http://www.mideastweb.org/Middle-East-Encyclopedia/shahada.htm). Mideastweb.org. 2008-11-14. . Retrieved 2011-05-04.

References

- *Encyclopædia Britannica Online*. Encyclopædia Britannica, Inc..
- *Encyclopædia Iranica*. Center for Iranian Studies, Columbia University. ISBN 1-56859-050-4.
- Martin, Richard C.. *Encyclopaedia of Islam and the Muslim world; vol.1*. MacMillan. ISBN 0-02-865604-0.
- Corbin, Henry (1993 (original French 1964)). *History of Islamic Philosophy, Translated by Liadain Sherrard, Philip Sherrard*. London; Kegan Paul International in association with Islamic Publications for The Institute of Ismaili Studies. ISBN 0-7103-0416-1.
- Dakake, Maria Massi (2008). *The Charismatic Community: Shi'ite Identity in Early Islam*. SUNY Press. ISBN 0-7914-7033-4.
- Holt, P. M.; Bernard Lewis (1977a). *Cambridge History of Islam, Vol. 1*. Cambridge University Press. ISBN 0-521-29136-4.
- Lapidus, Ira (2002). *A History of Islamic Societies* (2nd ed.). Cambridge University Press. ISBN 978-0-521-77933-3.
- Momen, Moojan (1985). *An Introduction to Shi'i Islam: The History and Doctrines of Twelve*. Yale University Press. ISBN 0-300-03531-4.
- Sachedina, Abdulaziz Abdulhussein (1988). *The Just Ruler (al-sultān Al-ʻādil) in Shīʻite Islam: The Comprehensive Authority of the Jurist in Imamite Jurisprudence*. Oxford University Press US. ISBN 0-19-511915-0.
- Tabatabaei, Sayyid Mohammad Hosayn; Seyyed Hossein Nasr (translator) (1979). *Shi'ite Islam*. Suny press. ISBN 0-87395-272-3.

Further reading

- Peter J. Chelkowski (ed.), *Eternal Performance: Taziyah and Other Shiite Rituals* (Salt lake City (UT), Seagull Books, 2010) (Seagull Books - Enactments).
- Corbin, Henry (1993). *History of Islamic Philosophy, translated by Liadain Sherrard and Philip Sherrard*. Kegan Paul International in association with Islamic Publications for The Institute of Ismaili Studies. ISBN 0-7103-0416-1.
- Halm, Heinz (2004). *Shi'ism*. Edinburgh University Press. ISBN 0-7486-1888-0.
- Halm, Heinz (2007). *The Shi'ites: A Short History*. Markus Wiener Pub. ISBN 1-55876-437-2.
- Lalani, Arzina R. (2000). *Early Shi'i Thought: The Teachings of Imam Muhammad Al-Baqir*. I.B.Tauris. ISBN 1-86064-434-1.
- Momen, Moojan (1985). *An Introduction to Shi'i Islam: The History and Doctrines of Twelver Shi'ism*. Yale University Press. ISBN 0-300-03499-7.
- Shirazi, Sultanu'l-Wa'izin. *Peshawar Nights, A Transcript of a Dialogue between Shia and Sunni scholars* (http://www.al-islam.org/peshawar/index.html). Ansariyan Publications. ISBN 978-964-438-320-5.
- Nasr, Seyyed Hossein; Hamid Dabashi (1989). *Expectation of the Millennium: Shiʻism in History*. SUNY Press. ISBN 0-88706-843-X.
- Rogerson, Barnaby (2007). *The Heirs of Muhammad: Islam's First Century and the Origins of the Sunni Shia split*. Overlook Press. ISBN 1-58567-896-1.
- Wollaston, Arthur N. (2005). *The Sunnis and Shias*. Kessinger Publishing. ISBN 1-4254-7916-2.
- Moosa, Matti (1988). *Extremist Shiites: The Ghulat Sects*. Syracuse University Press. ISBN 0-8156-2411-5.

External links

- Early Troubles (http://shiabook.blogspot.co.uk/2011_01_01_archive.html)
- YaHusain.com, Shia Website with informative lectures in English & Urdu (http://www.yahusain.com/)
- Islamic - Shia Website (http://www.roshd.org/)
- Al-Islam.org, A Digital Islamic Library (http://www.al-islam.org/)
- The Shiapedia, an online Shia encyclopedia (http://theshiapedia.com/index.php?title=Main_Page)
- Shiite Muslim Ashura (http://pachian.ir/post-41.aspx)
- Murajaat - A Shi'i/Sunni debate (http://www.al-islam.org/murajaat/index.htm)
- Patheos Library – Shi'a Islam (http://www.patheos.com/Library/Shia-Islam.html)
- shiasource.com (http://www.shiasource.com/)
- Imam Al-Khoei Foundation (http://www.al-khoei.org/) (Twelver)
- Official Website of Nizari Ismaili (http://www.theismaili.org/) (Ismaili)
- Official Website of Alavi Bohra (http://www.alavibohra.org/) (Ismaili)
- Dawoodi Bohra (http://www.mumineen.org/) (Ismaili)
- The Institute of Ismaili Studies (http://www.iis.ac.uk/home.asp?l=en) (Ismaili)
- Shia (http://www.dmoz.org/Society/Religion_and_Spirituality/Islam/Shia/) at the Open Directory Project
- Institute for Interreligious Dialogue (http://www.iid.org.ir/IIDE/), Tehran
- al-shia.org (http://www.al-shia.org/html/eng/index.php) Ahlulbayt Global Informations Center
- Shiite Muslims on Tasua - Photos (http://ara-photos.net/en/1391/09/14/reuters-news-agency-photos-mourning-shiite-muslims/)

Zand dynasty

Zandiyeh dynasty	
← ← 1750–1794 →	
 Flag	 Coat of arms
Capital	Shiraz
Languages	Persian (official)
Government	Monarchy
Shah	
- 1750–1779	Karim Khan Zand (first)
- 1789–1794	Lotf Ali Khan Zand (last)
History	
- Zandiyeh dynasty begins	1750
- Qajar dynasty begins	1794

The **Zandiyeh dynasty** (Zand) (Persian: سلسله زندیه), was a dynasty led by Karim Khan Zand that ruled southern and central Iran in the 18th century.

History

Karim Khan Zand

Monument of Karim Khan in Shiraz

The dynasty was founded by Karim Khan, chief of the Zand tribe, which is a tribe of Kurds[1] [2] [3] [4] [5] [6] [7] or Lurs according to David Yerushalmi.[8] He became one of Nader Shah's generals.[9] Nader Shah moved the Zand tribe from their home in Lakestan to the eastern steppes of Khorasan. After Nader's death, the Zand tribe, under the guidance of Karim Khan, went back to their original land.[10] After Adil Shah was made king Karim Khan and his soldiers defected from the army and along with Ali Morad Khan Bakhtiari and Abolfath Khan Haft Lang, two other local chiefs, became a major contender but was challenged by several adversaries.[11] Abolfath Khan was the Prime Minister, Karim Khan became the army chief commander and Ali Morad Khan became the regent.[11] Karim Khan declared Shiraz his capital. He gained control of central and southern parts of Iran. In order to add legitimacy to his claim, Karim Khan placed the infant Shah Ismail III, the grandson of the last Safavid king in 1757 on the throne. Ismail was a figurehead king and real power was vested in Karim Khan. Karim Khan chose to be the military commander and Alimardan Khan was the civil administrator. Soon enough Karim Khan managed to eliminate his partner as well as the puppet king and in 1760, founded his own dynasty. He refused to accept the title of the king and instead named himself *The Advocate of the People*.

By 1760, Karim Khan had defeated all his rivals and controlled all of Iran except Khorasan, in the northeast, which was ruled by Shah Rukh. His foreign campaigns against Azad Khan in Azerbaijan and against the Ottomans in Mesopotamia brought Azerbaijan and the province of Basra into his control. But he never stopped his campaigns against his arch-enemy, Mohammad Hassan Khan Qajar, the chief of the Ghovanloo Qajars. The latter was finally defeated by Karim Khan and his sons, Agha Mohammad Khan and Hossein Quoli Khan, were brought to Shiraz as hostages.

Karim Khan's monuments in Shiraz include the famous Arg of Karim Khan, Vakil Bazaar, and several mosques and gardens. He is also responsible for building of a palace in the town of Tehran, the future capital of the Qajar dynasty.

Decline and Fall

Karim Khan's death in 1779 left his territory vulnerable to threats from his enemies. His son and successor Abu al-Fath was an incompetent ruler who was heavily influenced by his half uncle (and Karim Khan's commander), Zaki Khan. Other rulers such as Ali Morad and Jafar Khan also failed to follow the policies of Karim Khan and soon enough, the country was under attack from all sides.

The biggest enemies of the Zands, the Qajar chiefs, led by the former hostage, Agha Mohammad Khan, were advancing fast against the declining kingdom. Finally, in 1789, Lotf Ali Khan, a grand-nephew of Karim Khan, declared himself the new king. His reign (until 1794) was spent mostly in war with the Qajar khan. He was finally captured and brutally killed in the fortress of Bam, putting an effective end to the Zand Dynasty.

Politically, it is also important that the Zands, especially Karim Khan, chose to call themselves *Vakilol Ro'aya* (*Advocate of the People*) instead of kings. Other than the obvious propaganda value of the title, it can be a reflection of the popular demands of the time, expecting rulers with popular leanings instead of absolute monarchs who were totally detached from the population, like the earlier Safavids.

After the Islamic Revolution of 1979 the Zand was the only dynasty whose names on public places and monuments were not removed by the new Republican government. Interestingly, a union was formed between the Zands and the Qajars in so far as Karim Khan's grand daughter, Bolour Khanum Zandieh married Mohammad Shah Qajar and bore him two daughters, Princess Ozra and Princess Effat ed-Dowleh.[12]

Culture

Zand era art sample, dated 1790.

The Zand era was an era of relative peace and economic growth for the country.[13] Many territories that were once captured by the Ottomans in the late Safavid era were retaken, and Iran was once again a coherent and prosperous country. After Iranian painting reached its height at the end of the 17th century, a special school of painting took shape during the Zand era in the 17th and 18th centuries.[14] The art of this era is remarkable and, despite the short length of the dynasty, a distinct Zand art had the time to emerge. Many Qajar artistic traits were copied from the Zand examples.

In foreign policy, Karim Khan attempted to revive the Safavid era trade by allowing the British to establish a trading post in the port of Bushehr. This opened the hands of the British East India company in Iran and increased their influence in the country.[15] The taxation system was reorganized in a way that taxes were levied fairly. The judicial system was fair and generally humane. Capital punishment was rarely implemented.[16]

Zand Benevolent

Esteemed academic, John Perry—the leading English authority on Karim Khan Zand's era—writes of a forward thinking and notably popular leader, Karim Khan Zand, who he described as a man "before his time" and who—by opening up international trade, employing a fair fiscal system and showing respect for existing religious institutions—succeeded in creating a peaceful and prosperous state in a particularly turbulent epoch of history.[17]

The Zand Benevolent Trust (dedicated to humanitarism and charity) has been set up by a number of Karim Khan Zand's descendants, including Nazanin Khajeh-Noori and Michael-Mehrdod Khan-e Zand Khajeh-Noori (aka Michael Khajeh-Noori), both great grand children of Princess Bolour Khanum Zandieh, the grand daughter of Karim Khan Zand, who married the Qajar King, Mohammad Shah Qajar.[12] The Zand Benevolent Trust is a global charity dedicated to bringing hope and relief to children and the vulnerable.[18]

Order of Zand

The Most Noble Order of Zand, founded in 2009, has been set up to, among other things, bestow companionship each year on a handful of unsung heroes to the honorary Most Noble Order of the Zand (which may include a cash award) . True to the spirit of the man described by Sir John Malcolm as "though born of an inferior rank, obtained power without crime, and who exercised it with a moderation that, for the times in which he lived, was as singular as his humanity and justice" nominees for consideration for selection are taken from a list proposed by any member of the public from any country in the world.

George Hosking, Founder, Chief Executive and Research Co-ordinator of The WAVE Trust (Worldwide Alternatives to ViolencE) has been one of the first to be bestowed with companion membership of the order in recognition of his remarkable efforts and reflecting the enormity of his contribution to the humane development of mankind and the creation of a truly peaceful society for us and our children to live in. George Hosking has been singularly key in founding and progressing The WAVE Trust's work to prevent violence and child abuse by tackling it at its roots and making headway at a government and international level and the path to making compassion, empathy and peace the norm rather than an idealistic goal paid lip service by the many but heed by the few.

Rulers / Kings of the Zand dynasty

- Karim Khan, 1750–1779 دنز ناخ میر
- Abol Fath Khan, 1779 دنز ناخ حتفلوبا
- Mohammad Ali Khan, 1779 دنز ناخدمحم
- Sadiq Khan, 1779–1782 دنز ناخ قداص
- Ali Murad Khan, 1782–1785 دنز ناخ دارمیلع
- Jafar Khan, 1785–1789 دنز ناخ رفعج
- Lotf Ali Khan, 1789–1794 دنز ناخ یلعفطل

Vakeel mosque, Shiraz. *Karim-khan Zand*, and *Lotf-ali Khan Zand* are remembered well by the people of Shiraz.

Family tree

Bodaq Khan Zand		Agha Beygom		Inaq Khan Zand
Allah Morad Khan	Agha Beygom		**Karim Khan** 1750–1779	**Sadiq Khan** 1779–1781
Koda Morad Khan	**Ali Murad Khan** 1782–1785	**Abol Fath Khan** 1779	**Mohammad Ali Khan** 1779	**Jafar Khan Zand** 1785–1789
Sayed Murad Khan 1789				**Lotf Ali Khan** 1789–1794

External links

- http://www.iranchamber.com/history/zand/zand.php [19]
- http://www.iranologie.com/history/history13.html [20]
- http://www.friesian.com/iran.htm#zand [21]
- http://www.zandbenevolent.com [22]

See also

- History of Iran
- List of kings of Persia
- Karim Khan
- Persian name
- List of Shi'a Muslims dynasties

References

[1] *A fourth pretender was Karim Khan, son of Aymak of the Zand, a section of Lak tribe*, Sir Percy Molesworth Sykes, *A History of Persi*, Macmillan and co., limited, 1930, p. 277. (http://books.google.com/books?id=y6BCAAAAIAAJ&q="A+fourth+pretender+was+Karim+Khan,+son+of+Aymak+of+the+Zand"&dq="A+fourth+pretender+was+Karim+Khan,+son+of+Aymak+of+the+Zand"&hl=en&ei=b_RKTr-IK4jqmAWpmNHuBw&sa=X&oi=book_result&ct=result&resnum=1&ved=0CC0Q6AEwAA)

[2] *One of the contenders for power was Karim Khan Zand, a member of the Lak tribe near Shiraz*, William Marsden, Stephen Album, *Marsden's Numismata orientalia illustrata*, Attic Books, 1977, ISBN 978-0-915018-16-1, p. 158. (http://books.google.com/books?id=zDtmAAAAMAAJ&q="One+of+the+contenders+for+power+was+Karim+Khan+Zand,+a+member+of+the+Lak+tribe+near+Shiraz,"&dq="One+of+the+contenders+for+power+was+Karim+Khan+Zand,+a+member+of+the+Lak+tribe+near+Shiraz,"&hl=en&ei=vfVKTuL8J6H8mAXfp8TxBw&sa=X&oi=book_result&ct=result&resnum=1&ved=0CCkQ6AEwAA)

[3] Karim Khan, the founder of the Zand dynasty of Persia that succeeded the Afsharids, was himself born to a family of these Lak deportees (of the Zand tribe), Mehrdad R. Izady, *The Kurds: A Concise Handbook*, Taylor & Francis, 1992, ISBN 978-0-8448-1727-9, p. 12. (http://books.google.com/books?id=I9mr6OgLjBoC&pg=PA12&dq="Karim+Khan,+the+founder+of+the+Zand+dynasty+of+Persia+that+succeeded+the+Afsharids,+was+himself+born+to+a+family+of+these+Lak+deportees+(of+the+Zand+tribe).+"&hl=en&ei=MfdKTpKjO8r9mAXQjtWSCA&sa=X&oi=book_result&ct=result&resnum=1&ved=0CCkQ6AEwAA#v=onepage&q="Karim Khan, the founder of the Zand dynasty of Persia that succeeded the Afsharids, was himself born to a family of these Lak deportees (of the Zand tribe)."&f=false)

[4] *Kurdish leader, Karim Khan Zand,...*, Wadie Jwaideh, *The Kurdish National Movement: Its Origins and Development*, Syracuse University Press, 2006, ISBN 978-0-8156-3093-7, p. 17. (http://books.google.com/books?id=FCbspX-dGPYC&pg=PA17&dq="Kurdish+leader,+Karim+Khan+Zand,+who+for+nearly"&hl=en&ei=9flKTu6BGcL-mAW206zxBw&sa=X&oi=book_result&ct=result&resnum=1&ved=0CCwQ6AEwAA#v=onepage&q="Kurdish leader, Karim Khan Zand, who for nearly"&f=false)

[5] Lokman I. Meho, Kelly L. Maglaughlin, *Kurdish Culture and Society: An Annotated Bibliography*, Greenwood Publishing Group, 2001. ISBN 978-0-313-31543-5, p. 308. (http://books.google.com/books?id=sl4PIeyWriUC&pg=PA308&dq="Karim+Khan+Zand,+the+founder+of+the+Kurdish+Zand+dynasty"&hl=en&ei=QftKTviXOLD2mAWLiPSICA&sa=X&oi=book_result&ct=result&resnum=1&ved=0CCwQ6AEwAA#v=onepage&q="Karim Khan Zand, the founder of the Kurdish Zand dynasty"&f=false)

[6] *...the bulk of the evidence points to their being one of the northern Lur or Lak tribes, who may originally have been immigrants of Kurdish origin.*, Peter Avery, William Bayne Fisher, Gavin Hambly, Charles Melville (ed.), *The Cambridge History of Iran: From Nadir Shah to the Islamic Republic*, Cambridge University Press, 1991, ISBN 978-0-521-20095-0, p. 64. (http://books.google.com/books?id=H20Xt157iYUC&pg=PA64&dq="the+bulk+of+the+evidence+points+to+their+being+one+of+the+northern+Lur+or+Lak+tribes,+who+may+originally+have+been+immigrants+of+Kurdish+origin.+"&hl=en&ei=Qv1KTtz2K4mOmQX07_X5Bw&sa=X&oi=book_result&ct=result&resnum=1&ved=0CCkQ6AEwAA#v=onepage&q="the bulk of the evidence points to their being one of the northern Lur or Lak tribes, who may originally have been immigrants of Kurdish origin."&f=false)

[7] Jwaideh, Wadie (2006). *The Kurdish national movement: its origins and development* (http://books.google.no/books?id=FCbspX-dGPYC&pg=PA17&dq=Karim+KhanZand+Kurdish#v=onepage&q=Karim Khan Kurdish&f=false). .

[8] *Muhammad Karim Khan, of the Zand clan of the Lur tribe, suc- ceeded in imposing his authority on parts of the defunct Safavid empire*, David Yeroushalmi, *The Jews of Iran in The Nineteenth Century: Aspects of History, Community, and Culture*, BRILL, 2009, ISBN 978-90-04-15288-5, p. xxxix. (http://books.google.com/books?id=XYlGS3s3zTQC&pg=PR39&dq="Muhammad+Karim+Khan,+of+the+Zand+clan+of+the+Lur+tribe,+suc-+ceeded+in+imposing+his+authority+on+parts+of+the+defunct+Safavid+empire"&hl=en&ei=_gJLTsPHGYPEmAXM_oiPCA&sa=X&oi=book_result&ct=result&resnum=1&ved=0CCsQ6AEwAA#v=onepage&q="Muhammad Karim Khan, of the Zand clan of the Lur tribe, suc- ceeded in imposing his authority on parts of the defunct Safavid empire"&f=false)

[9] History of Iran: Zand Dynasty (http://www.iranchamber.com/history/zand/zand.php)
[10] http://www.iranologie.com/history/history13.html
[11] History of Iran (http://www.farhangsara.com/history_afshar.htm)
[12] http://www.qajarpages.org/mohammadshahchildren.html
[13] http://www.bookrags.com/wiki/Zand_dynasty
[14] New Page 1 (http://www2.irib.ir/worldservice/englishRADIO/IRAN/iranart.htm)
[15] Iran & Iranian History – Afshar and Zand:: Iran Visitor (http://www.iranvisitor.com/index.php?cID=417&pID=1159)
[16] History of Iran – Kajar Dynasty – InfoHub (http://www.infohub.com/forums/showthread.php?t=2812)
[17] Makers of the Muslim World Series -Karim Khan Zand, by John R. Perry, Oneworld Publications October 2006, ISBN 1-85168-435-2
[18] http://www.zandbeloved.org
[19] http://www.iranchamber.com/history/zand/zand.php
[20] http://www.iranologie.com/history/history13.html
[21] http://www.friesian.com/iran.htm#zand
[22] http://www.zandbenevolent.com

Qajar dynasty

Sublime State of Persia

Dowlat-e Elliye-ye Irân

←  **1785–1925**  →

←   →

Flag

Coat of arms

Anthem
Salâmati-ye Shâh
(Well-being of the King)

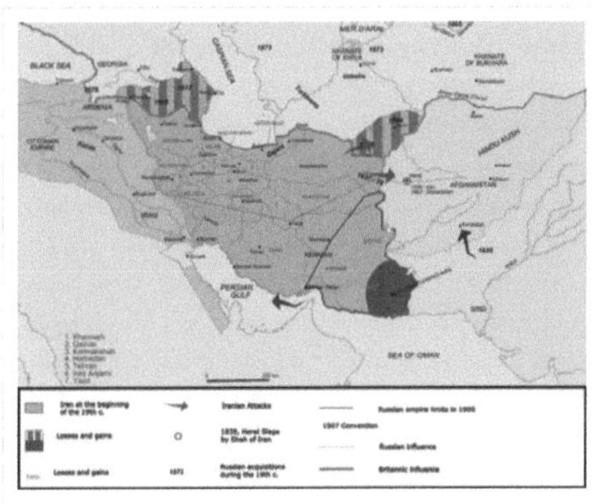

Map of Iran under the Qajar dynasty in the 19th century.

Capital		Tehran
Languages		Persian (court literature, administrative, cultural, official),[1] [2]
Government		Absolute monarchy (1785-1906) Constitutional monarchy (1906-1925)
Shah, Mirza		
-	1794-1797	Mohammad Khan Qajar (first)
-	1909-1925	Ahmad Shah Qajar (last)
Prime Minister		
-	1906	Mirza Nasrullah Khan (first)

-	1923-1925	Reza Pahlavi (last)
History		
-	Qajar dynasty begins	1785
-	Treaty of Gulistan	1813
-	Treaty of Turkmenchay	1828
-	Constitutional Revolution	1906
-	Pahlavi dynasty begins	1925
Currency		Toman
Today part of		

The **Qajar dynasty** (Qajar) (Persian: سلسله قاجاریه or دودمان قاجار; also anglicized as Ghajar or Kadjar) is a Persianized[3] Iranian royal family of Turkic origin,[4] [5] [6] [7] [8] which ruled Persia (Iran) from 1785 to 1925.[3] [9] The Qajar family took full control of Iran in 1794, deposing Lotf 'Ali Khan, the last of the Zand dynasty, and re-asserted Persian sovereignty over parts of the Caucasus. In 1796, Mohammad Khan Qajar seized Mashhad which was under Durrani suzerainty,[10] putting an end to the Afsharid dynasty, and Mohammad Khan was formally crowned as shah.[11]

Origins

The Qajar (or Ghajar) rulers were members of the Karagoz of the Qajars, originally the members of the Qarapapaqs of the larger Oghuz peoples.[4] [5] [6] [7] Qajars first settled during the Mongol period in the vicinity of Azerbaijan and were among the seven Qizilbash tribes that supported the Safavids.[12] The Safavids "left Arran (present-day Republic of Azerbaijan) to local Turkic speaking khans",[13] and, "in 1554 Ganja was governed by Shahverdi Soltan Ziyadoglu Qajar, whose family came to govern Karabakh in southern Arran".[14]

Qajars filled a number of diplomatic missions and governorships in the 16-17th centuries for the Safavids. The Qajars were resettled by Shah Abbas I throughout Persia. The great number of them also settled in Astarabad (present-day Gorgan, Iran) near the south-eastern corner of the Caspian Sea,[5] and it would be this branch of Qajars that would rise to power. The immediate ancestor of Qajars, Shah Qoli Khan Qajar Quvanlu (also spelled Ghovanloo) of the Quvanlus of Ganja, married into the Quvanlu Qajars of Astarabad. His son, Fath Ali Khan Qajar (born c. 1685-1693) was a renowned military commander during the rule of the Safavid shahs Husayn and Tahmasp II. He was killed on the orders of Tahmasp Qoli Khan Afshar (Nader Shah) in 1726. Fath Ali Khan's son Mohammad Hassan Khan Qajar (1722-1758) was killed at the behest of Karim Khan Zand, and was the father of Agha Mohammad Khan and Hossein Qoli Khan (Jahansouz Shah) Qajar (father of "Baba Khan," the future Fath Ali Shah Qajar).

Within 126 years between the demise of the Safavid state and the rise of Nasir al-Din Shah, the Qajars evolved from a shepherd-warrior tribe with strongholds in northern Persia into a Persian dynasty with all the trappings of a Perso-Islamic monarchy.[3]

Rise to power

"Like virtually every dynasty that ruled Persia since the 11th century, the Qajars came to power with the backing of Turkic tribal forces, while using educated Persians in their bureaucracy".[15] In 1779, after brutally murdering Mohammad Karim Khan Zand, the Zand dynasty ruler of southern Persia, Agha Mohammad Khan, the leader of the Qajar tribe, set out to reunify Iran. Agha Mohammad Khan was known as one of the cruelest kings, even by the 18th century Iranian standards.[5] In his quest for power, he razed cities, massacred entire populations, and blinded some 20,000 men in the city of Kerman because the local populace had chosen to defend the city against his siege.[5]

The Qajar armies were composed of a small number of Turkoman bodyguards and Georgian slaves.[16] By 1794, Agha Mohammad Khan had eliminated all his rivals, including Lotf 'Ali Khan, the last of the Zand dynasty. He reestablished Persian control over the territories in the Caucasus. Agha Mohammad established his capital at Tehran, a village near the ruins of the ancient city of Rayy. In 1796 he was formally crowned as shah. In 1797, Agha Mohammad was assassinated in Shusha, the capital of Karabakh khanate, and was succeeded by his nephew, Fath Ali Shah Qajar.

War with Russia

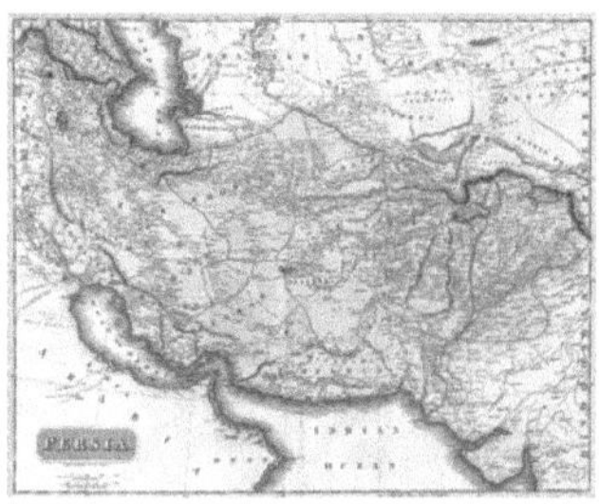

1814 map of Persia.

In 1803, under Fath Ali Shah, Qajars set out to fight against the Russian Empire, in what was known as Russo-Persian War of 1804-1813, due to concerns about the Russian expansion into Caucasus which was an Iranian domain, although some of the Khanates of the Caucasus were considered independent or semi-independent by the time of Russian expansion in 19th century,[17] this period marked the first major economic and military encroachments on Iranian interests during the colonial era. Qajar army suffered a major military defeat in the war and under the terms of the Treaty of Gulistan in 1813, Iran recognized Russian annexation of Georgia and most of the Caucasus region. The second Russo-Persian War of the late 1820s ended even more disastrously for Qajar Iran with temporary occupation of Tabriz and the signing of Treaty of Turkmenchay in 1828, acknowledging Russian sovereignty over the entire South Caucasus, the area north of the Aras River.

Fath Ali Shah's reign saw increased diplomatic contacts with the West and the beginning of intense European diplomatic rivalries over Iran. His grandson Mohammad Shah, who fell under the Russian influence and made two unsuccessful attempts to capture Herat, succeeded him in 1834. When Mohammad Shah died in 1848 the succession passed to his son Nasser-e-Din, who proved to be the ablest and most successful of the Qajar sovereigns.

Development and decline

Mullahs in the royal presence. The painting style is distinctly Qajar.

During Nasser-e-Din Shah's reign, Western science, technology, and educational methods were introduced into Persia and the country's modernization was begun. Nasser ed-Din Shah tried to exploit the mutual distrust between Great Britain and Russia to preserve Persia's independence, but foreign interference and territorial encroachment increased under his rule. He contracted foreign loans to finance expensive trips to Europe. These trips were part of a strategy to put Persia on the map as an independent, ancient but civilized state . Although the trips in this field were rather successful, he was not able to prevent Britain and Russia from encroaching into regions of traditional Persian influence. In 1856, during the Anglo-Persian War, Britain prevented Persia from reasserting control over Herat. The city had been part of Persia in Safavid times, but Herat had been under non-Persian rule since the mid-18th century. Britain also extended its control to other areas of the Persian Gulf during the 19th century. Meanwhile, by 1881, Russia had completed its conquest of present-day Turkmenistan and Uzbekistan, bringing Russia's frontier to Persia's northeastern borders and severing historic Persian ties to the cities of Bukhara and Samarqand. Several trade concessions by the Persian government put economic affairs largely under British control. By the late 19th century, many Persians believed that their rulers were beholden to foreign interests.

A Zoroastrian family in Qajar Iran

Mirza Taghi Khan Amir Kabir, was the young prince Nasser-e-Din's advisor and constable. With the death of Mohammad Shah in 1848, Mirza Taqi was largely responsible for ensuring the crown prince's succession to the throne. When Nasser ed-Din succeeded to the throne, Amir Nezam was awarded the position of prime minister and the title of Amir Kabir, the Great Ruler.

At that time, Persia was nearly bankrupt. During the next two and a half years Amir Kabir initiated important reforms in virtually all sectors of society. Government expenditure was slashed, and a distinction was made between the private and public purses. The instruments of central administration were overhauled, and Amir Kabir assumed responsibility for all areas of the bureaucracy. Foreign interference in Persia's domestic affairs was curtailed, and foreign trade was encouraged. Public works such as the bazaar in Tehran were undertaken. Amir Kabir issued an edict banning ornate and excessively formal writing in government documents; the beginning of a modern Persian prose style dates from this time.

A former Persian Legation in Washington, D.C.

One of the greatest achievements of Amir Kabir was the building of Dar ol Fonoon, the first modern university in Persia and the Middle East. Dar-ol-Fonoon was established for training a new cadre of administrators and acquainting them with Western techniques. Amir Kabir ordered the school to be built on the edge of the city so it could be expanded as needed. He hired French and Russian instructors as well as Persians to teach subjects as different as Language, Medicine, Law, Geography, History, Economics, and Engineering. Unfortunately, Amir Kabir did not live long enough to see his greatest monument completed, but it still stands in Tehran as a sign of a great man's ideas for the future of his country.

These reforms antagonized various notables who had been excluded from the government. They regarded the Amir Kabir as a social upstart and a threat to their interests, and they formed a coalition against him, in which the queen mother was active. She convinced the young shah that Amir Kabir wanted to usurp the throne. In October 1851 the shah dismissed him and exiled him to Kashan, where he was murdered on the shah's orders. Through his marriage to Ezzat od-Doleh, Amir Kabir had been the brother-in-law of the shah.

Constitutional Revolution

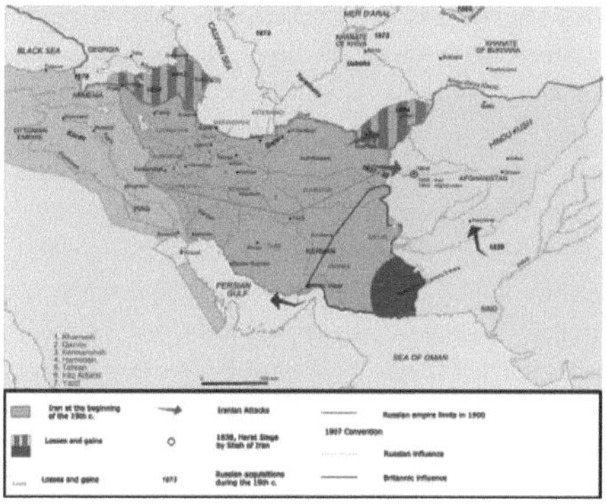

Persia in 19th and 20th centuries.

When Nasser al-Din Shah Qajar was assassinated by Mirza Reza Kermani in 1896, the crown passed to his son Mozaffar-e-din. Mozaffar-e-din Shah was moderate and kind, but was a relatively ineffective ruler. Royal extravagances coincided with an inadequate ability to secure state revenue which further exacerbated the financial woes of the Qajar. In response the Shah procured two large loans from Russia (in part to fund personal trips to Europe.) Public anger mounted as the Shah sold off concessions - such as road building monopolies, authority to collect duties on imports, etc. - to European interested in return for generous payments to the Shah and his officials. Popular demand to curb arbitrary royal authority in favor of rule of law increased as concern regarding growing foreign penetration and influence heightened.

Qajar-era currency bill featuring a depiction of Nasser al-Din Shah Qajar.

The shah's failure to respond to protests by the religious establishment, the merchants, and other classes led the merchants and clerical leaders in January 1906 to take sanctuary from probable arrest in mosques in Tehran and outside the capital. When the shah reneged on a promise to permit the establishment of a "house of justice", or consultative assembly, 10,000 people, led by the merchants, took sanctuary in June in the compound of the British legation in Tehran. In August the shah, through the issue of a decree promised a constitution. In October an elected assembly convened and drew up a constitution that provided for strict limitations on royal power, an elected parliament, or Majles, with wide powers to represent the people, and a government with a cabinet subject to confirmation by the Majles. The shah signed the constitution on December 30, 1906, but refusing to forfeit all of his power to the Majles, attached a caveat that made his signature on all laws required for their enactment. He died five days later. The Supplementary Fundamental Laws approved in 1907 provided, within limits, for freedom of press, speech, and association, and for security of life and property. The hopes for constitutional rule were not realized, however.

Mozaffar-e-din Shah's son Mohammad Ali Shah (reigned 1907–1909), who, through his mother, was also the grandson of Prime-Minister Amir Kabir (see before), with the aid of Russia, attempted to rescind the constitution and abolish parliamentary government. After several disputes with the members of the Majlis, in June 1908 he used his Russian-officered Persian Cossacks Brigade to bomb the Majlis building, arrest many of the deputies, and close down the assembly. Resistance to the shah, however, coalesced in Tabriz, Isfahan, Rasht, and elsewhere. In July 1909, constitutional forces marched from Rasht to Tehran lead by Mohammad Vali Khan Sepahsalar Khalatbari Tonekaboni, deposed the Shah, and re-established the constitution. The ex-shah went into exile in Russia. Mohammad Ali Shah died in San Remo, Italy in April 1925. As fate would have it, every future Shah of Iran would also die in exile.

On 16 July 1909, the Majles voted to place Mohammad Ali Shah's 11 year old son, Ahmad Shah on the throne. Although the constitutional forces had triumphed, they faced serious difficulties. The upheavals of the Constitutional Revolution and civil war had undermined stability and trade. In addition, the ex-shah, with Russian support, attempted to regain his throne, landing troops in July 1910. Most serious of all, the hope that the Constitutional Revolution would inaugurate a new era of independence from the great powers ended when, under the Anglo-Russian Agreement of 1907, Britain and Russia agreed to divide Persia into spheres of influence. The Russians were to enjoy exclusive right to pursue their interests in the northern sphere, the British in the south and east; both powers would be free to compete for economic and political advantage in a neutral sphere in the center. Matters came to a head when Morgan Shuster(also spelled Schuster), a United States administrator hired as treasurer general by the Persian government to reform its finances, sought to collect taxes from powerful officials who were Russian protégés and to send members of the treasury gendarmerie, a tax department police force, into the Russian zone. When in December 1911 the Majlis unanimously refused a Russian ultimatum demanding Shuster's dismissal, Russian troops, already in the country, moved to occupy the capital. To prevent this, on 20 December, Bakhtiari chiefs and their troops surrounded the Majles building, forced acceptance of the Russian ultimatum, and shut down the assembly, once again suspending the constitution.

Fall of the dynasty

Soltan Ahmad Shah was born 21 January 1898 in Tabriz, and succeeded to the throne at age 11. However, the occupation of Persia during World War I by Russian, British, and Ottoman troops was a blow from which Ahmad Shah never effectively recovered.

In February 1921, Reza Khan, commander of the Persian Cossack Brigade, staged a coup d'état, becoming the effective ruler of Iran. In 1923, Ahmad Shah went into exile in Europe. Reza Khan induced the Majles to depose Ahmad Shah in October 1925, and to exclude the Qajar dynasty permanently. Reza Khan was subsequently proclaimed Shah as *Reza Shah Pahlavi*, reigning from 1925 to 1941,

Ahmad Shah died on 21 February 1930 in Neuilly-sur-Seine, France.

Shahs of Persia, 1794-1925

	Name	Portrait	Title	Born-Died	Entered office	Left office
1	Mohammad Khan Qajar	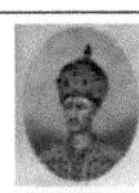	Shahanshah	1742 - 1797	20 March 1794	17 June 1797
2	Fat'h-Ali Shah Qajar		Shahanshah	1772 - 1834	17 June 1797	23 October 1834
3	Mohammad Shah Qajar		Shah	1808 - 1848	23 October 1834	5 September 1848
4	Naser al-Din Shah Qajar		Shahanshah	1831 - 1896	5 September 1848	1 May 1896

5	Mozaffar al-Din Shah Qajar		Shahanshah and Sultan	1853 - 1907	1 May 1896	3 January 1907
6	Mohammad Ali Shah Qajar		Shahanshah	1872 - 1925	3 January 1907	16 July 1909
7	Ahmad Shah Qajar		Sultan	1898 - 1930	16 July 1909	15 December 1925

Qajar Royal Family

The Qajar Imperial Family in exile is currently headed by the eldest descendant of Mohammad Ali Shah, Soltan Mohammad Ali Mirza Qajar, while the Heir Presumptive to the Qajar throne is Mohammad Hassan Mirza II, the grandson of Mohammad Hassan Mirza, Soltan Ahmad Shah's brother and heir. Mohammad Hassan Mirza died in England in 1943, having proclaimed himself shah in exile in 1930 after the death of his brother in France.

Today, the descendants of the Qajars often identify themselves as such and hold reunions to stay socially acquainted through the *Kadjar Family Association*.[18]

Qajar dynasty since 1925

Heads of the Qajar Imperial Family

The headship of the Imperial Family is inherited by the eldest male descendant of Mohammad Ali Shah.

- Soltan Ahmad Shah Qajar (1925-1930)
- Fereydoun Mirza (1930-1975)
- Soltan Hamid Mirza (1975-1988)
- Soltan Mahmoud Mirza (1988)
- Soltan Ali Mirza Qajar (1988–2011)
- Soltan Mohammad Ali Mirza (2011–present)

Heirs Presumptive of the Qajar dynasty

The Heir Presumptive is the Qajar heir to the Persian throne.

- Soltan Ahmad Shah Qajar (1925-1930)
- Mohammad Hassan Mirza (1930-1943)
- Fereydoun Mirza (1943-1975)
- Soltan Hamid Mirza (1975-1988)
- Mohammad Hassan Mirza II (1988–)

Notable members of Qajar family

Politics

- Mohammed Mosaddeq, Iranian Prime Minister, a qajar descendant on his maternal side
- Prince Abdol Hossein Mirza Farmanfarma, Iranian prime minister
- Amir Abbas Hoveyda, Iranian economist and politician, prime minister of Iran from 1965 to 1977, a qajar descendant on his maternal side
- Ali Amini, Iranian Prime Minister
- Prince Iraj Eskandari, Iranian communist politician
- Princess Maryam Farman Farmaian (b.1914-d.2008) Iranian communist politician, founder of the women's section of the Tudeh Party of Iran
- Ardeshir Zahedi (b. 1928-) Iranian diplomat, qajar descendant on his maternal side

Mohammad Mosaddeq

Social work

- Princess Sattareh Farmanfarmaian, Iranian Social work pioneer
- Prince Sabbar Farmanfarmaian, health minister in Mosaddeq cabinet

Business

- Princess Fakhr-od-dowleh

Women rights

- Princess Mohtaram Eskandari, intellectual and pioneering figures in Iranian women's movement

Iraj Mirza

Literature

- Prince Iraj Mirza (b. 1874-d. 1926), Iranian poet and translator
- Princess Lobat Vala (b. 1930), Iranian poet and campaigner for the Women Liberation
- Nader Naderpour, Iranian poet, a Qajar descendant on his maternal side
- Fereydoun Moshiri, Iranian poet, a Qajar descendant on his maternal side
- Shahrnush Parsipur, Iranian novelist, a Qajar descendant on her maternal side

Music

- Gholam-Hossein Banan, Iranian musician and singer, Qajar descendant on his maternal side

Gholam-Hossein Banan

Popular culture

- Marjane Satrapi, Iranian graphic novelist, Qajar descendant on her maternal side

Entertainment

- Princess Shireen Fathi, an American-Iranian performer, most notably as a dancer in the Persian Parade held annually in New York city, a Qajar descendant on her paternal side

Fereydoun Moshiri

See also

- Iran
- History of Iran
- List of kings of Persia
- Mirza Kouchek Khan
- Abdolhossein Teymourtash
- Qajar art
- List of Shi'a Muslims dynasties

References

[1] Homa Katouzian, "State and Society in Iran: The Eclipse of the Qajars and the Emergence of the Pahlavis", Published by I.B.Tauris, 2006. pg 327: "In post-Islamic times, the mother-tongue of Iran's rulers was often Turkic, but Persian was almost invariably the cultural and administrative language"

[2] Homa Katouzian, "Iranian history and politics", Published by Routledge, 2003. pg 128: "*Indeed, since the formation of the Ghaznavids state in the tenth century until the fall of Qajars at the beginning of the twentieth century, most parts of the Iranian cultural regions were ruled by Turkic-speaking dynasties most of the time. At the same time, the official language was Persian, the court literature was in Persian, and most of the chancellors, ministers, and mandarins were Persian speakers of the highest learning and ability*"

[3] Abbas Amanat, The Pivot of the Universe: Nasir Al-Din Shah Qajar and the Iranian Monarchy, 1831–1896, I.B.Tauris, pp 2–3

[4] Genealogy and History of Qajar (Kadjar) Rulers and Heads of the Imperial Kadjar House (http://www.kadjarfamily.org/articles_11.cfm)

[5] Cyrus Ghani. *Iran and the Rise of the Reza Shah: From Qajar Collapse to Pahlavi Power*, I.B. Tauris, 2000, ISBN 1-86064-629-8, p. 1

[6] William Bayne Fisher. *Cambridge History of Iran*, Cambridge University Press, 1993, p. 344, ISBN 0-521-20094-6

[7] Dr Parviz Kambin, A History of the Iranian Plateau: Rise and Fall of an Empire, Universe, 2011, p.36, online edition (http://books.google.com/books?id=tZkSLozhXxAC&printsec=frontcover&hl=en#v=onepage&q).

[8] Jamie Stokes, Anthony Gorman, Encyclopedia of the Peoples of Africa and the Middle East, 2010, p.707, Online Edition (http://books.google.com/books?id=stl97FdyRswC&printsec=frontcover&hl=en), *The Safavid and Qajar dynasties, rulers in Iran from 1501 to 1722 and from 1795 to 1925 respectively, were Turkic in origin.*

[9] Choueiri, Youssef M., *A companion to the history of the Middle East*, (Blackwell Ltd., 2005), 231,516.

[10] H. Scheel; Jaschke, Gerhard; H. Braun; Spuler, Bertold; T Koszinowski; Bagley, Frank (1981). *Muslim World* (http://books.google.com.pk/books?id=VNgUAAAAIAAJ). Brill Archive. p. 370. ISBN 9004061967, 9789004061965. . Retrieved 28 September 2012.

[11] Qajar Dynasty on Encyclopædia Britannica (http://www.britannica.com/eb/article-9062092/Qajar-Dynasty)

[12] Encyclopedia Iranica. The Qajar Dynasty. Online Edition (http://www.iranica.com/newsite/articles/v13f3/v13f3001e.html)

[13] K. M. Röhrborn, Provinzen und Zentralgewalt Persiens im 16. und 17. Jahrhundert, Berlin, 1966, p. 4

[14] Encyclopedia Iranica. Ganja. Online Edition (http://www.iranica.com/articles/v10f3/v10f372.html)

[15] Nikki R. Keddie. "The Iranian Power Structure and Social Change 1800-1969: An Overview", *International Journal of Middle East Studies*, Vol. 2, No. 1. (January, 1971), p. 4

[16] Ira Marvin Lapidus. *A History of Islamic Societies*, Cambridge University Press, 2002, ISBN 0-521-77933-2, p. 469.

"*Even when rulers on the plateau lacked the means to effect suzerainty beyond the Aras, the neighboring Khanates were still regarded as* ***Iranian dependencies***. Naturally, it was those Khanates located closes to the province of Azarbaijan which most frequently experienced attempts to re-impose Iranian **suzerainty**: the Khanates of Erivan, Nakhchivan and Qarabagh across the Aras, and the cis-Aras Khanate of Talish, with its administrative headquarters located at Lankaran and therefore very vulnerable to pressure, either from the direction of Tabriz or Rasht. Beyond the Khanate of Qarabagh, **the Khan of Ganja** and the Vali of Gurjistan (ruler of the Kartli-Kakheti kingdom of south-east Georgia), although less accessible for purposes of coercion, **were also regarded as the Shah's vassals, as were the Khans of Shakki and Shirvan, north of the Kura river**. The contacts between Iran and the Khanates of Baku and Qubba, however, were more tenuous and consisted mainly of maritime commercial links with Anzali and Rasht. The effectiveness of these somewhat haphazard assertions of suzeiranty dependend on the ability of a particular Shah to make his will felt, and the determination of the local khans to evade obligations they regarded as onerous." **The Cambridge history of Iran By William Bayne Fisher, Published by Cambridge University Press, 1991, pp. 145-146**

[18] "Qajar People" (http://www.kadjarfamily.org/). Qajars. . Retrieved 31 October 2012.

External links

- The Qajar (Kadjar) Pages (http://www.qajarpages.org/)
- Qajar Family Website (http://www.kadjarfamily.org/)
- Royal Ark-Qajar Website (http://www.4dw.net/royalark/Persia/qajar13.htm/) by Christopher Buyers
- Royal Ark-Qajar Website (http://www.4dw.net/royalark/Persia/qajar11.htm) by Christopher Buyers
- Some Photos of Qajar Family Members (http://www.worldisround.com/articles/255115/index.html)

Battle of Chaldiran

The **Battle of Chaldiran** or **Chaldoran** (Persian: چالدران , Turkish: *Çaldıran*) occurred on 23 August 1514 and ended with a victory for the Ottoman Empire over the Safavid Empire. As a result, the Ottomans gained immediate control over eastern Anatolia and northern Iraq. The battle, however, was just the beginning of 41 years of destructive war between the two empires that only ended in 1555 with the Treaty of Amasya. While the Ottomans often had the upper hand, the Persians for the most part held their ground. All Safavid losses in Shia-dominated metropolitan regions of Persia, such as Azerbaijan, Luristan and Kirmanshahan, proved temporary, being recovered from the Ottomans soon after each battle. The loss of Iraq, as well as Kurdish and Armenian eastern Anatolia, however, became permanent.

At Chaldiran, the Ottomans had a larger, better equipped army numbering 60,000 to 200,000, while the Kizilbash Turcomans numbered some 40,000 to 80,000. Shah Ismail I, who was wounded and almost captured in the battle, retired to his palace and withdrew from government administration[1] after his wives were captured by Selim I,[2] with at least one married off to one of Selim's statesmen.[3] The battle is one of major historical importance because it not only negated the idea that the Murshid of the Shia-Qizilbash was infallible,[4] but it also fully defined the Ottoman-Safavid borders and led Kurdish chiefs to assert their authority and switch their allegiance from the Safavids to the Ottomans.[5]

Background

After Selim I's successful struggle against his brothers for the throne of the Ottoman Empire, he was free to turn his attention to the internal unrest he believed was stirred up by the Shia Kizilbash, who had sided with other members of the Dynasty against him and had been semi-officially supported by Bayezid II. Selim now feared that they would incite the population against his rule in favor of Shah Isma'il leader of the Shia Safavids, and by some of his supporters believed to be family of the Prophet. Selim secured a jurist opinion that described Isma'il and the Kizilbash as "unbelievers and heretics" enabling him to undertake extreme measures on his way eastward to pacify the country.[6] In response, Shah Isma'il accused Sultan Selim of aggression against fellow Muslims, violating religious sexual rules and shedding innocent blood.[7]

When Selim started his march east, the Safavids were invaded in the east by the Uzbek state recently brought to prominence by Abu 'l-Fath Muhammad, who had fallen in battle against Isma'il only a few years before. To avoid the prospect of fighting a war on two fronts, Isma'il employed a scorched earth policy against Selim in the west.[8]

The terrain of eastern Anatolia and the Caucuses is extremely rough and combined with the difficulty in supplying the army in light of Isma'il's scorched earth campaign while marching against Muslims, Selim's army was discontented. The Janissaries even fired their muskets at the Sultan's tent in protest at one point. When Selim learned of the Safavid army forming at Chaldiran, he quickly moved to engage Isma'il in part to stifle the discontent of his army.[9]

Battle

The Ottomans deployed heavy artillery and thousands of Janissaries equipped with gunpowder weapons behind a barrier of carts. The Safavids used cavalry to engage the Ottoman forces. The Safavids attacked the Ottoman wings in an effort to avoid the Ottoman artillery positioned at the center. However, the Ottoman artillery was highly maneuverable and the Safavids suffered disastrous losses.[10] The advanced Ottoman weaponry was the deciding factor of the battle as the Safavid forces, who only had traditional weaponry, were decimated. The Safavids also suffered from poor planning and ill-disciplined troops unlike the Ottomans.[11]

Battle of Chaldiran

Aftermath

Following their victory the Ottomans temporarily captured the Safavid capital of Tabriz, which they evacuated quickly. The Shia defeat at Chaldiran brought an end to the Shia uprisings in Ottoman Empire. After two of his wives were captured by Selim[12] Ismail was heartbroken and resorted to drinking alcohol.[13] Ismail did not participate in government affairs,[14] as his aura of invincibility was shattered.

After the defeat at Chaldiran, however, the Safavids made drastic domestic changes. Ismail's son, Tahmasp I deployed cannons in subsequent battles.[15]

After the battle of Chaldiran, Selim I would then throw his forces southward to fight the Mamluk Sultanate in the Ottoman–Mamluk War (1516–1517).

Persian manuscript showing Shah Isma'il in the battle with an Ottoman commander

Battlefield

The site of the battle is near Chala Ashaqi village, around 6 km west of the town of Siyah Cheshmeh, south of Maku, north of Qareh Ziyaeddin. A large brick dome was built at the battlefield site in 2003 along with a statue of Seyid Sadraddin, one of the main Safavid commanders.

Quotes

After the battle, Selim referring to Ismail believed that his adversary was:

> Always drunk to the point of losing his mind and totally neglectful of the affairs of the state.[16]

See also

- Shi'a–Sunni relations

References

[1] Moojan Momen, *An Introduction to Shi'i Islam: The History and Doctrines of Twelver Shi'ism*, (Yale University Press, 1985), 107.
[2] *The Cambridge History of Iran*, ed. William Bayne Fisher, Peter Jackson, Laurence Lockhart, 224
[3] Leslie P. Peirce, *The Imperial Harem: Women and Sovereignty in the Ottoman Empire*, (Oxford University Press, 1993), 37.
[4] *The Cambridge History of Iran*, ed. William Bayne Fisher, Peter Jackson, Laurence Lockhart, 359.
[5] Martin Sicker, *The Islamic World in Ascendancy: From the Arab conquests to the Siege of Vienna*, (Praeger Publishers, 2000), 197.
[6] Caroline Finkel, *Osman's Dream*, (Basic Books, 2006), 104. .
[7] Caroline Finkel, *Osman's Dream*, 105.
[8] Caroline Finkel, *Osman's Dream*, 105
[9] Caroline Finkel, *Osman's Dream*, 106.
[10] Andrew James McGregor, *A Military History of Modern Egypt: From the Ottoman Conquest to the Ramadan War*, (Greenwood Publishing, 2006), 17.
[11] Gene Ralph Garthwaite, *The Persians*, (Blackwell Publishing, 2005), 164.
[12] *The Cambridge history of Iran*, ed. William Bayne Fisher, Peter Jackson, Laurence Lockhart, pg. 224.
[13] *The Cambridge history of Islam*, Part 1, ed. Peter Malcolm Holt, Ann K. S. Lambton, Bernard Lewis, pg. 401
[14] Elton L. Daniel, *The History of Iran*, (ABC-CLIO, 2012), 86
[15] *Gunpowder and Firearms in the Mamluk Sultanate Reconsidered*, Robert Irwin, **The Mamluks in Egyptian and Syrian politics and society**, ed. Michael Winter and Amalia Levanoni, (Brill, 2004), 127
[16] Rudi Matthee, *The Pursuit of Pleasure: Drugs and Stimulants in Iranian history, 1500-1900*, (Princeton University Press, 2005), 77

Ottoman–Safavid War (1532–1555)

The **Ottoman–Safavid War of 1532–1555** was fought between the Ottoman Empire of Suleiman the Magnificent and the Persian Safavid Empire of Tahmasp I.

Background

The war was triggered by territorial disputes between the two empires, especially when the Bey of Bitlis decided to put himself under Persian protection.[1] Also, Tahmasp had the governor of Baghdad, a sympathiser of Suleiman, assassinated.

On the diplomatic front, Persia had been engaged in discussions with the Habsburgs for the formation of a Habsburg-Persian alliance that would attack the Ottoman Empire on two fronts.[1]

Campaign of the Two Iraqs (First campaign, 1532–1534)

The Ottomans, first under the Grand Vizier Ibrahim Pasha, and later joined by Suleiman himself, successfully attacked Safavid Iraq, recaptured Bitlis, and proceeded to capture Tabriz and then Baghdad in 1534.[1] Tahmasp remained elusive as he kept retreating ahead of the Ottoman troops, adopting a scorched earth strategy.

Second campaign (1548–1549)

Attempting to defeat the Shah once and for all, Suleiman embarked upon a second campaign in 1548–1549. Again, Tahmasp adopted a scorched earth policy, laying waste to Armenia. Meanwhile, the French king Francis I, enemy of the Habsburgs, and Suleiman the Magnificent were moving forward with a Franco-Ottoman alliance, formalized in 1536, that would counterbalance the Habsburg threat. In 1547, when Suleiman attacked Persia, France sent its ambassador Gabriel de Luetz, to accompany him in his campaign.[2] Gabriel de Luetz gave military advice to Suleiman, as when he advised on artillery placement during the Siege of Van.[2] Suleiman made gains in Tabriz and

Iranian conquered Armenia, secured a lasting presence in the province of Van, and took some forts in Georgia.

Third campaign (1553–1555)

In 1553 Suleiman began his third and final campaign against the Shah, in which he first lost and then regained Erzurum. Ottoman territorial gains were secured by the Peace of Amasya in 1555. Suleiman returned Tabriz, but kept Baghdad, lower Mesopotamia, the mouths of the Euphrates and Tigris, and part of the Persian Gulf coast.

Due to his heavy commitment in Persia, Suleiman was only able to send limited naval support to France in the Franco-Ottoman Invasion of Corsica (1553).

Notes

[1] *The Cambridge history of Islam* by Peter Malcolm Holt,Ann K. S. Lambton, Bernard Lewis p.330 (http://books.google.com/books?id=j15MBH-FlwkC&pg=PA330)

[2] *The Cambridge history of Iran* by William Bayne Fisher p.384*ff*

French ambassador Gabriel de Luetz d'Aramon participated in the Ottoman campaign.

The walled city of Van, which Gabriel de Luetz d'Aramon helped conquer.

Ottoman–Safavid War (1578–1590)

The **Ottoman–Safavid War (1578–1590)** was a war between Safavid Persia under Mohammed Khodabanda and later Abbas I, and the Ottoman Empire under Murad III. It began in 1577-1578 and ended in 1590.

The Ottomans started the war, with the objective of conquering Azerbaijan and the Caucasus.[1] The war lasted for years, but the Ottomans took Tiflis in 1578, Kars and Tabriz in 1585, and Georgia became a tributary vassal of the Ottoman Empire.[2] As a result, the Ottomans effectively came to rule over Azerbaijan and the Caucasus as far as the Caspian Sea.[3]

A peace of Istanbul was concluded on 21 March 1590, in which Iran confirmed these Ottoman conquests, promised to end Shiite propaganda in Ottoman territories and persecution of Sunnis in her own lands.[3]

The advent of the Ottoman-Safavid war temporarily deflected Ottoman interest from European affairs, where the Ottoman Empire had been active with the Franco-Ottoman alliance and the support of the Dutch Revolt, in an interesting episode of mutually-supportive relations between Islam and Protestantism.

See also

- Lala Mustafa Pasha's Caucasian campaign
- Battle of Çıldır
- Battle of Torches
- Treaty of Istanbul (1590)

Notes

[1] *The Encyclopedia of world history* Peter N. Stearns, p.352 (http://books.google.com/books?id=MziRd4ddZz4C&pg=PA352)

[2] *Islam* by Gerhard Endress, p.194 (http://books.google.com/books?id=LsAGIZLZcRUC&pg=PA194)

[3] Stearns, p.352

Selim I

Selim I	
Sultan of the Ottoman Empire **Custodian of the Two Holy Mosques** **Muslim Caliph** **Amir al-Mu'minin**	

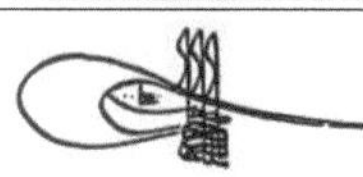	
Period	Growth of the Ottoman Empire
Coronation	1512
Full Name	Selim I
Birthplace	Amasya
Predecessor	Bayezid II
Successor	Suleiman I
Consort	Ayşe Hafsa Sultan
Offspring	Suleiman I Hatice Sultan Beyhan Sultan Şah Sultan
Royal House	House of Osman
Dynasty	Ottoman Dynasty
Father	Bayezid II
Valide Sultan	Gülbahar Sultan

Selim I, **Yavuz Sultân Selim Khan**, **Hâdim-ül Haramain-ish Sharifain** *(Servant of the Holy Cities of Mecca and Medina)* (Ottoman Turkish: سليم اوّل, Modern Turkish: *I.Selim*), nicknamed **Yavuz** "the Stern" or "the Steadfast", but often rendered in English as "the Grim" (October 10, 1465/1466/1470 – September 22, 1520), was the Sultan of the Ottoman Empire from 1512 to 1520.[1] His reign is notable for the enormous expansion of the Empire, particularly his conquest between 1516-1517 of the entire Mamluk Sultanate of Egypt, which included all of Sham, Hejaz, and Egypt itself. With the heart of the Arab World now under their control, the Ottomans became the dominant power in the region, and in the Islamic world. Upon conquering Egypt, Selim took the title of Caliph of Islam, being the first Ottoman sultan to do so. He was also granted the title of "Khâdim ül Haramain ish Sharifain" *(Servant of the Holy Cities of Mecca and Medina)*, by the Sharif of Mecca in 1517.

Selim's reign represented a sudden change in the expansion policy of the empire, which was working mostly against the West and the Beyliks before his reign.[2] On the eve of his death in 1520, the Ottoman Empire spanned almost 1 billion acres (trebling during Selim's reign).

Life

Turkish miniature of Sultan Selim I

Born in Amasya, Selim dethroned his father Bayezid II (1481–1512) in 1512. Bayezid's death followed immediately thereafter.[3] Selim put his brothers (Şehzade Ahmet and Şehzade Korkut) and nephews to death upon his accession in order to eliminate potential pretenders to the throne. This fratricidal policy was motivated by bouts of civil strife that had been sparked by the antagonism between Selim's father Beyazid and his uncle Cem Sultan, and between Selim himself and his brother Ahmet. His biological mother was Gül-Bahār Khātûn,[4] who had never acquired the title of *Valide Khātûn* since she had died before Selim's accession to the Ottoman throne. According to another theory, Selim was the biological son of A'ishā *(Ayşe)* Khātûn I[5] [6] who died at Trebizond on 1505 and was the daughter of *Alaüddevle Bozkurt Bey*, the eleventh ruler of the Dulkadirids centered around Elbistan in *Kahramanmaraş*.

Selim I was described as being tall, having very broad shoulders and a long mustache. He was skilled in politics and was said to be fond of fighting.[7]

Conquest of the Middle East

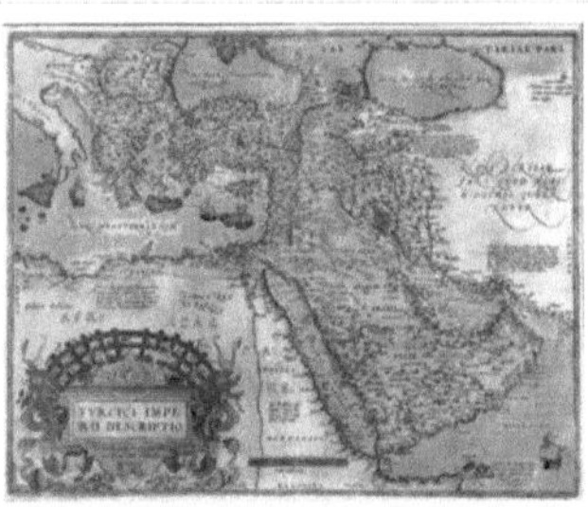

Outline of the Ottoman Empire, from the *Theatro d'el Orbe de la Tierra* de Abraham Ortelius, Anvers, 1602, updated from the 1570 edition.

Safavid Empire

Battle of Chaldiran.

For Selim, one of the first challenges as Sultan was the growing tension between himself and Shah Ismail who had recently brought the Safavids to power and had switched the state religion from Sunni Islam to the adherence of the Twelver Shia Islam. By 1510, Ismail had conquered west part of Iran[8] and was of a great threat to his Sunni Muslim neighbors to the west. In 1514, Selim I attacked Ismā'il's kingdom to stop the spread of Shiism into Ottoman dominions. Selim and Ismā'il had been exchanging a series of belligerent letters prior to the attack. Selim I defeated Ismā'il at the Battle of Chaldiran in 1514.[9] Ismā'il's army was more mobile and their soldiers were better prepared but the Ottomans prevailed due in large part to their efficient modern army, and possession of artillery, black powder and muskets. Ismā'il was wounded and almost captured in battle, and Selim I entered the Iranian capital of Tabriz in triumph on September 5,[10] but did not linger. A mutiny among his troops fearing a counter attack and entrapment by the fresh Safavid forces called in from the interior, forced the triumphant Ottomans to withdraw prematurely. This allowed Ismā'il to recover quickly. The Battle of Chaldiran, was of historical significance, in which the reluctancy showed by Shah Ismail to accept the advantages of modern firearms and the importance of artillery was decisive.[11] After the battle, Selim referring to Ismail stated that his adversary was: "Always drunk to the point of losing his mind and totally neglectful of the affairs of the state.[12]

Syria, Palestine, Egypt and the Arabian Peninsula

Selim then conquered the Mamluk Sultanate of Egypt, defeating the Mamluk Egyptians first at the Battle of Marj Dabiq, and then at the Battle of Ridanieh. This led to the Ottoman annexation of the entire sultanate, from Syria and Palestine in Sham, to Hejaz in the Arabian Peninsula, and ultimately Egypt itself. This permitted him to extended Ottoman power to the Muslim holy cities of Mecca and Medina, hitherto under Egyptian rule. Rather than style himself the *Hakim ul Haremeyn*, or *The Ruler of The Two Holy Shrines*, he accepted the more pious title *Khadim ul Haremeyn*, or *The Servant of The Two Holy Shrines*.[3] [13]

After the conquest of Egypt and the Holy Cities in 1517, Selim induced Al-Mutawakkil III (1509–17), the last in the line of Abbasid caliphs who resided in Cairo since 1261 as nominal rulers legitimizing the *de facto* rule of the Mamluk sultans over the Mamluk Sultanate,[14] to formally surrender the title of Caliph and its emblems, the sword and the mantle of Muhammad.[2] These are kept in the Topkapı Palace Museum at Istanbul, Turkey.

Death

After his return from his Egyptian campaign, Selim began to prepare for an expedition which is believed to be against Hungary. This campaign was cut short when he was overwhelmed by sickness and subsequently died in the ninth year of his reign. He was about fifty-five years of age. It is said that Selim succumbed to sirpence, a skin infection which he developed during his long campaigns on horseback. (Sirpence was an anthrax infection sometimes seen among leatherworkers and others who worked with livestock). Some historians claim that he was poisoned by the doctor tending to his infection[1] and some historians claim that the disease he suffered from was skin cancer. He died at Corlu, Tekirdağ.

Selim I on his deathbed.

Titles

After claiming the Caliphate, Selim assumed the title *Malik ul-Barreyn, wa Khakan ul-Bahrayn, wa Kasir ul-Jayshayn, wa Khadim ul-Haramayn* - that is, *King of the Two Lands (continents Europe and Asia), Khagan of the Two Seas (Mediterranean and Indian Seas), Conqueror of the Two Armies (European and Safavid armies), and Servant of the Two Holy Shrines (Mecca and Medina).* This title alludes to his dominions in Europe and Asia (namely, Balkan, Anatolia, and much of the Fertile Crescent), his control over the Mediterranean and Black seas, his defeat of both the Mamluk and Safavid armies, and his guardianship of the shrines of Mecca and Medina.

Personality

By most accounts, Selim had a fiery temper and had very high expectations of his subordinates. Several of his viziers were executed for various reasons. A famous anecdote relates how another vizier playfully asked the Sultan for some preliminary notice of his doom so that he might have time to put his affairs in order. The Sultan laughed and replied that indeed he had been thinking of having the vizier killed, but had no one fit to take his place, otherwise he would gladly oblige. Lord Kinross in his history of the Ottomans reports that life at Sultan Selim's court was full of opportunities, and there were always plenty of applicants to the highest offices, regardless of the risks.

Selim was one of the empires most successful and respected rulers, he was energetic and very hard working. Accordingly, his court was dynamic, with the rewards as great as the risks. During his eight years of ruling he didn't have time to rest. Although he was a leader, he was also very humble and modest. His reign was short, but may have prepared the Ottoman empire for its zenith under the achievements of his son.[15] A popular legend has it that Selim had filled the royal treasury to the brink and locked it with his own seal. He decreed that "he who will fill the treasury more than this, may use his seal to lock it." The treasury remained locked with Selim's seal until the collapse of the Empire 400 years later.

Selim was also a distinguished poet who wrote both Turkish and Persian verse under the nickname *mahlas Selimi*; collections of his Persian poetry are extant today.[15] In one of his poems, he wrote;

> *A carpet is large enough to accommodate two sufis, but the world is not large enough for two Kings.*
>
> — Yavuz Sultan Selim

Foreign Relations

Relations with the Shah Ismail

While marching into Persia in 1514, Selim's troops suffered from the scorched-earth tactics of Shah Ismail. The Sultan hoped to lure Ismail into an open battle before his troops starved to death, and began writing insulting letters to the Shah, accusing him of cowardice:

> *They, who by perjuries seize scepters ought not to skulk from danger, but their breast ought, like the shield, to be held out to encounter peril; they ought, like the helm, to affront the foeman's blow.*

Ismail responded to Selim's third message, quoted above, by sending an envoy to deliver a letter accompanied by a box of opium. The Shah's letter insultingly implied that Selim's prose must have been the work of an unqualified writer on drugs. Selim was known to occasionally use opium, and grew enraged by the Shah's denigration of his literary talent. He ordered the Persian envoy to be torn to pieces.[16]

Relations with Babur

Babur's early relations with the Ottomans were initially troubled because the Ottoman Sultan Selim I provided Babur's arch rival Ubaydullah Khan with powerful Matchlocks and Cannons to counter the influence of the Safavids.[17] In the year 1507, when ordered to accept Selim I as his rightful suzerain Babur refused, and gathered Qizilbash servicemen in order to counter the forces of Ubaydullah Khan during the Battle of Ghazdewan. In the year 1513, Ottoman Sultan Selim I reconciled with Babur (probably fearing that he would join the Safavids), dispatched Ustad Ali Quli the artilleryman and Mustafa Rumi the Matchlock marksman and many other Ottoman Turks, in order to assist Babur in his conquests. Thenceforth this particular assistance proved to be the basis of future Mughal-Ottoman relations.[18]

References

[1] Yavuz Sultan Selim Biography (http://www.sevgi.k12.tr/~ottomanempire/ingosmanli/Sultans/yavuz_sultan_selim_biography.htm) Retrieved on 2007-09-16

[2] The Rise of the Turks and the Ottoman Empire (http://www.shsu.edu/~his_ncp/Turkey2.html) Retrieved on 2007-09-16

[3] The Classical Age, 1453-1600 (http://www.turizm.net/turkey/history/ottoman2.html) Retrieved on 2007-09-16

[4] Yavuz Bahadıroğlu, Resimli Osmanlı Tarihi, Nesil Yayınları *(Ottoman History with Illustrations, Nesil Publications)*, 15th Ed., 2009, page 157, ISBN 978-975-269-299-2

[5] "Yavuz Sultan Selim Han" (http://www.kultur.gov.tr/EN/belge/2-1280/yavuz-sultan-selim-han.html). Republic of Turkey Ministry of Culture and Tourism. . Retrieved 2009-02-06.

[6] "Mother Of Yavuz Sultan Selim" (http://www.osmanli.org.tr/osmanlisultanlari-5-218.html). Osmanlı Araştırmaları Vakfı *(Ottoman Research Foundation)*. .

[7] "Sultan Selim the Excellent" (http://www.ottomanonline.net/sultans/9.html). Ottomanonline.net. . Retrieved 2012-03-20.

[8] BBC, (LINK (http://www.bbc.co.uk/religion/religions/islam/history/safavidempire_1.shtml))

[9] Michael Axworthy *Iran: Empire of the Mind* (Penguin, 2008) p.133

[10] The later Crusades, 1274-1580: from Lyons to Alcazar Door Norman Housley, page 120, 1992

[11] "Morgan, David. "Shah Isma'il and the Establishment of Shi'ism"" (http://coursesa.matrix.msu.edu/~fisher/hst373/readings/morgan.html). Coursesa.matrix.msu.edu. . Retrieved 2012-03-20.

[12] The pursuit of pleasure: drugs and stimulants in Iranian history, 1500-1900 By Rudolph P. Matthee, pg. 77

[13] Yavuz Sultan Selim Government (http://www.sevgi.k12.tr/~ottomanempire/ingosmanli/Sultans/yavuz_sultan_selim_government.htm) Retrieved on 2007-09-16

[14] Thompson, J., *A History of Egypt*, AUC Press 2008, p. 194; Vatikiotis, P.J., *The History of Modern Egypt*, The Johns Hopkins University Press, 1992, p.20

[15] *Necdet Sakaoğlu, Bu Mülkün Sultanları, pg.127*

[16] Crider, Elizabeth Fortuato (1969). *The Foreign Relations of the Ottoman Empire Under Selim I, 1512-1520* (http://rave.ohiolink.edu/etdc/view?acc_num=osu1224163441)(Master's Thesis). Ohio State University, 1969, page 20. Retrieved on 2011-04-12

[17] http://books.google.com/books?id=uB1uAAAAMAAJ&q=Ubaydullah+Khan#search_anchor

[18] http://books.google.com/books?id=uB1uAAAAMAAJ&q=Babur&source=gbs_word_cloud_r&cad=6

External links

- "Selim". *Encyclopædia Britannica* (11th ed.). 1911.

Ottoman–Safavid War (1603–1618)

The **Ottoman–Safavid War** was a war between Safavid Persia under Abbas I of Persia and the Ottoman Empire under Sultan Ahmed I. It began in 1603 and ended with Safavid victory in 1618.

Course

Abbas I had recently undertaken a major reform of his army through the English gentleman of fortune Robert Shirley and the shah's favorite ghulam and chancellor Allahverdi Khan and thus opened in 1603, with the first Safavid pitched victory won in 1604. This forced the Ottomans to return the territory they had seized from Persia, including Baghdad. He then won another victory at Basra in 1605, soon after extending his empire beyond the Euphrates; and forcing the Ottomans to cede Shirvan and Kurdistan to Persia in 1611. Hostilities ceased momentarily in 1614 with the Persian army at its peak. Hostilities later resumed and Abbas completely defeated the allied Ottoman-Tartar force near Sultanieh in 1618, bringing the war to an end on terms highly favourable to the Persians.

See also

- Treaty of Nasuh Pasha
- Treaty of Serav
- List of conflicts in the Middle East

Ottoman–Persian War (1821–1823)

The **Ottoman-Qajar War** was fought between Qajar Empire and the Ottoman Empire from 1821 to 1823.

After severe losses in Georgia during the Russo-Persian War of 1804–13, Crown Prince Abbas Mirza (1783–1833) of Persia vowed to modernize his armies. He sent Persians to England to learn Western military techniques and invited British officers to train his troops, especially newly introduced infantry. Ottoman incursions into Iran's Azerbaijan region under the pretense of punishing rebellious tribesmen led to declaration of war in 1821. Mohammed Ali Mirza defeated the governor of Baghdad's invasion of Persia and then besieged Baghdad, the siege was lifted due to his death. In retaliation, Abbas Mirza attacked eastern Anatolia (Turkey) in 1821. Ottoman and Persian forces met in the Lake Van region. The war culminated in Persian victory at the Battle of Erzurum in 1821, where Abbas Mirza's army routed a superior Ottoman force. Peace did not come until the Treaty of Erzurum two years later; both sides recognized the previous borders, with no territorial changes.

Ottoman–Safavid War (1623–1639)

The **Ottoman–Safavid War of 1623–1639** was the last of a series of conflicts fought between the Ottoman Empire and Safavid Persia, then the two major powers of the Near East, over control of Mesopotamia. After initial Persian success in capturing Baghdad and most of modern Iraq, the war became a stalemate, as the Persians were unable to press further into the Ottoman Empire, and the Ottomans themselves were distracted by wars in Europe and weakened by internal turmoil. Eventually, the Ottomans were able to recover Baghdad, and the signing of the Treaty of Zuhab ended the war in an Ottoman victory, with Mesopotamia remaining thenceforth in Ottoman hands, until lost in the aftermath of World War I.

Background

Starting in 1514, for over a century the Ottoman Empire and Savafid Persia were engaged in almost constant warfare over control of the Caucasus and Mesopotamia. The two states were the greatest powers of the Middle East, and the rivalry was further fueled by dogmatic differences: the Ottomans were Sunnis, while the Safavids were staunchly Shia Muslims of the Qizilbash sect, and seen as heretics by the Ottomans.[1]

After the Battle of Chaldiran eliminated Safavid influence in Anatolia, during the war of 1532–1555 the Ottomans conquered Arab Iraq, taking Baghdad in 1534 and securing recognition of their gains by the Treaty of Amasya in 1555.[2] Peace lasted for two decades before another war began in 1578. It ended with the Treaty of Istanbul in 1590, with a clear Ottoman victory: the Ottomans occupied Georgia, Yerevan, and even the former Safavid capital, Tabriz. The Persians were hard pressed, as the Ottoman advances were combined with an attack by the Shaybanids into Persian Khorasan.[3]

The new Persian Shah, Abbas I, reorganized his army, raising the new *ghulam* infantry in imitation of the Janissaries,[4] and bided his time. In 1603, he launched an offensive that retook Tabriz, Azerbaijan and Georgia. The Ottomans, distracted by wars with the Habsburg Monarchy in Europe, failed to offer effective resistance.[5] After a successful conclusion of the War against the Mughals, feeling encouraged by the internal turmoil within the Ottoman Empire that followed the murder of Sultan Osman II, Abbas resolved to attack the Ottoman possessions in Iraq.[5]

The war

The Shah's opportunity came with a series of rebellions in the Ottoman Empire: Abaza Mehmed Pasha, the governor of Erzurum, rose in rebellion, while Baghdad had been since 1621 in the hands of an officer of the Janissaries, the *Subashi* Bakr, and his followers.[6] [7] Bakr had sought his recognition as the local pasha from the Porte, but the Sultan had ordered Hafız Ahmed Pasha, the governor of Diyarbakir, to intervene.[7] Bakr then turned to Abbas, who sent troops to Bakr's aid. To forestall a Persian capture of Baghdad, Hafız Ahmed quickly restored relations with Bakr, who in turn returned to the Ottoman allegiance. In response, the Persians besieged Baghdad, and took it on 14 January 1624, with the aid of Bakr's son, Muhammad.[7] [8] The fall of the city was followed by the massacre of a large part of its Sunni inhabitants, as the Shah endeavored to transform Baghdad into a purely Shiite city.[4]

The fall of Baghdad was a major blow to Ottoman prestige. Ottoman garrisons and the local tribes began to defect, and the Persians soon captured most of Iraq, including the cities of Kirkuk and Mosul and the Shia holy shrines of Najaf and Karbala, which the Shah visited.[5] [9] In 1625, Hafız Ahmed Pasha, now Grand Vizier, marched to retake Baghdad. Despite a "scorched earth" policy ordered by the Shah, the Ottoman army reached Baghdad and invested it in November on three sides.[9] The Ottoman assaults on the city managed to penetrate the outer fortifications, but failed to take the city before the arrival of a relief army under Shah Abbas. The Ottomans then withdrew within their strongly fortified camp, and continued to prosecute the siege.[9] In response, Abbas decided to intercept Ottoman supply convoys. This strategy bore fruit: the Ottomans were forced to risk an attack on the Persian army, which was repulsed with heavy losses, and on 4 July 1626, the Ottoman army lifted the siege and withdrew to Mosul.[7] [10]

In 1629, the Ottomans, under the new and capable Grand Vizier Gazi Ekrem Khüsrev Pasha, and having secured peace with the Habsburgs, mustered their forces for another offensive.[11] A severe winter, including floods, made operations in central Iraq impossible, however, and Khüsrev turned his army east, invading Persia proper. On 4 May 1630, he routed the Persians under Zainal Khan Begdeli Shamlu in battle at Mahidasht near Kermanshah and proceeded to sack the city of Hamadan.[7] [12] Khüsrev Pasha then turned back towards Baghdad and besieged it in November. However the siege had to be lifted soon, as the onset of another heavy winter threatened his lines of communication.[12] [13] In the wake of his withdrawal, the Persians re-established their control of Iraq, and subdued the rebellious Kurdish populations. The next few years saw constant raiding and skirmishes, without either side claiming any decisive advantage. Shah Safi sent a peace delegation to the Ottoman court, but the new Grand Vizier, Tabanıyassi Mehmed Pasha rejected its demands.[12] The Caucasian front of the Persians flared up again in 1633, when the restless Georgian kingdoms of Kartli and Kakheti, which, under the rule of King Teimuraz, defied Safavid sovereignty. In 1634, Rustam Khan, a Georgian convert, was sent by the Shah to subdue them. Teimuraz was defeated, but managed to escape to safety in Imereti. He would however manage to restore himself on the throne of Kakheti in 1638, and even win Persian recognition of this fact.[14]

In 1635, in a conscious effort to emulate his warrior predecessors, Sultan Murad IV himself took up the leadership of the army. The Ottomans took Revan (on 8 August) and plundered Tabriz.[12] [15] The victorious Sultan returned in triumph to Constantinople, but his victories were short-lived: in the spring of the next year, Shah Safi retook Revan and defeated an Ottoman army.[16] [17] Renewed Persian peace proposals failed, and in 1638, Murad IV himself led an army against Baghdad. The city fell in December after a siege of 39 days, effectively restoring Ottoman control over Iraq, and peace negotiations began soon after.[16] [17]

Aftermath

The Treaty of Zuhab, concluded on 17 May 1639, finally settled the Ottoman–Persian frontier, with Yerevan remaining Persian, and Iraq ceded to the Ottomans. Mesopotamia, which had formed an important part of various Persian empires from the time of the Achaemenids, was irrevocably lost.[16] The peace established a permanent equilibrium of power in the region, and despite future conflicts and minor adjustments, the frontier postulated by the treaty remains to this day the western border of Iran with Iraq and Turkey.[16] [18]

See also

- Capture of Baghdad (1638)

References

[1] Finkel (2006), pp. 104–105
[2] Finkel (2006), pp. 125, 135
[3] Holt, Lambton & Lewis (1978), p. 338
[4] Faroqhi (2006), p. 47
[5] Holt, Lambton & Lewis (1978), p. 339
[6] Finkel (2006), pp. 203–205
[7] Cooper (1979), p. 631
[8] Finkel (2006), p. 205
[9] Savory (2007), p. 89
[10] Savory (2007), p. 90
[11] Roemer (1989), p. 283
[12] Roemer (1989), p. 284
[13] Cooper (1979), pp.631–632
[14] Roemer (1989), p. 286
[15] Finkel (2006), pp. 215–216
[16] Roemer (1989), p. 285
[17] Finkel (2006), p. 217

[18] Cooper (1979), p. 634

Sources

- Cooper, J. P. (1979). *The New Cambridge Modern History, Volume IV: The Decline of Spain and the Thirty Years War, 1609–48/59*. CUP Archive. ISBN 0521297134.
- Faroqhi, Suraiya (2006). *The Cambridge History of Turkey: The Later Ottoman Empire, 1603–1839*. Cambridge University Press. ISBN 9780521620956.
- Finkel, Caroline (2006). *Osman's Dream: The Story of the Ottoman Empire 1300–1923*. London: John Murray. ISBN 978-0-7195-6112-2.
- Holt, P. M.; Lambton, Ann K. S.; Lewis, Bernard (1978). *The Central Islamic Lands from Pre-Islamic Times to the First World War*. Cambridge: Cambridge University Press. ISBN 0521291356.
- Newman, Andrew J. (2006). *Safavid Iran: Rebirth of a Persian Empire*. I.B.Tauris. ISBN 9781860646676.
- Roemer, H. R. (1986). "The Safavid Period". *The Cambridge History of Iran, Vol. 6: The Timurid and Safavid Periods*. Cambridge: Cambridge University Press. pp. 189–350. ISBN 0521200946.
- Savory, Roger (2007). *Iran Under the Safavids*. Cambridge: Cambridge University Press. ISBN 9780521042512.

Suleiman the Magnificent

Suleiman the Magnificent
Suleiman the Lawgiver

Sultan of the Ottoman Empire
Custodian of the Two Holy Mosques
Islamic Caliph
Amir al-Mu'minin

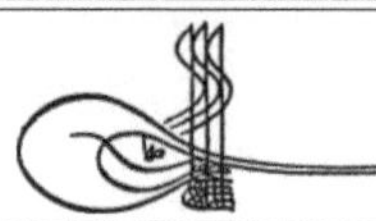

Period	Growth of the Ottoman Empire
Coronation	30 September 1520
Full Name	Suleiman bin Selim Khan
Born	6 November 1494
Birthplace	Trabzon
Died	c. 5 September 1566 (aged 71)
Place of death	Szigetvár, Hungary
Buried	Süleymaniye Mosque, Constantinople[1] [2] (present day Istanbul)
Predecessor	Selim I
Successor	Selim II
Consort	Hürrem Sultan
Wives	Haseki Hürrem Sultan Mahidevran Sultan

Offspring	[3] Şehzade Mahmud (1511–1521) Şehzade Mustafa (1515–1553) Şehzade Murad (1519–1521) Şehzade Mehmed (1521–1543) Şehzade Abdullah (1522-1524) Mihrimah Sultan (1522–1578) Selim II (1524–1574) Raziye Sultan (1525-?) Şehzade Beyazıt (1525–1561) Şehzade Mehmed (1526-1533) Şehzade Cihangir (1531–1553) Şehzade Orhan (1543-1562) Şehzade Osman (1545-1562) Şehzade Orhan (1554-1562) Şehzade Mehmed (died when he was 3 years old) Fatma Sultan Şehzade Ahmed Unknown name (1521-1522)
Royal House	House of Osman
Dynasty	Ottoman Dynasty
Father	Selim I
Valide Sultan	Ayşe Hafsa Sultan

Suleiman I (pron.: /ˌsʊlɪˈmɑːn/; Ottoman Turkish: سلطان سليمان اول, *Sultān Suleimān-i evvel* or قانونى سلطان سليمان, *Kānūnī Sultān Suleimān*, Modern Turkish: *I. Süleyman* (Turkish pronunciation: [sylejˈmɑn]) or *Kanuni Sultan Süleyman*; 6 November 1494 – c. 5 September 1566) was the tenth and longest-reigning Emperor, Sultan of the Ottoman Empire, from 1520 to his death in 1566. He is known in the West as **Suleiman the Magnificent**[4] and in the East, as "**The Lawgiver**" (Turkish: *Kanuni*; Arabic: القانونى, *al-Qānūnī*), for his complete reconstruction of the Ottoman legal system. Suleiman became a prominent monarch of 16th century Europe, presiding over the apex of the Ottoman Empire's military, political and economic power. Suleiman personally led Ottoman armies to conquer the Christian strongholds of Belgrade, Rhodes, and most of Hungary before his conquests were checked at the Siege of Vienna in 1529. He annexed most of the Middle East in his conflict with the Safavids and large swathes of North Africa as far west as Algeria. Under his rule, the Ottoman fleet dominated the seas from the Mediterranean to the Red Sea and the Persian Gulf.[5]

At the helm of an expanding empire, Suleiman personally instituted legislative changes relating to society, education, taxation, and criminal law. His canonical law (or the *Kanuns*) fixed the form of the empire for centuries after his death. Not only was Suleiman a distinguished poet and goldsmith in his own right; he also became a great patron of culture, overseeing the golden age of the Ottoman Empire's artistic, literary and architectural development.[6] He spoke five languages: Ottoman Turkish, Arabic, Chagatai, Persian and Serbian.

In a break with Ottoman tradition, Suleiman married a harem girl, Roxelana, who became Hürrem Sultan; her intrigues as queen in the court and power over the Sultan made her quite renowned. Their son, Selim II, succeeded Suleiman following his death in 1566 after 46 years of rule.

Early life

Suleiman was born in Trabzon along the coast of the Black Sea, probably on 6 November 1494.[7] His mother was Valide Sultan Aishe Hafsa Sultan or Hafsa Hatun Sultan, who died in 1534. At the age of seven, he was sent to study science, history, literature, theology, and military tactics in the schools of the Topkapı Palace in Constantinople(now İstanbul) As a young man, he befriended Pargalı Ibrahim, a slave who later became one of his most trusted advisers.[8] From the age of seventeen, young Suleiman was appointed as the governor of first Kaffa (Theodosia), then Sarukhan (Manisa) with a brief tenure at Adrianople (now Edirne).[9] Upon the death of his father, Selim I (1465–1520), Suleiman entered Constantinople and acceded to the throne as the tenth Ottoman Sultan. An early description of Suleiman, a few weeks following his accession, was provided by the Venetian envoy Bartolomeo Contarini: "He is twenty-six years of age, tall, but wiry, and of a delicate complexion. His neck is a little too long, his face thin, and his nose aquiline. He has a shade of a moustache and a small beard; nevertheless he has a pleasant mien, though his skin tends to pallor. He is said to be a wise Lord, fond of study, and all men hope for good from his rule."[10] Some historians claim that in his youth Suleiman had an admiration for Alexander the Great.[11] [12] He was influenced by Alexander's vision of building a world empire that would encompass the east and the west, and this created a drive for his subsequent military campaigns in Asia and in Africa, as well as in Europe.

Military campaigns

Conquests in Europe

Upon succeeding his father, Suleiman began a series of military conquests, eventually suppressing a revolt led by the Ottoman-appointed governor of Damascus in 1521. Suleiman soon made preparations for the conquest of Belgrade from the Kingdom of Hungary—something his great-grandfather Mehmed II had failed to achieve. Its capture was vital in eliminating the Hungarians who, following the defeats of the Serbs, Bulgarians, Byzantines and Albanians, remained the only formidable force who could block further Ottoman gains in Europe. Suleiman encircled Belgrade and began a series of heavy bombardments from an island in the Danube. With a garrison of only 700 men, and receiving no aid from Hungary, Belgrade fell in August 1521.[13]

Battle between the Turks and Christians, 16th century

Suleiman as a young man

News of the conquest of one of Christendom's major strongholds spread fear across Europe. As the ambassador of the Holy Roman Empire to Constantinople was to note, "The capture of Belgrade was at the origin of the dramatic events which engulfed Hungary. It led to the death of King Louis, the capture of Buda, the occupation of Transylvania, the ruin of a flourishing kingdom and the fear of neighbouring nations that they would suffer the same fate..."[14]

The road to Hungary and Austria lay open, but Suleiman diverted his attention to the Eastern Mediterranean island of Rhodes, the home base of the Knights Hospitaller. In the summer of 1522, taking advantage of the navy he inherited from his father, Suleiman dispatched an armada of some 400 ships while personally leading an army of 100,000 across Asia Minor to a point opposite the island.[15] Following a siege of five months with brutal encounters, Rhodes capitulated and Suleiman allowed the Knights of Rhodes to depart. They eventually formed their new base in Malta.

As relations between Hungary and the Ottoman Empire deteriorated, Suleiman resumed his campaign in Eastern Europe and on 29 August 1526, he defeated Louis II of Hungary (1506–26) at the Battle of Mohács. In its wake, Hungarian resistance collapsed and the Ottoman Empire became the pre-eminent power in Eastern Europe.[16] Upon encountering the lifeless body of King Louis, Suleiman is said to have lamented: "I came indeed in arms against him; but it was not my wish that he should be thus cut off while he scarcely tasted the sweets of life and royalty."[17] [18] While Suleiman was campaigning in Hungary, Turkmen tribes in central Anatolia revolted under the leadership of Kalender Çelebi.

Under Charles V and his brother Ferdinand, the Habsburgs reoccupied Buda and took Hungary. As a result, in 1529, Suleiman once again marched through the valley of the Danube and regained control of Buda and in the following autumn laid siege to Vienna. It was to be the Ottoman Empire's most ambitious expedition and the apogee of its drive towards the West. With a reinforced garrison of 16,000 men,[19] the Austrians inflicted upon Suleiman his first defeat, sowing the seeds of a bitter Ottoman-Habsburg rivalry which lasted until the 20th century.[20] A second attempt to conquer Vienna failed in 1532, with Suleiman retreating before reaching the city. In both cases, the Ottoman army was plagued by bad weather (forcing them to leave behind essential siege equipment) and was hobbled by overstretched supply lines.[21]

By the 1540s a renewal of the conflict in Hungary presented Suleiman with the opportunity to avenge the defeat suffered at Vienna. Some Hungarian nobles proposed that Ferdinand, who was ruler of neighbouring Austria and tied to Louis II's family by marriage, be King of Hungary, citing previous agreements that the Habsburgs would take the Hungarian throne if Louis died without heirs.[22] However, other nobles turned to the nobleman John Zápolya who, being supported by Suleiman, remained unrecognized by the Christian powers of Europe.

King John Sigismund of Hungary with Suleiman in 1556.

Ottoman Siege of Esztergom (1543).

In 1541 the Habsburgs once again engaged in conflict with the Ottomans, attempting to lay siege to Buda. With their efforts repulsed, and more Habsburg fortresses captured as a result,[23] Ferdinand and his brother Charles V were forced to conclude a humiliating five-year treaty with Suleiman. Ferdinand renounced his claim to the Kingdom of Hungary and was forced to pay a fixed yearly sum to the Sultan for the Hungarian lands he continued to control. Of more symbolic importance, the treaty referred to Charles V not as 'Emperor', but in rather plainer terms as the 'King of Spain', leading Suleiman to consider himself the true 'Caesar'.[24]

With his main European rivals subdued, Suleiman had assured the Ottoman Empire a powerful role in the political landscape of Europe.

Ottoman–Safavid War

As Suleiman stabilized his European frontiers, he now turned his attention to the ever present threat posed by the Shi'a Safavid dynasty of Persia. Two events in particular were to precipitate a recurrence of tensions. First, Shah Tahmasp had the Baghdad governor loyal to Suleiman killed and replaced with an adherent of the Shah, and second, the governor of Bitlis had defected and sworn allegiance to the Safavids.[25] As a result, in 1533, Suleiman ordered his Grand Vizier Pargalı Ibrahim Pasha to lead an army into Asia where he retook Bitlis and occupied Tabriz without resistance. Having joined Ibrahim in 1534, Suleiman made a push towards Persia, only to find the Shah sacrificing territory instead of facing a pitched battle, resorting to harassment of the Ottoman army as it proceeded along the harsh interior.[26] When in the following year Suleiman and Ibrahim made a grand entrance into Baghdad, its commander surrendered the city, thereby confirming Suleiman as the leader of the Islamic world and the legitimate successor to the Abbasid Caliphs.[27]

Miniature depicting Suleiman marching with an army in Nakhchivan, summer 1554

Suleiman the Magnificent on horseback. Painting by Hans Eworth.

Attempting to defeat the Shah once and for all, Suleiman embarked upon a second campaign in 1548–1549. As in the previous attempt, Tahmasp avoided confrontation with the Ottoman army and instead chose to retreat, using scorched earth tactics in the process and exposing the Ottoman army to the harsh winter of the Caucasus.[26] Suleiman abandoned the campaign with temporary Ottoman gains in Tabriz and the Urmia region, a lasting presence in the province of Van, and some forts in Georgia.[28] In 1553 Suleiman began his third and final campaign against the Shah. Having initially lost territories in Erzurum to the Shah's son, Suleiman retaliated by recapturing Erzurum, crossing the Upper Euphrates and laying waste to parts of Persia. The Shah's army continued its strategy of avoiding the Ottomans, leading to a stalemate from which neither army made any significant gain. In 1554, a settlement was signed which was to conclude Suleiman's Asian campaigns. It included the return of Tabriz, but secured Baghdad, lower Mesopotamia, the mouths of the river Euphrates and Tigris, as well as part of the Persian Gulf.[29] The Shah also promised to cease all raids into Ottoman territory.[30]

Campaigns in the Indian Ocean

Ottoman fleet in the Indian Ocean in the 16th century.

Ottoman ships had been sailing in the Indian Ocean, since the year 1518. Ottoman Admirals such as: Hadim Suleiman Pasha, Seydi Ali Reis[31] and Kurtoğlu Hızır Reis are known to have voyaged to the Mughal imperial ports of Thatta, Surat and Janjira. The Mughal Emperor Akbar, himself is known to have exchanged six documents with Suleiman the Magnificent.[31] [32] [33]

In the Indian Ocean, Suleiman led several naval campaigns against the Portuguese in an attempt to remove them and reestablish trade with India. Aden in Yemen was captured by the Ottomans in 1538, in order to provide an Ottoman base for raids against Portuguese possessions on the western coast of modern Pakistan and India.[34] Sailing on to India, the Ottomans failed against the Portuguese at the Siege of Diu in September 1538, but then returned to Aden where they fortified the city with 100 pieces of artillery.[34] [35] From this base, Sulayman Pasha managed to take control of the whole country of Yemen, also taking Sa'na.[34] Aden rose against the Ottomans however and invited the Portuguese instead, so that the Portuguese were in control of the city until its seizure by Piri Reis in the Capture of Aden (1548).

With its strong control of the Red Sea, Suleiman successfully managed to dispute control of the Indian trade routes to the Portuguese and maintained a significant level of trade with the Mughal Empire of South Asia throughout the 16th century.[36] His admiral Piri Reis led an Ottoman fleet in the Indian Ocean, achieving the Capture of Muscat in 1552.

In 1564, Suleiman received an embassy from Aceh (modern Indonesia), requesting Ottoman support against the Portuguese. As a result an Ottoman expedition to Aceh was launched, which was able to provide extensive military support to the Acehnese.[37]

Mediterranean and North Africa

Barbarossa Hayreddin Pasha defeats the Holy League under the command of Andrea Doria at the Battle of Preveza in 1538

Having consolidated his conquests on land, Suleiman was greeted with the news that the fortress of Koroni in Morea (the modern Peloponnese) had been lost to Charles V's admiral, Andrea Doria. The presence of the Spanish in the Eastern Mediterranean concerned Suleiman, who saw it as an early indication of Charles V's intention to rival Ottoman dominance in the region. Recognizing the need to reassert the navy's preeminence in the Mediterranean, Suleiman appointed an exceptional naval commander in the form of Khair ad Din, known to Europeans as Barbarossa. Once appointed admiral-in-chief, Barbarossa was charged with rebuilding the Ottoman fleet, to such an extent that the Ottoman navy equalled in number those of all other Mediterranean countries put together.[38] In 1535 Charles V won an important victory against the Ottomans at Tunis, which together with the war against Venice the following year, led Suleiman to accept proposals from Francis I of France to form an alliance against Charles.[25] In 1538, the Spanish fleet was defeated by Barbarossa at the Battle of Preveza, securing the eastern Mediterranean for the Turks for 33 years until the defeat at the Battle of Lepanto in 1571.

Francis I (left) and Suleiman the Magnificent (right) initiated a Franco-Ottoman alliance from the 1530s.

East of Morocco, huge territories in North Africa were annexed. The Barbary States of Tripolitania, Tunisia, and Algeria became autonomous provinces of the Empire, serving as the leading edge of Suleiman's conflict with Charles V, whose attempt to drive out the Turks failed in 1541.[39] The piracy carried on thereafter by the Barbary pirates of North Africa can be seen in the context of the wars against Spain. For a short period Ottoman expansion secured naval dominance in the Mediterranean. Ottoman navies also controlled the Red Sea, and held the Persian Gulf until 1554, when their ships were defeated by the navy of the Portuguese Empire. The Portuguese had taken Ormus (in the Strait of Hormuz) in 1515 and would continue to vie with Suleiman's forces for control of Aden, in present-day Yemen.

The Siege of Malta in 1565: Arrival of the Turkish fleet, by Matteo Perez d'Aleccio

In 1542, facing a common Habsburg enemy, Francis I sought to renew the Franco-Ottoman alliance. As a result, Suleiman dispatched 100 galleys[40] under Barbarossa to assist the French in the western Mediterranean. Barbarossa pillaged the coast of Naples and Sicily before reaching France where Francis made Toulon the Ottoman admirals naval headquarters. The same campaign had seen Barbarossa attack and capture Nice in 1543. By 1544, a peace between Francis I and Charles V had put a temporary end to the alliance between France and the Ottoman Empire.

Elsewhere in the Mediterranean, when the Knights Hospitallers were re-established as the Knights of Malta in 1530, their actions against Muslim navies quickly drew the ire of the Ottomans who assembled another massive army in order to dislodge the Knights from Malta. The Ottomans invaded in 1565, undertaking the Great Siege of Malta, which began on May 18 and lasted until September 8, and is portrayed vividly in the frescoes of Matteo Perez d'Aleccio in the Hall of St. Michael and St. George. At first it seemed that this would be a repeat of the battle on Rhodes, with most of Malta's cities destroyed and half the Knights killed in battle; but a relief force from Spain entered the battle, resulting in the loss of 30,000 Ottoman troops and the victory of the local Maltese citizenry.[41]

Administrative reforms

While Sultan Suleiman was known as "the Magnificent" in the West, he was always *Kanuni* Suleiman or "The Lawgiver" to his own Ottoman subjects. As the historian Lord Kinross notes, "Not only was he a great military campaigner, a man of the sword, as his father and great-grandfather had been before him. He differed from them in the extent to which he was also a man of the pen. He was a great legislator, standing out in the eyes of his people as a high-minded sovereign and a magnanimous exponent of justice".[42] The overriding law of the empire was the Shari'ah, or Sacred Law, which as the divine law of Islam was outside of the Sultan's powers to change. Yet an area of distinct law known as the *Kanuns* (canonical legislation) was dependent on Suleiman's will alone, covering areas such as criminal law, land tenure and taxation.[43] He collected all the judgments that had been issued by the nine Ottoman Sultans who preceded him. After eliminating duplications and choosing between contradictory statements, he issued a single legal code, all the while being careful not to violate the basic laws of Islam.[44] It was within this framework that Suleiman, supported by his Grand Mufti Ebussuud, sought to reform the legislation to adapt to a rapidly changing empire. When the Kanun laws attained their final form, the code of laws became known as the *kanun-i Osmani*, or the "Ottoman laws". Suleiman's legal code was to last more than three hundred years.[45]

A bas-relief of Suleiman adorning the interior of the United States House of Representatives. It is one of 23 commemorating famous lawmakers throughout history.

Suleiman gave particular attention to the plight of the rayas, Christian subjects who worked the land of the Sipahis. His Kanune Raya, or "Code of the Rayas", reformed the law governing levies and taxes to be paid by the rayas, raising their status above serfdom to the extent that Christian serfs would migrate to Turkish territories to benefit from the reforms.[46] The Sultan also played a role in protecting the Jewish subjects of his empire for centuries to come. In late 1553 or 1554, on the suggestion of his favorite doctor and dentist, the Spanish Jew Moses Hamon, the Sultan issued a *firman* formally denouncing blood libels against the Jews.[47] Furthermore, Suleiman enacted new criminal and police legislation, prescribing a set of fines for specific offenses, as well as reducing the instances requiring death or mutilation. In the area of taxation, taxes were levied on various goods and produce, including animals, mines, profits of trade, and import-export duties. In addition to taxes, officials who had fallen into disrepute were likely to have their land and property confiscated by the Sultan.

Education was another important area for the Sultan. Schools attached to mosques and funded by religious foundations provided a largely free education to Muslim boys in advance of the Christian countries of the time.[48] In his capital, Suleiman increased the number of *mektebs* (primary schools) to fourteen, teaching boys to read and write as well as the principles of Islam. Young men wishing further education could proceed to one of eight *medreses* (colleges), whose studies included grammar, metaphysics, philosophy, astronomy, and astrology.[48] Higher *medreses* provided education of university status, whose graduates became *imams* or teachers. Educational centers were often one of many buildings surrounding the courtyards of mosques, others included libraries, refectories, fountains, soup kitchens and hospitals for the benefit of the public.

Cultural achievements

Tughra of Suleiman the Magnificent.

Under Suleiman's patronage, the Ottoman empire entered the golden age of its cultural development. Hundreds of imperial artistic societies (called the *Ehl-i Hiref*, "Community of the Talented") were administered at the Imperial seat, the Topkapı Palace. After an apprenticeship, artists and craftsmen could advance in rank within their field and were paid commensurate wages in quarterly annual installments. Payroll registers that survive testify to the breadth of Suleiman's patronage of the arts, the earliest of documents dating from 1526 list 40 societies with over 600 members. The *Ehl-i Hiref* attracted the empire's most talented artisans to the Sultan's court, both from the Islamic world and recently conquered territories in Europe, resulting in a blend of Islamic, Turkish and European cultures.[49] Artisans in service of the court included painters, book binders, furriers, jewellers and goldsmiths. Whereas previous rulers had been influenced by Persian culture (Suleiman's father, Selim I, wrote poetry in Persian), Suleiman's patronage of the arts had seen the Ottoman Empire assert its own artistic legacy.[50]

Suleiman himself was an accomplished poet, writing in Persian and Turkish under the nom de plume *Muhibbi* (Lover). Some of Suleiman's verses have become Turkish proverbs, such as the well-known *Everyone aims at the same meaning, but many are the versions of the story*. When his young son Mehmed died in 1543, he composed a moving chronogram to commemorate the year: *Peerless among princes, my Sultan Mehmed.*[51] [52] In addition to Suleiman's own work, many great talents enlivened the literary world during Suleiman's rule, including Fuzuli and Baki. The literary historian E. J. W. Gibb observed that "at no time, even in Turkey, was greater encouragement given to poetry than during the reign of this Sultan".[51] Suleiman's most famous verse is:

> The people think of wealth and power as the greatest fate,
> But in this world a spell of health is the best state.
> What men call sovereignty is a worldly strife and constant war;
> Worship of God is the highest throne, the happiest of all estates.[53]

Süleymaniye Mosque in Istanbul, built by Mimar Sinan, Suleiman's chief architect.

Suleiman also became renowned for sponsoring a series of monumental architectural developments within his empire. The Sultan sought to turn Constantinople into the center of Islamic civilization by a series of projects, including bridges, mosques, palaces and various charitable and social establishments. The greatest of these were built by the Sultan's chief architect, Mimar Sinan, under whom Ottoman architecture reached its zenith. Sinan became responsible for over three hundred monuments throughout the empire, including his two masterpieces, the Süleymaniye and Selimiye mosques—the latter built in Adrianople (now Edirne) in the reign of Suleiman's son Selim II. Suleiman also restored the Dome of the Rock in Jerusalem and the Jerusalem city walls (which are the current walls of the Old City of Jerusalem), renovated the Kaaba in Mecca, and constructed a complex in Damascus.[54]

Personal life

Wives

- Fülane Hatun: Suleiman's first wife, and his real name is unknown. Mahmud's mother who was born in 1512. Records at that time but not a real name Fulani name comes as a tribute for ladies. Suleiman "Fulani" is defined as three mistress. One of them is his son Sultan Mahmud's mother, Fatima gave birth in 1514 and the other "Fulani" There is nicknamed haseki a favorite. Other "Fulani babe," he was born the same year as defined Sultan'la Mihrimah 1522 lady's mother's name is not listed today, and some sources record that the scene of the effective period of the past, but as the son of the Sultan for the Hürrem twin Hürrem Sultan Mihrimah's Prince Abdullah confirmed that another woman to have given his mother, his mother's lady.
- Gülfem Babe: Suleiman, the opinions of Hürrem in the direction of the Sultan is the next or previous haseki, the concubine wife of origin. Gülfem Hatun who died in 1561 or 1562, a baby who was born in 1521 and that same year his son Murad, who died October 12th 1521 is believed that she was the mother.
- Mahidevran Sultan (in some sources as the name goes Gulbahar): born around the year 1500 and the current origin Mahidevran Sultan Mustafa brought to the world in 1515. Mustafa Suleiman followed by the October 6 1553. in Bursa also began to live. She died on February 3rd 1581, in Bursa.
- Hürrem Sultan: Starboard lordship or estimated 1520's accession to the throne after the entry into harem concubine origin Hürrem, 1521, Mehmed, 1522 Mihrimah'ı, 1524 Selim, Bayezid 1525, 1531 in the Cihangir district, brought to the world. She died on April 15 in 1558.

Relationship with Hürrem Sultan

Suleiman was infatuated with Hürrem Sultan, a harem girl from Ruthenia, then part of Poland. In the West foreign diplomats, taking notice of the palace gossip about her, called her "Russelazie" or "Roxelana", referring to her Ruthenian (Ukrainian) origins.[55] The daughter of an Orthodox priest,[29] she was captured by Tatars from Crimea, sold as a slave in Constantinople, and eventually rose through the ranks of the Harem to become Suleiman's favourite. Breaking with two centuries of Ottoman tradition,[29] a former concubine had thus become the legal wife of the Sultan, much to the astonishment of observers in the palace and the city.[56] He also allowed Hürrem Sultan to remain with him at court for the rest of her life, breaking another tradition—that when imperial heirs came of age, they would be sent along with the imperial concubine who bore them to govern remote provinces of the Empire, never to return unless their progeny succeeded to the throne.[57]

Under his pen name, Muhibbi, Suleiman composed this poem for Roxelana:

> "Throne of my lonely niche, my wealth, my love, my moonlight.
> My most sincere friend, my confidant, my very existence, my Sultan, my one and only love.
> The most beautiful among the beautiful...
> My springtime, my merry faced love, my daytime, my sweetheart, laughing leaf...
> My plants, my sweet, my rose, the one only who does not distress me in this world...
> My Constantinople, my Caraman, the earth of my Anatolia
> My Badakhshan, my Baghdad and Khorasan
> My woman of the beautiful hair, my love of the slanted brow, my love of eyes full of mischief...
> I'll sing your praises always
> I, lover of the tormented heart, Muhibbi of the eyes full of tears, I am happy."[58]

Pargalı Ibrahim Pasha

Pargalı Ibrahim Pasha was the boyhood friend of Suleiman. Ibrahim was originally a Christian Greek from Parga, Epirus,[59] [60] and when he was young was educated at the Palace School under the devshirme system. Suleiman made him the royal falconer, then promoted him to first officer of the Royal Bedchamber.[61] Ibrahim Pasha rose to Grand Vizier in 1523 and commander-in-chief of all the armies. Suleiman also conferred upon Ibrahim Pasha the honor of *beylerbey* of Rumelia, granting Ibrahim authority over all Turkish territories in Europe, as well as command of troops residing within them in times of war. According to a 17th century chronicler, Ibrahim had asked Suleiman not to promote him to such high positions, fearing for his safety; to which Suleiman replied that under his reign no matter what the circumstance, Ibrahim would never be put to death.[62]

The Ottoman Sultan Suleiman the Magnificent, awaits the arrival of his Grand Vizier Pargalı Ibrahim Pasha at Buda, in the year 1529.

Yet Ibrahim eventually fell from grace with the Sultan. During his thirteen years as Grand Vizier, his rapid rise to power and vast accumulation of wealth had made Ibrahim many enemies among the Sultan's court. Reports had reached the Sultan of Ibrahim's impudence during a campaign against the Persian Safavid empire: in particular his adoption of the title *serasker sultan* was seen as a grave affront to Suleiman.[63]

Suleiman's suspicion of Ibrahim was worsened by a quarrel between the latter and the Minister of Finance Iskender Chelebi. The dispute ended in the disgrace of Chelebi on charges of intrigue, with Ibrahim convincing Suleiman to sentence the Minister to death. Before his death however, Chelebi's last words were to accuse Ibrahim of conspiracy against the Sultan.[63] These dying words convinced Suleiman of Ibrahim's disloyalty,[63] and on 15 March 1536 Ibrahim was executed.

Succession

Suleiman's two Haseki Sultans had borne him eight sons, four of whom survived past the 1550s. They were Mustafa, Selim, Beyazıt, and Cihangir. Of these, only Mustafa, the eldest, was not Hürrem Sultan's son, but rather Mahidevran Gülbahar Sultan's ("Rose of Spring"), and therefore preceded Hürrem's children in the order of succession. Hürrem was aware that should Mustafa become Sultan her own children would be strangled. Yet Mustafa was recognised as the most talented of all the brothers and was supported by Pargalı İbrahim Pasha, who was by this time Suleiman's Grand Vizier. The Austrian ambassador Busbecq would note "Suleiman has among his children a son called Mustafa, marvellously well educated and prudent and of an age to rule, since he is 24 or 25 years old; may God never allow a Barbary of such strength to come near us", going on to talk of Mustafa's "remarkable natural gifts".[64] Hürrem is usually held at least partly responsible for the intrigues in nominating a successor. Although she was Suleiman's wife, she exercised no official public role. This did not, however, prevent Hürrem from wielding powerful political influence. Since the Empire lacked, until the reign of Ahmed I, any formal means of nominating a successor, successions usually involved the death of competing princes in order to avert civil unrest and rebellions. In attempting to avoid the execution of her sons, Hürrem used her influence to eliminate those who supported Mustafa's accession to the throne.[53]

Thus in power struggles apparently instigated by Hürrem,[61] Suleiman had Ibrahim murdered and replaced with her sympathetic son-in-law, Rüstem Pasha. By 1552, when the campaign against Persia had begun with Rüstem appointed commander-in-chief of the expedition, intrigues against Mustafa began. Rüstem sent one of Suleiman's most trusted men to report that since Suleiman was not at the head of the army, the soldiers thought the time had come to put a younger prince on the throne; at the same time he spread rumors that Mustafa had proved receptive to the idea. Angered by what he came to believe were Mustafa's plans to claim the throne, the following summer upon return from his campaign in Persia, Suleiman summoned him to his tent in the Ereğli valley,[65] stating he would "be able to clear himself of the crimes he was accused of and would have nothing to fear if he came".[66]

Mustafa was confronted with a choice: either he appeared before his father at the risk of being killed; or, if he refused to attend, he would be accused of betrayal. In the end, Mustafa chose to enter his father's tent, confident that the support of the army would protect him. Busbecq, who claims to have received an account from an eyewitness, describes Mustafa's final moments. As Mustafa entered his father's tent, Suleiman's Eunuchs attacked Mustafa, with the young prince putting up a brave defence. Suleiman, separated from the struggle only by the linen hangings of the tent, peered through the chamber of his tent and "directed fierce and threatening glances upon the mutes, and by menacing gestures sternly rebuked their hesitation. Thereupon, the mutes in their alarm, redoubling their efforts, hurled Mustafa to the ground and, throwing the bowstring round his neck, strangled him."[67]

Cihangir is said to have died of grief a few months after the news of his half-brother's murder.[68] The two surviving brothers, Beyazıt and Selim, were given command in different parts of the empire. Within a few years, however, civil war broke out between the brothers, each supported by his loyal forces.[69] With the aid of his father's army, Selim defeated Beyazıt in Konya in 1559, leading the latter to seek refuge with the Safavids along with his four sons. Following diplomatic exchanges, the Sultan demanded from the Safavid Shah that Beyazıt be either extradited or executed. In return for large amounts of gold, the Shah allowed a Turkish executioner to strangle Beyazıt and his four sons in 1561,[68] clearing the path for Selim's succession to the throne seven years later. On 5 or 6 September 1566,[70] Suleiman, who had set out from Constantinople to command an expedition to Hungary, died before an Ottoman victory at the Battle of Szigetvár in Hungary.[71]

Legacy

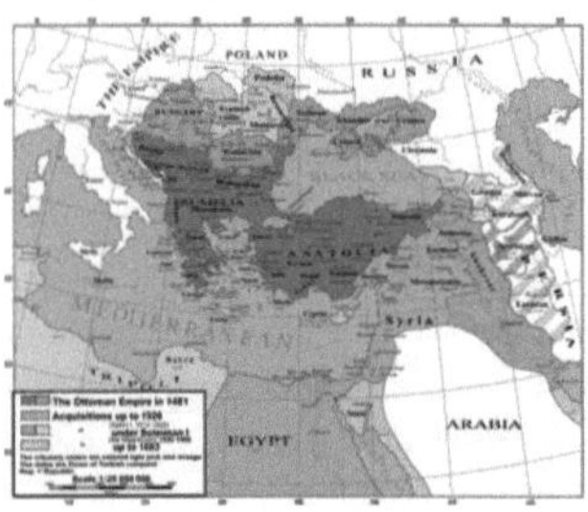

Suleiman I's conquests were followed by continuous territorial expansion until the Empire's peak in 1683.

At the time of Suleiman's death, the Ottoman Empire was one of the world's foremost powers.[72] Suleiman's conquests had brought under the control of the Empire the major Muslim cities (Mecca, Medina, Jerusalem, Damascus, and Baghdad), many Balkan provinces (reaching present day Croatia and Austria), and most of North Africa. His expansion into Europe had given the Ottoman Turks a powerful presence in the European balance of power. Indeed, such was the perceived threat of the Ottoman Empire under the reign of Suleiman that Austria's ambassador Busbecq warned of Europe's imminent conquest: "On [the Turks'] side are the resources of a mighty empire, strength unimpaired, habituation to victory, endurance of toil, unity, discipline, frugality and watchfulness... Can we doubt what the result will be?...When the Turks have settled with Persia, they will fly at our throats supported by the might of the whole East; how unprepared we are I dare not say."[73]

Even thirty years after his death "Sultan Solyman" was quoted by the English playwright William Shakespeare as a military prodigy in *The Merchant of Venice* (Act 2, Scene 1).

Türbe (tomb) of Sultan Süleyman at Süleymaniye Mosque.

Suleiman's legacy was not, however, merely in the military field. The French traveler Jean de Thévenot a century later bears witness to the "strong agricultural base of the country, the well being of the peasantry, the abundance of staple foods, and the pre-eminence of organization in Suleiman's government".[74] The administrative and legal reforms which earned him the name Law Giver ensured the Empire's survival long after his death, an achievement which "took many generations of decadent heirs to undo".[75]

Through his personal patronage, Suleiman also presided over the Golden Age of the Ottoman Empire, representing the pinnacle of the Ottoman Turks' cultural achievement in the realm of architecture, literature, art, theology and philosophy.[6] [76] Today the skyline of the Bosphorus, and of many cities in modern Turkey and the former Ottoman provinces, are still adorned with the architectural works of Mimar Sinan. One of these, the Süleymaniye Mosque, is the final resting place of Suleiman and Hürrem Sultan: they are buried in separate domed mausoleums attached to the mosque.

Notes

[1] *The Encyclopædia Britannica*, Vol.7, Edited by Hugh Chisholm, (1911), 3; *Constantinople, the capital of the Turkish Empire...*

[2] Britannica, Istanbul (http://concise.britannica.com/ebc/article-9368294/Istanbul):*When the Republic of Turkey was founded in 1923, the capital was moved to Ankara, and Constantinople was officially renamed Istanbul in 1930.*

[3] (http://tr.wikipedia.org/wiki/I._SÃ¼leyman)

[4] Merriman.

[5] Mansel, 61.

[6] Atıl, 24.

[7] Clot, 25.

[8] Barber, Noel (1973). *The Sultans*. New York: Simon & Schuster. pp. 36. ISBN 0-7861-0682-4.

[9] Clot, 28.

[10] Kinross, 175.

[11] Lamb, 14.

[12] Barber, 23.

[13] Imber, 49.

[14] Clot, 39.

[15] Kinross, 176.

[16] Kinross, 187.

[17] Severy, 580

[18] Embree, Suleiman The Magnificent (http://www.ccds.charlotte.nc.us/History/MidEast/04/embree/embree.htm).

[19] Turnbull, Stephen (2003). *The Ottoman Empire 1326 – 1699*. New York: Osprey Publishing. pp. 50.

[20] Imber, 50.

[21] Labib, 444.

[22] Imber, 52.

[23] Imber, 53.

[24] Imber, 54.

[25] Imber, 51.

[26] Sicker, 206.

[27] Clot, 93.

[28] 1548–49 (http://www.bartleby.com/67/794.html)

[29] Kinross, 236.

[30] 1553–55 (http://www.bartleby.com/67/795.html)

[31] Azmi Özcan (1997). *Pan-Islamism: Indian Muslims, the Ottomans and Britain, 1877-1924* (http://books.google.com/books?id=s04pus5jBNwC&pg=PA11). BRILL. pp. 11–. ISBN 978-90-04-10632-1. . Retrieved 30 September 2012.
[32] http://jis.oxfordjournals.org/content/7/1/32.extract
[33] Naimur Rahman Farooqi (1989). *Mughal-Ottoman relations: a study of political & diplomatic relations between Mughal India and the Ottoman Empire, 1556-1748* (http://books.google.com/books?id=uB1uAAAAMAAJ). Idarah-i Adabiyat-i Delli. . Retrieved 30 September 2012.
[34] *The history of Aden, 1839-72* by Zaka Hanna Kour p.2 (http://books.google.com/books?id=KrsmNygcbNgC&pg=PA2)
[35] *An economic and social history of the Ottoman Empire* by Halil İnalcik p.326 (http://books.google.com/books?id=Ovg_RQlklU4C&pg=PA326)
[36] *History of the Ottoman Empire and modern Turkey* by Ezel Kural Shaw p.107 (http://books.google.com/books?id=UVmsI0P9RDUC&pg=PA107)
[37] *Cambridge illustrated atlas, warfare: Renaissance to revolution, 1492-1792* by Jeremy Black p.17 (http://books.google.com/books?id=neUKEvaYPZYC&pg=PA17)
[38] Clot, 87.
[39] Kinross, 227.
[40] Kinross, 53.
[41] The History of Malta (http://www.malta.com/about-malta/history-of-malta.html)
[42] Kinross, 205.
[43] Imber, 244.
[44] Greenblatt, 20.
[45] Greenblatt, 21.
[46] Kinross, 210.
[47] Mansel, 124.
[48] Kinross, 211.
[49] Atıl, The Golden Age of Ottoman Art (http://www.saudiaramcoworld.com/issue/198704/the.golden.age.of.ottoman.art.htm), 24–33.
[50] Mansel, 70.
[51] Halman, Suleyman the Magnificent Poet (http://web.archive.org/web/20060309091926/http://www.byegm.gov.tr/yayinlarimiz/NEWSPOT/1999/JulyAug/N6.htm)
[52] Muhibbî (Kanunî Sultan Süleyman)(**Turkish**)[[Category:Articles with Turkish language external links (http://www.turkcebilgi.net/kim-kimdir/m/muhibbi-kanuni-sultan-suleyman-31612.html)]] In Turkish the chronogram reads شهزادەلر گزيدەسى سلطان محمدم(*Şehzadeler güzidesi Sultan Muhammed'üm*), in which the Arabic Abjad numerals total 955, the equivalent in the Islamic calendar of 1543 AD.
[53] Mansel, 84.
[54] Atıl, 26.
[55] Ahmed, 43.
[56] Mansel, 86.
[57] Imber, 90.
[58] A 400 Year Old Love Poem (http://www.womeninworldhistory.com/sample-10.html)
[59] Margaret Rich Greer, Walter Mignolo, Maureen Quilligan. the Black Legend: the discourses of religious and racial difference in the Renaissance empires. (http://books.google.com/books?id=T4_rFZLGF1IC&pg=PA41&dq=Parga+ibrahim+Pasha+greek&hl=el&ei=sCVdTIaoA8_-sQb1gNG4Bw&sa=X&oi=book_result&ct=result&resnum=1&ved=0CC0Q6AEwADgU#v=snippet&q="Ibrahim Pasha, his intimate and grand vezir, a Greek from Parga in Epirus"&f=false"Rereading), University of Chicago Press, 2007. ISBN 978-0-226-30722-0, p. 41: "Ibrahim Pasha, his intimate and grand vezir, a Greek from Parga in Epirus"
[60] Willem Frederik Bakker. Byzantina et Neohellenica Neerlandica. (http://books.google.com/books?id=hAUVAAAAIAAJ&pg=PA312&dq=Parga+ibrahim+Pasha+greek&hl=el&ei=sCVdTIaoA8_-sQb1gNG4Bw&sa=X&oi=book_result&ct=result&resnum=3&ved=0CDoQ6AEwAjgU#v=onepage&q="Ibrahim was the son of a Greek fisherman from the Epirot town of Parga"&f=false"Studia) BRILL, 1972. ISBN 978-90-04-03552-2 ,p. 312
[61] Mansel, 87.
[62] Clot, 49.
[63] Kinross, 230.
[64] Clot, 155.
[65] Ünal, Tahsin (1961). Anıt. pp. 9–22.
[66] Clot, 157.
[67] Kinross, 239.
[68] Mansel, 89.
[69] Kinross, 240.
[70] Yapp, Suleiman I (http://uk.encarta.msn.com/encyclopedia_761575054_2/Suleiman_I.html)
[71] Imber, 60.
[72] Clot, 298.

[73] Lewis, 10.
[74] Ahmed, 147.
[75] Lamb, 325.
[76] Russell, The Age of Sultan Suleyman (http://query.nytimes.com/gst/fullpage.html?res=9B0DE7D6163EF935A15752C0A961948260).

References

Printed Sources

- Ahmed, Syed Z (2001). *The Zenith of an Empire : The Glory of the Suleiman the Magnificent and the Law Giver*. A.E.R. Publications. ISBN 978-0-9715873-0-4.
- Atıl, Esin (1987). *The Age of Sultan Süleyman the Magnificent*. Washington, D.C.: National Gallery of Art. ISBN 978-0-89468-098-4.
- Atıl, Esin (July/August 1987). "The Golden Age of Ottoman Art" (http://www.saudiaramcoworld.com/issue/198704/the.golden.age.of.ottoman.art.htm). *Saudi Aramco World* (Houston, Texas: Aramco Services Co) **38** (4): 24–33. ISSN 1530-5821. Retrieved 2007-04-18.
- Barber, Noel (1976). *Lords of the Golden Horn : From Suleiman the Magnificent to Kamal Ataturk*. London: Pan Books. ISBN 978-0-330-24735-1.
- Clot, André (1992). *Suleiman the Magnificent : The Man, His Life, His Epoch*. London: Saqi Books. ISBN 978-0-86356-126-9.
- Garnier, Edith *L'Alliance Impie* Editions du Felin, 2008, Paris ISBN 978-2-86645-678-8 Interview (http://www.neopodia.mobi/20090519-histoire-renaissance-alliance-impie-francois-1er-ier-ottoman-soliman-suleyman-charles-quint-forces-en-presence)
- Greenblatt, Miriam (2003). *Süleyman the Magnificent and the Ottoman Empire*. New York: Benchmark Books. ISBN 978-0-7614-1489-6.
- Imber, Colin (2002). *The Ottoman Empire, 1300–1650 : The Structure of Power*. New York: Palgrave Macmillan. ISBN 978-0-333-61386-3.
- Kinross, Patrick (1979). *The Ottoman centuries: The Rise and Fall of the Turkish Empire*. New York: Morrow. ISBN 978-0-688-08093-8.
- Labib, Subhi (November 1979). "The Era of Suleyman the Magnificent: Crisis of Orientation". *International journal of Middle East studies* (London: Cambridge University Press) **10** (4): 435–451. ISSN 0020-7438.
- Lamb, Harold (1951). *Suleiman, the Magnificent, Sultan of the East*. Garden City, N.Y.: Doubleday. ISBN 1-4067-7271-2. OCLC 397000.
- Levey, Michael (1975). *The World of Ottoman Art*. Thames & Hudson. ISBN 0-500-27065-1.
- Lewis, Bernard (2002). *What Went Wrong? : Western Impact and Middle Eastern Response*. London: Phoenix. ISBN 978-0-7538-1675-2.
- Mansel, Phillip (1998). *Constantinople : City of the World's Desire, 1453–1924*. New York: St. Martin's Griffin. ISBN 978-0-312-18708-8.
- Merriman, Roger Bigelow (1944). *Suleiman the Magnificent, 1520–1566*. Cambridge, Massachusetts: Harvard University Press. ISBN 1-4067-7272-0. OCLC 784228.
- Severy, Merle (November 1987). "The World of Süleyman the Magnificent". *National geographic* (Washington, D.C.: National Geographic Society) **172** (5): 552–601. ISSN 0027-9358.
- Sicker, Martin (2000). *The Islamic World In Ascendancy : From the Arab Conquests to the Siege of Vienna*. Westport, Connecticut: Praeger. ISBN 978-0-275-96892-2.
- "Suleiman The Lawgiver" (http://www.saudiaramcoworld.com/issue/196402/.suleiman.the.lawgiver..htm). *Saudi Aramco World* (Houston, Texas: Aramco Services Co) **15** (2): 8–10. March/April 1964. ISSN 1530-5821. Retrieved 2007-04-18.

On-line sources

- "1548-49" (http://www.bartleby.com/67/794.html). *The Encyclopedia of World History*. 2001. Retrieved 2007-04-18.
- "1553-55" (http://www.bartleby.com/67/795.html). *The Encyclopedia of World History*. 2001. Retrieved 2007-04-18.
- "A 400 Year Old Love Poem" (http://www.womeninworldhistory.com/sample-10.html). *Women in World History Curriculum Showcase*. Retrieved 2007-04-18.
- Embree, Mark (2004). "Suleiman The Magnificent" (http://web.archive.org/web/20060930054118/http://www.ccds.charlotte.nc.us/History/MidEast/04/embree/embree.htm). Archived from the original (http://www.ccds.charlotte.nc.us/History/MidEast/04/embree/embree.htm) on September 30, 2006. Retrieved 2007-04-18.
- Halman, Talat (1988). "Suleyman the Magnificent Poet" (http://web.archive.org/web/20060309091926/http://www.byegm.gov.tr/yayinlarimiz/NEWSPOT/1999/JulyAug/N6.htm). Archived from the original (http://www.byegm.gov.tr/yayinlarimiz/NEWSPOT/1999/JulyAug/N6.htm) on 2006-03-09. Retrieved 2007-04-18.
- "The History of Malta" (http://web.archive.org/web/20070501155037/http://www.malta.com/about-malta/history-of-malta.html). 2007. Archived from the original (http://www.malta.com/about-malta/history-of-malta.html) on May 1, 2007. Retrieved 2007-04-27.
- "Muhibbî (Kanunî Sultan Süleyman)" (http://www.turkcebilgi.net/kim-kimdir/m/muhibbi-kanuni-sultan-suleyman-31612.html) (in Turkish). *Türkçe Bilgi—Kim kimdir?*. Retrieved 2008-01-13.
- Russell, John (2007-01-26). "The Age of Sultan Suleyman" (http://query.nytimes.com/gst/fullpage.html?res=9B0DE7D6163EF935A15752C0A961948260). New York Times. Retrieved 2007-08-09.
- Yapp, Malcolm Edward (2007). "Suleiman I" (http://uk.encarta.msn.com/encyclopedia_761575054_2/Suleiman_I.html). *Microsoft Encarta*. Retrieved 2008-04-17.

Further reading

- Alum Bati, "Harem Secrets", Trafford, 2008, ISBN=1-4251-5750-5
- Bridge, Anthony (1983). *Suleiman the Magnificent, Scourge of Heaven*. New York: F. Watts. ISBN 0-88029-169-9. OCLC 9853956.
- Downey, Fairfax Davis. *The Grand Turke, Suleyman the Magnificent, sultan of the Ottomans*. New York: Minton, Balch & Company. OCLC 25776191.
- Hooker, Richard. "The Ottomans: Suleyman" (http://www.wsu.edu:8080/~dee/OTTOMAN/SULEYMAN.HTM). Retrieved 2007-09-02.
- Lybyer, Albert Howe (1913). *The government of the Ottoman empire in the time of Suleiman the Magnificent* (http://archive.org/details/governmentofotto00lyby). Cambridge, Massachusetts: Harvard University Press. ISBN 0-404-14681-3. OCLC 1562148.
- André Clot *Soliman Le Magnifique*, Fayard, Paris, 1983, 469 p. ISBN 2-213-01260-1

Murad III

Murad III	
His Imperial Majesty Grand Sultan, Commander of the Faithful and Successor of the Prophet of the Lord of the Universe	

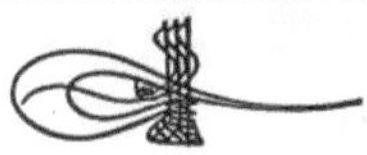	
Reign	1574–95
Period	Growth of the Ottoman Empire
Full Name	Murad III
Born	4 July 1546
Birthplace	Bozdağan or Manisa
Died	15/16 January 1595
Place of death	Topkapı Palace of Constantinople
Predecessor	Selim II
Successor	Mehmed III
Royal House	House of Osman
Dynasty	Ottoman Dynasty

Murad III (Ottoman Turkish: مراد ثالث *Murād-i s̱ālis̱*, Turkish:*III.Murat*) (4 July 1546 – 15/16 January 1595) was the Sultan of the Ottoman Empire from 1574 until his death.

Biography

Born in Bozdağan or Manisa, Murad III was the eldest son of sultan Selim II (1566–74), and succeeded his father in 1574. Murad began his reign by having his five younger brothers strangled.[1] His authority was undermined by the harem influences, more specifically, those of his mother and later of his favorite wife Safiye Sultan. The power had only been maintained under Selim II by the genius of the all-powerful Grand Vizier Mehmed Sokollu who remained in office until his assassination in October 1579. During his reign the northern borders with the Austro-hungarian empire were defended by the Bosniak kapetan Hasan Predojević. The reign of Murad III was marked by wars with Safavids and Habsburgs and Ottoman economic decline and institutional decay. The Ottomans also faced defeats during battles such as the Battle of Sisak.

Murad took great interest in the arts, particularly miniatures and books. He actively supported the court Society of Miniaturists, commissioning several volumes including the *Siyer-i Nebi*, the most heavily illustrated biographical work on the life of Muhammad, the *Book of Skills*, the *Book of Festivities* and the *Book of Victories*.[2] He had two large alabaster urns transported from Pergamon and placed on two sides of the nave in the Hagia Sophia in Constantinople and a large wax candle dressed in tin which was donated by him to the Rila monastery in Bulgaria is on display in the monastery museum.

From him descend all succeeding Sultans,[3] through his marriage to his maternal relative Valide Sultan Safiye Sultan, originally named Sofia Baffo, a Venetian noblewoman, mother of Mehmed III.

Numerous envoys and letters were exchanged between Elizabeth I and Sultan Murad III.[4] In one correspondence, Murad entertained the notion that Islam and Protestantism had "much more in common than either did with Roman Catholicism, as both rejected the worship of idols", and argued for an alliance between England and the Ottoman Empire.[5] To the dismay of Catholic Europe, England exported tin and lead (for cannon-casting) and ammunitions to the Ottoman Empire, and Elizabeth seriously discussed joint military operations with Murad III during the outbreak of war with Spain in 1585, as Francis Walsingham was lobbying for a direct Ottoman military involvement against the common Spanish enemy.[6] This diplomacy would be continued under Murad's successor Mehmed III, by both the sultan and Safiye Sultan alike.

Murad died in the Topkapı Palace of Constantinople[7] [8] in 1596.

In fiction

Orhan Pamuk's historical novel *Benim Adım Kırmızı* (*My Name is Red*, 1998) takes place at the court of Murad III, during nine snowy winter days of 1591, which the writer uses in order to convey the tension between East and West.

References

[1] Marriott, John Arthur. *The Eastern Question* (Clarendon Press, 1917), 96.
[2] Pamuk, Orhan. *My Name is Red*, Alfred A. Knopf, 2010. ISBN 978-0-307-59392-4
[3] See A. D. Alderson, The structure of the Ottoman dynasty [Oxford: Clarendon, 1956], Table XXXI et seq., for details.
[4] Karen Ordahl Kupperman. *The Jamestown project* (http://books.google.com/books?id=9VmYYEfyToQC&pg=PA39). p. 39. .
[5] Kupperman, p.40
[6] Kupperman, p.41
[7] *The Encyclopædia Britannica*, Vol.7, Edited by Hugh Chisholm, (1911), 3; *Constantinople, the capital of the Turkish Empire...*
[8] Britannica, Istanbul (http://concise.britannica.com/ebc/article-9368294/Istanbul):*When the Republic of Turkey was founded in 1923, the capital was moved to Ankara, and Constantinople was officially renamed Istanbul in 1930.*

External links

- Ancestry of Sultana Nur-Banu (Cecilia Venier-Baffo) (http://www.wargs.com/royal/venier.html)

[aged 48]

Mustafa I

Mustafa I	
Ottoman Sultan	
Caliph	

Reign	1622–23
Period	Growth of the Ottoman Empire
Full Name	Mustafa I
Born	1591
Died	January 1639 (aged 47–48)
Predecessor	Ahmed I
Successor	Murad IV
Royal House	House of Osman
Dynasty	Ottoman Dynasty

Mustafa I Deli (1591 – January 20, 1639) (Ottoman Turkish: مصطفی اول), son of Mehmed III, was the Sultan of the Ottoman Empire from 1617 to 1618 and from 1622 to 1623.

Biography

He was born in the Manisa Palace, as the younger brother of Ahmed I (1603–17). His mother was Valide Handan Sultan, an ethnic Greek originally named Helena.

He replaced as Sultan his brother Ahmed I, who died because of typhus at a young age, in 1617. Mustafa I was reported to be mentally retarded or at least neurotic and was never more than a tool of court cliques at the Topkapı Palace.[1] During the reign of his brother, he was confined to his room in virtual imprisonment, a system called Kafes, for fourteen years.

In 1618, after a short rule, he was deposed in favour of his young nephew Osman II (1618–22) and was sent back to Kafes. The conflict between the Janissaries and Osman II presented him with a second chance. After the assassination of Osman II in 1622 by Janisaries, he was placed back on the throne and held it for another year.[2] He had the participants in the coup against Osman II executed and believed that Osman II was still alive. He was seen searching for him throughout the palace, knocking on doors and crying out to his nephew to relieve him from the burden of sovereignty. His inability to rule led to deposition and confinement by Osman II's younger half-brother,

Murad IV (1623–40).[3]

He died sixteen years later in the Topkapı Palace of Constantinople.

Notes

[1] Imber, Colin. *The Ottoman Empire: The Structure of Power, 2nd ed.* New York: Palgrave Macmillan, 2009, pp. 66-68, 97-98. ISBN 0-230-57451-3.

[2] Imber. *The Ottoman Empire*, pp. 98-99.

[3] Imber. *The Ottoman Empire*, p. 99.

Osman II

Osman II	
Ottoman Sultan	
Caliph	

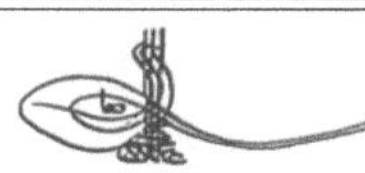	
Reign	1618–1622
Period	Growth of the Ottoman Empire
Full Name	Sultan Osman II
Born	November 3, 1604
Died	March 20, 1622 (aged 17)
Predecessor	Mustafa I (The Intestable)
Successor	Mustafa I (The Intestable)
Royal House	House of Osman
Dynasty	Ottoman Dynasty

Sultan **Osman II** or **Othman II** (commonly known as **Genç Osman** – meaning Osman the Young – in Turkish) (Ottoman Turkish عثمان ثانى *'Os̱mān-i s̱ānī*) (November 3, 1604 – May 20, 1622) was the Sultan of the Ottoman Empire from 1618 until his death on 20 May 1622. He married Aisha, born in 1607, without issue.

Biography

Osman II was born at Topkapı Palace, Constantinople, the son of Sultan Ahmed I (1603–17) and his wife Mâh-Firûze Hatice (Khadija) Valide Sultan, originally named Maria, a Greek.[1] At a young age his mother paid a lot of attention to his education, as a result of this Osman II was a known poet and had mastered many languages, including Arabic, Persian, Greek, Latin, Italian, and the court sign language. He ascended the throne at the early age of 14 as the result of a coup d'état against his uncle Mustafa I (The Intestable) (1617–18, 1622–23). Despite his youth, Osman II soon sought to assert himself as a ruler, and after securing the empire's eastern border by signing a peace treaty (treaty of Serav) with Safavid Iran, he personally led the Ottoman invasion of Poland during the Moldavian Magnate Wars. Forced to sign a peace treaty with the Poles after the Battle of Chotin (Chocim) (which was, in fact, a siege of Chotin defended by the Polish hetman Jan Karol Chodkiewicz) in September–October, 1621, Osman II returned home to Istanbul in shame, blaming the cowardice of the Janissaries and the insufficiency of his statesmen for his humiliation.

Probably the first Sultan to identify and attempt to tackle the Janissaries as a praetorian institution doing more harm than good to the modern empire, Osman II closed their coffee shops (the gathering points for conspiracies against the throne) and started planning to create a new, loyal and ethnic Turkic army consisting of Anatolian, Mesopotamian and Egyptian Turks and Turkmens. The result was a palace uprising by the janissaries, who promptly imprisoned the young sultan. When an executioner was sent to strangle him at Yedikule, Constantinople, Osman II refused to give in and started fighting the man and was only subdued when he was hit on his back with the rear end of an axe by one of his imprisoners. After that he was strangled with a bowstring. A combination of these stories is given by the French traveler Pouqueville, who writes that when the cord was thrown over his neck, Osman 'had the presence of mind to slip it with his hand, and knock down the principal executioner; on which his grand vizier seized him by the most sensible part of his body, when Osman fainted with pain, and was strangled.' [2]

Notes

[1] Shaw, Stanford Jay. *History of the Ottoman Empire and Modern Turkey*. Cambridge University Press. p. 191.

[2] Pouqueville, François Charles H.L.: *Travels through the Morea, Albania, and several other parts of the Ottoman Empire*, page 113-114, published 1806

Ahmed III

Ahmed III	
Ottoman Sultan	
Caliph	

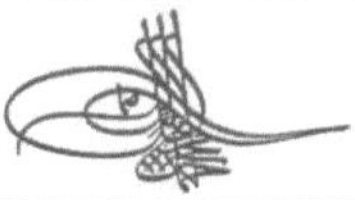	
Reign	December 28, 1703 – September 20, 1730
Period	Stagnation of the Ottoman Empire
Full Name	Ahmed III
Predecessor	Mustafa II
Successor	Mahmud I
Royal House	House of Osman
Dynasty	Ottoman Dynasty

Ahmed III (Ottoman Turkish: احمد ثالث *Aḥmed-i s̱ālis̱*) (December 30/31, 1673 – July 1, 1736) was Sultan of the Ottoman Empire and a son of Sultan Mehmed IV (1648–87). His mother was Mâh-Pâre Ummatullah (Emetullah) Râbi'a Gül-Nûş Valide Sultan, originally named Evmania Voria, who was an ethnic Greek.[1] [2] [3] [4] [5] [6] He was born at Hajioglupazari, in Dobruja. He succeeded to the throne in 1703 on the abdication of his brother Mustafa II (1695–1703). Nevşehirli Damat İbrahim Pasha and his daughter, Princess Hatice (wife of the former) directed the government from 1718 to 1730, a period referred to as the *Tulip Era*.

Biography

Portrait of Ahmed III by John Young

Ahmed III cultivated good relations with France, doubtless in view of Russia's menacing attitude - in fact, both his wives were Frenchwomen. He afforded refuge in Ottoman territory to Charles XII of Sweden (1682–1718) after the Swedish defeat at the hands of Peter I of Russia (1672–1725) in the Battle of Poltava of 1709. In 1710 he convinced the Ottoman Sultan Ahmed III to declare war against Russia, and the Ottoman forces under Baltacı Mehmet Pasha won a major victory at the Battle of Prut. In the aftermath, Russia returned Azov back to the Ottomans, agreed to demolish the fortress of Taganrog and others in the area, and to stop interfering in the affairs of the Polish-Lithuanian Commonwealth.

Miniature of Ahmed III.

Forced against his will into war with Russia, Ahmed III came nearer than any Ottoman sovereign before or since to breaking the power of his northern rival, whose armies his grand vizier Nevşehirli Damat İbrahim Pasha succeeded in completely surrounding at the Pruth River Campaign in 1711. The subsequent Ottoman victories against Russia enabled the Ottoman Empire to advance to Moscow, had the Sultan wished. However, this was halted as a report reached Constantinople[7][8] that the Safavids were invading the Ottoman Empire, causing a period of panic, turning the Sultan's attention away from Russia.

Sultan Ahmed III had become unpopular by reason of the excessive pomp and costly luxury in which he and his principal officers indulged; on September 20, 1730, a mutinous riot of seventeen janissaries, led by the Albanian Patrona Halil, was aided by the citizens as well as the military until it swelled into an insurrection in front of which the Sultan was forced to give up the throne.

Ahmed voluntarily led his nephew Mahmud I (1730–54) to the seat of sovereignty and paid allegiance to him as Sultan of the Empire. He then retired to the Kafes previously occupied by Mahmud and died at Topkapı Palace after six years of confinement.

Character of Ahmed III's rule

Sultan Ahmed III receives French ambassador Vicomte d'Andrezel at Topkapı Palace.

French ambassador Marquis de Bonnac being received by Sultan Ahmed III.

The reign of Ahmed III, which had lasted for twenty-seven years, although marked by the disasters of the Great Turkish War, was not unsuccessful. The recovery of Azov and the Morea, and the conquest of part of Persia, managed to counterbalance the Balkan territory ceded to the Habsburg Monarchy through the Treaty of Passarowitz, after the Ottoman Empire was defeated in Austro-Turkish War of 1716-18. In 1716, he sent an army of 33,000 men to capture Corfu from the Republic of Venice.

Ahmed III left the finances of the Ottoman Empire in a flourishing condition, which had remarkably been obtained without excessive taxation or extortion procedures. He was a cultivated patron of literature and art, and it was in his time that the first printing press authorized to use the Arabic or Turkish languages was set up in Constantinople, operated by Ibrahim Muteferrika (while the printing press had been introduced to Constantinople in 1480, all works published before 1729 were in Greek, Armenian, or Hebrew).

It was in this reign that an important change in the government of the Danubian Principalities was introduced: previously, the Porte had appointed Hospodars, usually native Moldavian and Wallachian boyars, to administer those provinces; after the Russian campaign of 1711, during which Peter the Great found an ally in Moldavia Prince Dimitrie Cantemir, the Porte began overtly deputizing Phanariote Greeks in that region, and extended the system to Wallachia after Prince Stefan Cantacuzino established links with Eugene of Savoy. The Phanariotes constituted a kind of *Dhimmi* nobility, which supplied the Porte with functionaries in many important departments of the state.

Relations with the Mughal Empire

Jahandar Shah

In the year 1712, the Mughal Emperor Jahandar Shah, a grandson of Aurangzeb sent gifts to the Ottoman Sultan Ahmad III and refereed to himself as the Ottoman Sultan's devoted admirer.[9]

Farrukhsiyar

The Mughal Emperor Farrukhsiyar a grandson of Aurangzeb, is also known to have sent a letter to the Ottomans bu this time it was received by the Grand Vizier Nevşehirli Damad Ibrahim Pasha providing a graphic description of the informing him of the efforts of the Mughal commander Syed Hassan Ali Khan Barha against the Rajput and Maratha rebellion.[10]

Marriages and issue

He married Aminā (Emine) Mihr-î-Mâh Sultan, and Râbi'a Sermi Sultan. By the first wife he had Mustafa III and by the second wife he had Abdul Hamid I.

In Fiction

Sultan Ahmed III

A passage in Voltaire's *Candide* has the book's eponymous main character meet the deposed Ahmed III while on a ship from Venice to Constantinople. The Sultan is in the company of five other deposed European monarchs, and he tells Candide, who initially doubts his credentials:

> I am not jesting, my name is Achmet III. For several years I was Sultan; I dethroned my brother; my nephew dethroned me; they cut off the heads of my viziers; I am ending my days in the old seraglio; my nephew, Sultan Mahmoud, sometimes allows me to travel for my health, and I have come to spend the Carnival at Venice" ([11]).

This episode was taken up by the modern Turkish writer Nedim Gürsel and made into the setting of his 2001 novel *Le voyage de Candide à Istanbul.*

In fact, there is no evidence of the deposed Sultan being allowed to make such foreign travels, nor did Voltaire (or Gürsel) assert that it had any actual historical foundation.

The Harem with the Sultan Ahmed III

See also

- Fountain of Ahmed III
- Ibrahim Muteferrika

References

[1] Freely, John (2001). *The lost Messiah.* Viking. p. 132. ISBN 0-670-88675-0. "He set up his harem there, his favourite being Rabia Giilniis Ummetiillah, a Greek girl from Rethymnon on Crete"

[2] Palmer, Alan (2009). *The decline and fall of the Ottoman Empire.* Barnes & Noble. p. 27. ISBN 1-56619-847-X. "Unusually, the twenty-nine year old Ahmed III was a brother, rather than a half- brother, of his predecessor; their Greek mother, Rabia"

[3] Bromley, J. S. (1957). *The New Cambridge Modern History.* University of California: University Press. p. 554. ISBN 0-521-22128-5. "the mother of Mustafa II and Ahmed III was a Greek"

[4] Sardo, Eugenio Lo (1999). *Tra greci e turchi: fonti diplomatiche italiane sul Settecento ottomano.* Consiglio nazionale delle ricerche. p. 82. ISBN 88-8080-014-0. "Their mother, a Greek, lady named Rabia Gülnûş, continued to wield influence as the Valide Sultan - mother of the reigning sultan"

[5] Library Information and Research Service (2005). *The Middle East.* Library Information and Research Service. p. 91. "She was the daughter of a Greek family and she was the mother of Mustafa II (1664-1703), and Ahmed III (1673-1736)."

[6] Baker, Anthony E - Freely, John (1993). *The Bosphorus.* Redhouse Press. p. 146. ISBN 975-413-062-0. "The Valide Sultan was born Evmania Voria, daughter of a Greek priest in a village near Rethymnon on Crete. She was captured by the Turks when they took Rethymnon in 1645."

[7] *The Encyclopædia Britannica,* Vol.7, Edited by Hugh Chisholm, (1911), 3; *Constantinople, the capital of the Turkish Empire...*

[8] Britannica, Istanbul (http://concise.britannica.com/ebc/article-9368294/Istanbul): *When the Republic of Turkey was founded in 1923, the capital was moved to Ankara, and Constantinople was officially renamed Istanbul in 1930.*
[9] http://books.google.com/books?id=uB1uAAAAMAAJ&q=Farrukhsiyar&source=gbs_word_cloud_r&cad=6
[10] *Mughal-Ottoman relations: a study of political & diplomatic relations ... - Naimur Rahman Farooqi* (http://books.google.com/books?id=uB1uAAAAMAAJ&q=Farrukhsiyar&source=gbs_word_cloud_r&cad=6). Books.google.com. . Retrieved 2012-04-29.
[11] http://www.archive.org/stream/candideotherroma00voltuoft/candideotherroma00voltuoft_djvu.txt

Sources

- This article incorporates text from the *History of Ottoman Turks* (1878)
- This article incorporates text from a publication now in the public domain: Chisholm, Hugh, ed. (1911). "Ahmed III" (http://www.1911encyclopedia.org/Ahmed_III). *Encyclopædia Britannica* (11th ed.). Cambridge University Press.

[aged 62]

Murad IV

Murad IV	
Ottoman Sultan	
Caliph	

Reign	1623–40
Period	Growth of the Ottoman Empire
Full Name	Sultan Murad IV
Born	July 26, 1612
Died	February 9, 1640 (aged 27)
Predecessor	Mustafa I
Successor	Ibrahim I
Royal House	House of Osman
Dynasty	Ottoman Dynasty

Murad IV Ghazi (Ottoman Turkish: مراد رابع *Murād-ı rābi'*) (July 26/27, 1612 – February 9, 1640) was the Sultan of the Ottoman Empire from 1623 to 1640, known both for restoring the authority of the state and for the brutality of his methods. Murad IV was born in Constantinople, the son of Sultan Ahmed I (1603–17) and the ethnic Greek[1] [2] [3] Valide Kösem Sultan (also known as Mahpeyker). Brought to power by a palace conspiracy in 1623, he succeeded his uncle Mustafa I (1617–18, 1622–23). He was only 11 when he took the throne.

Portrait of Murad IV.

Murad IV reign (1623-1640)

In the early years of Murad's reign, he was under the control of his relatives. During this period, peace and harmony in the Ottoman lands were completely lost and tyrants took control of the cities. His absolute rule started around 1632, when he took the authority and repressed all the tyrants, and he re-established the supremacy of Sultan.

Early Reign (1623-1632)

Murad IV was for a long time under the control of his relatives and during his early years as Sultan, his mother, Kösem Sultan, essentially ruled through him. The Empire fell into anarchy; the Safavid Empire invaded Iraq almost immediately, Northern Anatolia erupted in revolts, and in 1631 the Janissaries stormed the palace and killed the Grand Vizier, among others. Murad IV feared suffering the fate of his elder brother, Osman II (1618–22), and decided to assert his power.

Absolute Rule and Imperial Policies (1632-1640)

Murad IV tried to quell the corruption that had grown during the reigns of previous Sultans, and that had not been checked while his mother was ruling through proxy. He addressed this corruption with several policy changes, such as limiting wasteful spending.

Murad IV also banned alcohol, tobacco, and coffee in Constantinople.[4] He ordered execution for breaking this ban. He would patrol the streets and taverns of Constantinople in civilian clothes at night, policing the enforcement of his command. By prohibiting tobacco he assimilated the idles gathering in the taverns. He restored the judicial regulations by very strict punishments, including execution. Halil İnalcık reports that even though he was a ruthless supporter of alcohol prohibition, Murat IV was a habitual drinker.[5]

In a poem by the sultan under a nom de plume, he says "Even if the rivers became wine, they wouldn't fill my glass." In another poem he says "The wine is such a devil that I have to protect my people from it by drinking all of it".

Kösem Sultan (1589–1651) the mother of Murad IV who exercised a decisive influence on the sultan's decisions during his youth.

Relations with the Mughal Empire

In the year 1626, the Mughal Emperor Jahangir began to contemplate an alliance between the Ottomans, Mughals and Uzbeks against the Safavids, who had defeated the Mughals at Kandahar. He even wrote a letter to the Ottoman Sultan Murad IV, Jahangir's ambition however did not materialize due to his death in 1627. However, Jahangir's son and successor Shah Jahan perused the a goal of alliance with the Ottoman Empire.

While he was encamped in Baghdad, Murad IV is known to have met the Mughal Emperor Shah Jahan's ambassadors: Mir Zarif and Mir Baraka, who presented 1000 pieces of finely embroidered cloth and even armor. Murad IV gifted them with the finest weapons, saddles and Kaftans and ordered his forces to accompany the Mughals to the port of Basra, where they set sail to Thatta and finally Surat.[6]

Military campaigns

Militarily, Murad IV's reign is most notable for the war against Safavids in which Ottoman forces invaded Azerbaijan, occupied Revan, Tabriz and Hamadan, and recaptured Baghdad in 1638. The sultan had a famous quote about the fall of Baghdad: "Bağdat'ı almaya çalışmak, Bağdat'ın kendinden daha mı güzeldi ne" ("Trying to conquer Baghdad, was almost more beautiful than Baghad itself.").[7]

Murad IV himself commanded the Ottoman army in the last years of the war, and proved to be an outstanding field commander. He was the second Ottoman Sultan to command an army on the battlefield since death of Suleiman the Magnificent in 1566. During his campaign to Iran, he annihilated all rebels in Anatolia and restored order in the Empire.

The war was concluded by the Treaty of Kasrı Şirin in May 1639, which restored Mesopotamia to the Ottomans. After his return to Constantinople, he ordered respected statesmen of the Empire to prepare a new economic and political project to return to the Empire the old successful days.

Physical power

Murad IV was the last Warrior Sultan who led campaigns in front of his army and fought on the battlefield. His physical strength was phenomenal, which is described in detail on the books of Evliya Çelebi. He was especially known for his exceptional strength in wrestling - capable of fighting several opponents at the same time. His favorite weapon was a huge mace, weighing 60 kilograms (132 lbs), which he wielded effortlessly with a single hand.[8] Among his other favourite weapons are a longbow and a large two-handed broadsword weighing more than 50 kilograms (110 lbs). His weapons are today displayed at the Topkapı Palace Museum in Istanbul, intact and well preserved.

Architecture

Ottoman architectural concepts were incorporated into the design of the famous Taj Mahal, in the Mughal Empire by Shah Jahan.

Sultan Murad IV put emphasis on architecture and in his period many monuments were erected. Some of them are Meydanı Mosque, Bayram Pasha Dervish Lodge, Tomb, Fountain, Primary School, Konya Serefeddin Mosque.

The Mughal Emperor Shah Jahan had exchanged ambassadors with the Ottoman Sultan Murad IV, it was through these exchanges that he received Isa Muhammad Effendi and Ismail Effendi, two Turkish architects and students of the famous Koca Mimar Sinan Agha. Both of them later comprised among the Mughal team that would design and build the Taj Mahal.

Death

Murad IV died in İstanbul at the age of 27 in 1640. There are two separate claims on the cause of death. Western sources claim Murad IV, who had outlawed alcohol, died from cirrhosis of the liver. Ottoman sources claim that he died of gout.

On his deathbed, Murad IV ordered the execution of his mentally disabled brother, Ibrahim I (reigned 1640–48), which would have meant the end of the Ottoman line; but the order was not carried out.

Notes

[1] E. van Donzel, Islamic Desk Reference: Compiled from the Encyclopaedia of Islam, Brill Academic Publishers, p 219
[2] Robert Bator, Daily Life in Ancient and Modern Istanbul, Runestone Press, p 42
[3] Douglas Arthur Howard, The History of Turkey, Greenwood Press, p 195
[4] Hopkins, Kate (2006-03-24). "Food Stories: The Sultan's Coffee Prohibition" (http://www.accidentalhedonist.com/index.php/2006/03/24/food_stories_the_sultan_s_coffee_prohibi). . Retrieved 2006-09-12.
[5] İnalcık, Halil; Imber, Colin (1989). *The Ottoman Empire : the classical age, 1300-1600*. New Rochelle, N.Y.: Aristide D. Caratzas. p. 99. ISBN 0-89241-388-3.
[6] http://books.google.com/books?id=uB1uAAAAMAAJ&q=zarif#search_anchor
[7] *Akın Alıcı, Hayata Yön Veren Sözler*, 2004
[8] Tarih-i Na'imâ Vol 3 p 454

See also

- Ottoman-Commonwealth War (1633-1634)
- Koçi Bey
- Kösem Sultan

Mahmud I

Mahmud I

Ottoman Sultan

Caliph

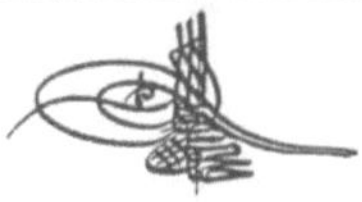

Reign	20 September 1730 – 13 December 1754
Period	Stagnation of the Ottoman Empire
Full Name	Mahmud I
Predecessor	Ahmed III
Successor	Osman III
Royal House	House of Osman
Dynasty	Ottoman Dynasty

Mahmud I (Ottoman Turkish: محمود اول, called **the Hunchback** (*Kambur*)(2 August 1696 – 13 December 1754) was the Sultan of the Ottoman Empire from 1730 to 1754.

Reign

He was born at the Edirne Palace, the son of Mustafa II (1695–1703); his mother was Valide Sultan Saliha Sabkati. Mahmud I was the older brother of Osman III (1754–57).

On 28 September 1730, Patrona Halil with a small group of fellow Janissaries aroused some of the citizens of Constantinople[1] [2] who opposed the reforms of Ahmet III.[3] Sweeping up more soldiers Halil led the riot to the Topkapı Palace and demanded the death of the grand vizer, Nevşehirli Damat İbrahim Pasha and the abdication of Ahmet III. Ahmet III acceded to the demands, had İbrahim Pasha strangled, and agreed to his nephew, Mahmud, becoming sultan.[3]

Mahmud I was recognized as sultan by the mutineers as well as by court officials but for some weeks after his accession the empire was in the hands of the insurgents. Halil rode with the new sultan to the Mosque of Eyub where the ceremony of girding Mahmud I with the Sword of Osman was performed; many of the chief officers were deposed and successors to them appointed at the dictation of the bold rebel who had served in the ranks of the Janissaries and who appeared before the sultan bare-legged and in his old uniform of a common soldier. A Greek butcher, named Yanaki, had formerly given credit to Halil and had lent him money during the three days of the insurrection. Halil showed his gratitude by compelling the Divan to make Yanaki Hospodar of Moldavia. However, Yanaki never took charge of this office.

The Khan of the Crimea assisted the Grand Vizier, the Mufti and the Aga of the Janissaries in putting down the rebellion. On 24 November 1731, Halil was strangled by the sultan's order[3] and in his presence, after a Divan in which Halil had dictated that war be declared against Russia. His Greek friend, Yanaki, and 7,000 of those who had supported him were also put to death. The jealousy which the officers of the Janissaries felt towards Halil, and their readiness to aid in his destruction, facilitated the exertions of Mahmud I's supporters in putting an end to the rebellion after it had lasted over a year.

The rest of Mahmud I's reign was dominated by wars with Persia, Austria and Russia.

Mahmud I entrusted government to his viziers and spent much of his time composing poetry.

He died at Topkapı Palace, Constantinople.

Relations with the Mughal Empire

Nadir Shah's devastating campaign against the Mughal Empire, created a void in the western frontiers of Persia, which was effectively exploited by the Ottoman Sultan Mahmud I, who initiated the Otttoman-Persian War (1743–1746), in which the Mughal Emperor Muhammad Shah closely cooperated with the Ottomans and their ambassador Haji Yusuf Agha, these relations between the two great empires continued until Muhammad Shah's death in 1748.[4]

Marriage

He married Haja 'Ali-Janab without issue.

Notes

[1] *The Encyclopædia Britannica*, Vol.7, Edited by Hugh Chisholm, (1911), 3; *Constantinople, the capital of the Turkish Empire...*

[2] Britannica, Istanbul (http://concise.britannica.com/ebc/article-9368294/Istanbul):*When the Republic of Turkey was founded in 1923, the capital was moved to Ankara, and Constantinople was officially renamed Istanbul in 1930.*

[3] Shaw, Stanford J. and Shaw, Ezel Kural (1976) *History of the Ottoman Empire and Modern Turkey, volume 1: Empire of the Gazis: the rise and decline of the Ottoman Empire, 1280-1808* Cambridge University Press, Cambridge, England, p. 240, ISBN 0-521-21280-4

[4] Farooqi, Naimur Rahman (1989). *Mughal-Ottoman relations: a study of political & diplomatic relations between Mughal India and the Ottoman Empire, 1556-1748.* Idarah-i Adabiyat-i Delli. ASIN: B0006ETWB8. See Google Books search (http://books.google.com/books?id=uB1uAAAAMAAJ&q=Muhammad+Shah&source=gbs_word_cloud_r&cad=6).

Sources

- Incorporates text from *History of Ottoman Turks* (1878)

[aged 58]

Mahmud II

Mahmud II	
Sultan of the Ottoman Empire	
Caliph	

Reign	1808–39
Period	Stagnation of the Ottoman Empire
Full Name	Mahmud II
Born	20 July 1789
Died	1 July 1839 (aged 49)
Predecessor	Mustafa IV
Successor	Abdülmecid I
Royal House	House of Osman
Dynasty	Ottoman Dynasty

Mahmud II (Ottoman Turkish: محمود ثانی *Mahmud-ı s̲ānī*) (20 July 1789 – 1 July 1839) was the 30th Sultan of the Ottoman Empire from 1808 until his death in 1839. He was born in the Topkapi Palace, Constantinople,[1] the posthumous son of Sultan Abdulhamid I. His reign is notable mostly for the extensive administrative, military and fiscal reforms he instituted, which culminated into the Decree of Tanzimat (Reorganization) that was carried out by his sons Abdülmecid I and Abdülaziz I.

Accession

His mother was Valide Sultan Naksh-i-Dil Haseki (who according to an unsubstantiated legend was a cousin of Napoleon's wife Josephine).[2] In 1808, Mahmud II's predecessor (and half-brother) Mustafa IV (1807–08) ordered his execution along with his cousin, the deposed Sultan Selim III (1789–1807), in order to defuse the rebellion. Selim III was killed, but Mahmud was safely kept hidden by his mother and was placed on the throne after the rebels deposed Mustafa IV. The leader of this rebellion, Alemdar Mustafa Pasha, then became Mahmud II's vizier.

There are many stories surrounding the circumstances of his attempted murder. A version by the 19th century Ottoman historian Cevdet Pasha gives the following account: one of his slaves, a Georgian girl named Cevri, gathered ashes when she heard the commotion in the palace surrounding the murder of Selim III. When the assassins approached the Harem chambers where Mahmud was staying, she was able to keep them away for a while by throwing ashes into their faces, temporary blinding them. This allowed Mahmud to escape through a window and climb onto the roof of the Harem. He apparently ran to the roof of the Third Court where other pages saw him and helped him come down with pieces of clothes that were quickly tied together as a ladder. By this time one of the leaders of the rebellion, Alemdar Mustafa Pasha arrived with his armed men and upon seeing the dead body of Selim III proclaimed Mahmud as *padishah*. The slave girl Cevri Kalfa was awarded for her bravery and loyalty and appointed *haznedar usta*, the chief treasurer of the imperial Harem, which was the second most important position in the hierarchy. A plain stone staircase at the *Altınyol* (Golden Way) of the Harem is called Staircase of Cevri (Jevri) Kalfa, since the events apparently happened around there and are associated with her.[3]

Reign overview

Mahmud Han bin Abdülhamid muzaffer daima

The stylized signature of Mahmud II was written in an expressive calligraphy. It reads: "Mahmud Khan son of Abdülhamid is forever victorious".

The vizier took the initiative in resuming reforms that had been terminated by the conservative coup of 1807 that had brought Mustafa IV to power. However, soon the vizier was killed by Ibrahim's army, and Mahmud II temporarily abandoned the reforms. Mahmud II's later reformation efforts were more successful.

During the early years of Mahmud II's reign, his governor of Egypt Mehmet Ali Paşa successfully reconquered the holy cities of Medina (1812) and Mecca (1813) from the Nejdi rebels.

His reign also marked the first breakaway from the Ottoman Empire, with Greece gaining its independence following a rebellion that started in 1821. In 1827 the combined British, French and Russian navies defeated the Ottoman Navy at the Battle of Navarino; in the aftermath, the Ottoman Empire was forced to recognize Greece with the Treaty of Constantinople in July 1832. This event, together with the occupation of the Ottoman province of Algeria by France in 1830, marked the beginning of the gradual break-up of the Ottoman Empire. Non-Turkish ethnic groups living in the empire's territories, especially in Europe, started their own independence movements.

Among Mahmud II's most notable acts during his reign was the abolition of the Janissary corps in 1826, permitting the establishment of a European-style conscript army, recruited largely from Turkish speakers of Rumelia and Asia Minor. Mahmud was also responsible for the subjugation of the Iraqi Mamluks by Ali Ridha Pasha in 1831. He ordered the execution of the renowned Ali Pasha of Tepelena. He sent his Grand Vizier to execute the Bosnian Muslim hero Husein Gradaščević and dissolute the Bosnia Eyalet.

He began preparations for the Tanzimat reforms in 1839. The Tanzimat marked the beginning of modernization in Turkey, and had immediate effects on social and legal aspects of life in the Empire, such as European style clothing, architecture, legislation, institutional organization and land reform.

He was concerned also for aspects of tradition. He made great efforts to revive the sport of archery. He ordered his archery student, Mustafa Kani, to write a book about the history, construction, and use of Turkish bows, from which comes most of what is now known of Turkish bowyery.[4]

Mahmud II died of tuberculosis - some say he was murdered - at the Esma Sultana Palace, Çamlıca, in 1839. His funeral was attended by crowds of people who came to bid the Sultan farewell. His son Abdülmecid succeeded him.

The mausoleum (*türbe*) of Sultan Mahmud II, located at Divan Yolu street.

Reforms

Legal reforms

Mahmud II before his clothing reform in 1826

Among his reforms are the edicts (or firmans), by which he closed the Court of Confiscations, and took away much of the power of the Pashas.

Previous to the first of the Firmans the property of all persons banished or condemned to death was forfeited to the crown; and a sordid motive for acts of cruelty was thus kept in perpetual operation, besides the encouragement of a host of vile Delators.

The second firman removed the ancient rights of Turkish governors to doom men to instant death by their will; the *Paşas*, the *Ağas*, and other officers, were enjoined that "they should not presume to inflict, themselves, the punishment of death on any man, whether Raya or Turk, unless authorized by a legal sentence pronounced by the Kadi, and regularly signed by the judge." Mahmud also created an appeal system by a criminal to one of the Kazasker (chief military judge) of Asia or Europe, and finally to the Sultan himself, if the criminal chose to persist in his appeal.

About the same time that Mahmud II ordained these changes, he personally set an example of reform by regularly attending the Divan, or state council, instead of secluding himself from the labors of state. The practice of the Sultan avoiding the Divan had been introduced as long ago as the reign of Suleiman I, and was considered as one of the causes of the decline of the Empire by a Turkish historian nearly two centuries before Mahmud II's time.

Mahmud II also addressed some of the worst abuses connected with the Vakifs, by placing their revenues under state administration. However, he did not venture to apply this vast mass of property to the general purposes of the government.

In his time the financial situation of the Empire was troubling, and certain social classes had long been under oppression under difficult taxes. In dealing with the complicated questions that therefore arose, Mahmud II is considered to have demonstrated the best spirit of the best of the Köprülüs. A Firman of February 22, 1834 abolished the vexatious charges which public functionaries, when traversing the provinces, had long been accustomed to take from the inhabitants. By the same edict all collection of money, except for the two regular half-yearly periods, was denounced as abuses. "No one is ignorant," said Sultan Mahmud II in this document, "that I am bound to afford support to all my subjects against vexatious proceedings; to endeavour unceasingly to lighten, instead of increasing their burdens, and to ensure peace and tranquility. Therefore, those acts of oppression are at once contrary to the will of God, and to my imperial orders."

Mahmud II after his clothing reform in 1826

The *haraç*, or capitation-tax, though moderate and exempting those who paid it from military service, had long been made an engine of gross tyranny through the insolence and misconduct of the government collectors. The Firman of 1834 abolished the old mode of levying it, and ordained that it should be raised by a commission composed of the *Kadı*, the Muslim governors, and the *Ayans*, or municipal chiefs of *Rayas* in each district. Many other financial improvements were effected. By another important series of measures, the administrative government was simplified and strengthened, and a large number of sinecure offices were abolished. Sultan Mahmud II provided a valuable personal example of good sense, and economy, organising the imperial household, suppressing all titles without duties, and all salaried officials without functions.

Military reforms

Mahmud II dealt effectively with the military fiefs, the "Tımar"s and the "Ziamet"s. These had been instituted to furnish the old effective military force, but had long ceased to serve this purpose. By attaching them to the public domains, Mahmud II materially strengthened the resources of the state, and put an end to a host of corruptions. One of the most resolute acts of his ruling was the suppression of the *Dere Beys*, the hereditary local chiefs (with power to nominate their successors in default of male heirs), which, in one of the worst abuses of the Ottoman feudal system, had made themselves petty princes in almost every province of the empire.

Mahmudiye (1829), built by the Imperial Naval Arsenal on the Golden Horn in Constantinople, was for many years the largest warship in the world. The 201 x 56 kadem (1 kadem = 37.887 cm) or 76.15 m × 21.22 m (ft × ft) ship of the line was armed with 128 cannons on 3 decks and carried 1,280 sailors on board (kadem, which translates as "foot", is often misinterpreted as equivalent in length to one imperial foot, hence the wrongly converted dimensions of "201 x 56 ft, or 62 x 17 m" in some sources.) She participated in numerous important naval battles, including the Siege of Sevastopol (1854–1855) during the Crimean War.

The reduction of these insubordinate feudatories was not effected at once, or without severe struggles and frequent insurrections. Mahmud II steadily persevered in this great measure and ultimately the island of Cyprus became the only part of empire in which power not emanating from the Sultan was allowed to be retained by *Dere Beys*.

One of his most notable achievement was the abolition (through use of military force, execution and exile, and banning of the Bektashi order) of the Janissary Corps, event known as The Auspicious Incident, in 1826 and the establishment of a modern Ottoman Army, named the Asakir-i Mansure-i Muhammediye (meaning 'Victorious Soldiers of Muhammad' in Ottoman Turkish).

Following the loss of Greece after the Battle of Navarino against the combined British-French-Russian flotilla in 1827, Mahmud II gave top priority to rebuilding a strong Ottoman naval force. The first steam ships of the Ottoman Navy were acquired in 1828. In 1829 the world's largest warship for many years, the 201 x 56 kadem (1 kadem = 37.887 cm) or 76.15 m × 21.22 m (ft × ft) ship of the line *Mahmudiye*, which had 128 cannons on 3 decks and carried 1,280 sailors on board, was built for the Ottoman Navy at the Imperial Naval Arsenal (Tersâne-i Âmire) on the Golden Horn in Constantinople (kadem, which translates as "foot", is often misinterpreted as equivalent in length to one imperial foot, hence the wrongly converted dimensions of "201 x 56 ft, or 62 x 17 m" in some sources.)

Marriages and issue

He married HH Valide Sultan Bezmiâlem, originally named Suzi (1807–1852), a Georgian or a Georgian Jew, and had:

- Abdülmecid I

He married HH Valide Sultan Pertevniyal, originally named Bezime (1812–1883),[5] [6] and had:

- Abdülaziz I

Wives

m. (first) Fatima Bash Kadin Effendi (d. February 1809).

m. (second) 'Ali-Janab Bash Kadin Effendi (d. 1839).

m. (third) Hajira Partav-Piyala Navifidan Bash Kadin Effendi (4 January 1793 - 25 December 1855).

m. (fourth) Misl-i-Nayab Kadin Effendi (d. 1825).

m. (fifth) Kamari Kadin Effendi (d. 1825).

m. (sixth) Abr-i-Raftar Kadin Effendi (d. 1825).

m. (seventh) Zarnigar Kadin Effendi (d. 1832), 4th Iqbal from 23 May 1826, later promoted to 7th Kadin Effendi, and 2nd Kadin Effendi.

m. (ninth) Ashub-i-Jahan Kadin Effendi (1793 - 10 June 1870).

m. (tenth) Vuzlat Kadin Effendi (d. 1830).

m. (eleventh) Nur-i-Tab Kadin Effendi (1810 - 2 January 1886).

m. (twelfth) 1811, Hajiya Khushair Kadin Effendi (d. at Mecca, 1859).

m. (thirteenth) Parviz Falak Kadin Effendi (d. 21 September 1863).

m. (fourteenth) Husn-i-Malika Khanum Effendi (1812 - October 1886), previously styled Bash Iqbal.

m. (fifteenth) Zayn-i-Falak Khanum Effendi (d. 20 December 1842), previously styled 2nd Iqbal.

m. (sixteenth) Labriz Falak Khanum Effendi (1810 - 9 February 1865), previously styled 4th Iqbal.

m. (seventeenth) 1826, Tiryal Khanum Effendi (1810 - 1883), previously styled 3rd Iqbal.

Sons

1) H.I.H. Prince (Shahzada) 'Abdu'l-Hamid. (6 April 1811 - 1815), son of 'Ali-Janab.

2) H.I.H. Prince (Vali Ahad-Shahzada) Sultan Murad Khan (25 December 1811 - 14 July 1812).

3) H.I.H. Prince (Shahzada) Bayezid (27 March 1812 - 25 June 1812).

4) H.I.H. Prince (Vali Ahad-Shahzada) Sultan 'Abdu'l-Hamid Khan (6 March 1813 - 20 April 1825) (s/o 'Aliyanab).

5) H.I.H. Prince (Shahzada) Osman (12 June 1813 - 10 April 1814), son of Partav-Piyala.

6) H.I.H. Prince (Shahzada) Ahmad (25 July 1814 - 16 July 1815).

7) H.I.H. Prince (Shahzada) Muhammad (26 August 1814 - 28 October 1814).

8) H.I.H. Prince (Shahzada) Muhammad. b. and died 4 August 1816.

9) H.I.H. Prince (Shahzada) Sulaiman (29 August 1817 - 14 December 1819).

10) H.I.H. Prince (Shahzada) Ahmad. b. 13 October 1819. He died before 24 December 1819.

11) H.I.H. Prince (Shahzada) Ahmad. b. 25 December 1819. He died young.

12) H.I.H. Prince (Shahzada) 'Abdu'llah. b. and d. 4 April 1820.

13) H.I.H. Prince (Shahzada) Mahmud (18 February 1822 - 23 October 1822), twin brother of Prince Muhammad.

14) H.I.H. Prince (Shahzada) Muhammad (18 February 1822 - 23 September 1822), twin brother of Prince Mahmud).

15) H.I.H. Prince (Shahzada) Ahmad (6 July 1822 - 9 April 1823).

16) H.M. Sultan (Shahzada) 'Abdu'l-Majid Khan I, Grand Sultan of Turkey (son of Bazm-i-Alam) - see below.

17) H.I.H. Prince (Shahzada) Ahmad (5 December 1823 - 1824).

18) H.I.H. Prince (Shahzada) 'Abdu'l-Hamid. (18 February 1827 - 15 November 1828).

19) H.M. Sultan 'Abdu'l-Aziz Khan I, Grand Sultan of Turkey, son of Partav-Nihal - see below.

20) H.I.H. Prince (Shahzada) Nizam ud-din (6 December 1835 - 24 February 1838).

21) H.I.H. Prince (Shahzada) Hafiz (1836 - 24 January 1839).

22) H.I.H. Prince (Shahzada) Kamal ud-din. He d. 18xx.

Daughters

1) H.I.H. Princess Fatma Sultana (4 February 1809 - 5 August 1809), daughter of Partav-Piyala. Said to have been the first child born to the dynasty for twenty years.

2) H.I.H. Princess Aisha Sultana (5 July 1809 - February 1810), daughter of Ashub-i-Jahan.

3) H.I.H. Princess Fatima Sultana (30 April 1810 - 7 May 1825), daughter of Partav-Piyala.

4) H.I.H. Princess Saliha Sultana (16 June 1811 - 5 February 1843), daughter of Ashub-i-Jahan. She married 24 May 1834 at the Nashatabad Palace, as second wife of Admiral H.H. Damad Gurcu Muhammad Halil Rifat Pasha (1795 - 4 March 1856), Ambassador to Russia 1829-1830, Naval C-in-C 1830-1832, 1843-1845, 1847-1848 and 1854-1855, Supreme C-in-C 1836-1838 and 1839-1840, Governor-General of Anatolia 1852. She died at the Nashatabad Palace, having had issue, one son and one daughter.

5) H.I.H. Princess Shah Sultana (22 May 1812 - September 1814).

6) H.I.H. Princess Mihrîmah Sultana (10 June 1812 - 3 July 1838), daughter of Khushair. She married 28 April 1836, Field Marshal H.H. Damad Muhammad Said Pasha (b. at Bursa, 1798; d. at Constantinople, December 1868), Supreme C-in-C 1838-1839 and 1846-1848, Naval C-in-C 1840-1841, sometime Minister for Commerce, son of Bursali Alyanak 'Ali Effendi. She had issue, one son.

7) H.I.H. Princess Amina Sultana (12 June 1813 - 20 June 1814), daughter of Partav-Piyala.

8) H.I.H. Princess Amina Sultana. Born 30 July 1814. She died young.

9) H.I.H. Princess Shah Sultana (14 October 1814 - 13 April 1817).

10) H.I.H. Princess Amina Sultana (7 January 1815 - 29 September 1816), daughter of Partav-Piyala. She died in a fire at the Sarai.

11) H.I.H. Princess Zainab Sultana (18 April 1815 - 8 January 1816) (d/o Khushair).

12) H.I.H. Princess Hamida Sultana. b. 14 July 1817. She d. before 1818.

13) H.I.H. Princess Jamila Sultana. b. 1818. She d. young.

14) H.I.H. Princess Hamida Sultana (4 July 1818 - 15 February 1819).

15) H.I.H. Princess Atiye Sultana (2 January 1824 - 11 August 1850), daughter of Parviz Falak). She m. 7 May 1840, as his second wife, Field Marshal H.H. Damad Ahmad Fathi Pasha (b. at Eyub, 1801; d. at Constantinople, 14 February 1858), Ambassador to Austria 1835-1836, at the Court of St James's 1838, and to France 1838, Marshal of the Artillery 1845-1852, posthumous son of Radosi Haji Hafiz Ahmad Agha, by his wife, Salaha Khanum. She had two daughters.

16) H.I.H. Princess Munira Sultana (16 October 1824 - 23 May 1825).

17) H.I.H. Princess Khadija Sultana (6 September 1825 - 19 December 1842), daughter of Parviz Falak. She died at the Bashiktash Palace.

18) H.I.H. Princess Adile Sultana (23 May 1826 - 12 February 1899), daughter of Zarnigar. Adopted by Navifidan in 1832. m. at the Nashatabad Palace, 12 June 1845, Grand Admiral H.H. Damad Haji Muhammad 'Ali Pasha 'Alioglu (b. at Hamshin, 1813; d. at Constantinople, 30 June 1868), Marshal of the Artillery 1844, Naval C-in-C 1845-1847, 1848-1849, 1851-1852, 1855-1858, 1858-1863 and 1866-1867, Supreme C-in-C 1849-1851, 1853-1854 and 1861-1863, and Grand Vazier 1852-1853, son of Haji Omar Agha. She had one son and two daughters.

19) H.I.H. Princess Fatima Sultana (20 July 1828 - 2 October 1830), daughter of Parviz Falak.

20) H.I.H. Princess Hayria Sultana (22 March 1831 - 1832).

21) H.I.H. Princess Hayria Sultana (12 January 1832 - 15 February 1833).

22) H.I.H. Princess Rafia Sultana (January 1836 - 24 January 1839).

In fiction

The 2006 historical detective novel *The Janissary Tree*, by Jason Goodwin, is set in 1836 Constantinople, with Mahmud II's modernising reforms (and conservative opposition to them) forming the background of the plot. The Sultan himself and his mother appear in several scenes.

The 1989 film *Intimate Power*, also known as *The Favorite*, is adapted from an historical fiction novel by Prince Michael of Greece. It portrays a legend about Aimée du Buc de Rivéry as a young captured French girl who, after spending years in an Ottoman harem, outlives two Sultans and protects Mahmud as his surrogate mother. Mahmud is a minor role in the film but is portrayed as both an adult and a child. The film concludes with a variation of his dramatic succession.

See also

- Peter the Great
- Atçalı Kel Mehmet Efe
- Sened-i İttifak
- Muhammad Ali
- Husein Gradaščević
- Ali Pasha
- Auspicious Incident

References

[1] Finkel, Caroline, *Osman's Dream*, (Basic Books, 2005), 57; "*Istanbul was only adopted as the city's official name in 1930..*".
[2] Christine Isom-Verhaaren, "Royal French Women in the Ottoman Sultans' Harem: The Political Uses of Fabricated Accounts from the Sixteenth to the Twenty-first Century", *Journal of World History*, vol. 17, No. 2, 2006
[3] Davis, Claire (1970). *The Palace of Topkapi in Istanbul*. New York: Charles Scribner's Sons. pp. 214–217. ASIN B000NP64Z2.
[4] Paul E Klopsteg. Turkish Archery and the Composite Bow. Chapter I, Background of Turkish Archery. Second edition, revised, 1947, published by the author, 2424 Lincolnwood Drive, Evanston, Ill.
[5] His profile in the Ottoman Web Site (http://www.osmanli700.gen.tr/english/sultans/32index.html)
[6] Daniel T. Rogers, "All my relatives: Valide Sultana Partav-Nihal" (http://freepages.genealogy.rootsweb.ancestry.com/~dantrogers/pafg1639.htm)

- Incorporates text from Edward Shepherd Creasy, *History of the Ottoman Turks; From the beginning of their empire to the present time* (1878).

Tahmasp I

Shah Tahmasp I	
Shahanshah of Persia	
 Painting of **Shah Tahmasp I** at Chehel Sotoon palace.	
Reign	23 May 1524 – 25 May 1576
Coronation	2 June 1524
Full name	'Abu'l Muzaffar 'Abu'l Fath Sultan Shah Tahmasp bin Shah Ismail al-Safavi al-Husayni al-Musavi[1]
Titles	Shahanshah, Sahib-i-Qiran, Sultan bar Salatin [2]
Born	22 February 1514
Birthplace	Shahabad
Died	14 May 1576 (aged 62)
Place of death	Qazvin
Buried	Ardebil
Predecessor	Ismail I
Successor	Ismail II
Royal House	Safavi
Father	Shah Ismail I
Mother	Tajlu Khanum
Religious beliefs	shia islam

Tahmasp or **Tahmasb I** (Persian: م ی بسامهت هاش) (22 February 1514 – 14 May 1576) was an influential Shah of Iran, who enjoyed the longest reign of any member of the Safavid dynasty. He was the son of Ismail I and Shah-Begi Khanum (known under the title Tajlu Khanum) of the Turcoman Mawsillu tribe.[3] [4]

He came to the throne aged ten in 1524 and during his minority was weak and came under the control of the Qizilbash, Turkic tribesmen who formed the backbone of Safavid power. The Qizilbash leaders fought among themselves for the right to be regents over Tahmasp. Upon adulthood, however, Tahmasp was able to reassert the power of the Shah and control the tribesmen.

His reign was marked by foreign threats, primarily from the Ottomans and the Uzbeks. In 1555, however, he regularized relations with the Empire through the Peace of Amasya. This peace lasted for 30 years, until it was broken in the time of Shah Mohammed Khodabanda.

He is also known for the reception he gave to the fugitive Mughal Emperor Humayun which is depicted in a painting on the walls of the Safavid palace of Chehel Sotoon. Humayun's return from Persia, accompanied by a large retinue of Persian noblemen and soldiers, signaled an important change in Mughal court culture, as the Central Asian origins of the dynasty were largely overshadowed by the influences of Persian art, architecture, language and literature.

One of Shah Tahmasp's more lasting achievements was his encouragement of the Persian rug industry on a national scale, possibly a response to the economic effects of the interruption of the Silk Road carrying trade during the Ottoman wars.

Biography

Regency 1524–1533

Tahmasp was just 10 years old when he succeeded his father Shah Isma'il I, the founder of Safavid rule in Iran. Too young to rule in his own right, Tahmasp came under the control of the Qizilbash. Some of the tribes recognised a Qizilbash leader, Div Sultan Rumlu, as regent (*atabeg*) to the shah, but others dissented and in 1526 a bloody civil war broke out among the differing factions. Div Sultan emerged victorious but his ally, Chuha Sultan Takkalu, turned against him and urged the shah to get rid of him. On 5 July 1527 as Div Sultan arrived for a meeting of the government, Tahmasp shot an arrow at him. When it failed to kill him, the shah's supporters finished him off.

Chuha Sultan now became regent. Iran's enemies, the Uzbeks, had taken advantage of the civil war to invade the north-eastern province of Khorasan. In 1528 Chuha Sultan and the shah marched with their army to reassert control of the region. Although they defeated the Uzbeks in a battle near Jam, Tahmasp was disgusted at the cowardice Chuha Sultan had displayed during the combat. Finally, in 1530/1, a quarrel broke out between members of the Takkalu and Shamlu Qizilbash factions and the Shamlus succeeded in killing Chuha Sultan. The Takkalus regained the advantage and some of them even tried to kidnap the shah. Tahmasp lost patience and ordered a general massacre of the Takkalu tribe. They never regained their influence in Iran.

The leader of the Shamlu faction, Husayn Khan, now assumed the regency but, in 1533, Tahmasp suspected Husayn Khan was plotting to overthrow him and had him put to death. Tahmasp was now old enough and confident enough to rule in his own right.[5]

Foreign threats 1533–1553

The 16th century Chehel Sotun pavillon in Qazvin. It's the last remain of the palace of Shah Tahmasp. it was heavily restored by the Qajars in the 19th century.

The discord in Iran had allowed its enemies, the Uzbek khans in the east and the Ottoman Empire in the west, to seize territory. The Ottomans were at the height of their power during the reign of Suleiman the Magnificent. They launched four invasions of Iran between 1533 and 1553. Since the Ottoman army possessed overwhelming numerical superiority, Tahmasp avoided pitched battle with them and resorted to alternative tactics.

In 1534, Suleiman invaded Iran with a force numbering 200,000 men and 300 pieces of artillery. Tahmasp could only field 7,000 men (of dubious loyalty) and a few cannons. The Ottomans seized the Safavid capital Tabriz, crossed Kurdistan and captured Baghdad. Tahmasp avoided direct confrontation with the Ottoman army, preferring to harass it then retreat, leaving scorched earth behind him. This scorched earth policy led to the loss of 30,000 Ottoman troops as they made their way through the Zagros mountains and Suleiman decided to abandon his campaign.[6]

Next, Suleiman tried to exploit the disloyalty of Tahmasp's brother Alqas Mirza, who was governor of the frontier province of Shirvan. Alqas had rebelled and, fearing his brother's wrath, he had fled to the Ottoman court. He persuaded Suleiman that if he invaded the Iranians would rise up and overthrow Tahmasp. In 1548, Suleiman and

Alqas entered Iran with a huge army but Tahmasp had already "scorched the earth" around Tabriz and the Ottomans could find few supplies to sustain themselves. Alqas penetrated further into Iran but the citizens of Isfahan and Shiraz refused to open their gates to him. He was forced to retreat to Baghdad where the Ottomans abandoned him as an embarrassment. Captured by the Iranians, his life was spared but he was condemned to spend the rest of his life in prison in the fortress of Qahqaha.[7]

During the final Ottoman invasion of Iran in 1553, Tahmasp seized the initiative and defeated Iskandar Pasha near Erzerum. He also captured one of Suleiman's favourites, Sinan Beg. This persuaded the sultan to come to terms at the Peace of Amasya in 1555. The treaty freed Iran from Ottoman attacks for three decades. Nevertheless, Tahmasp took the precaution of transferring his capital from Tabriz to Qazvin, which was further away from the border.[8]

Between 1540 and 1553, Tahmasp conducted military campaigns in the Caucasus region, capturing many Armenians, Georgians and Circassians. These would become an important new element in Iranian society.[9]

Royal refugees: Bayezid and Humayun

The Mughal Empire was Iran's eastern neighbour. In 1544, the Mughal emperor, Humayun, fled to Tahmasp's court after he had been overthrown by the rebel Sher Shah Suri (Sher Khan). Tahmasp insisted on the Sunni Humayun converting to Shi'ism before he would help him. Humayun reluctantly agreed and also gave Tahmasp the strategically important city of Kandahar in exchange for Iranian military assistance against the heirs of Sher Khan and his own rebellious brothers. By 1555, he had regained his throne.[10] [11]

Shah Tahmasp greets the exiled Humayun.

Humayun was not the only royal figure to seek refuge at Tahmasp's court. A dispute arose in the Ottoman Empire over who was to succeed the aged Suleiman the Magnificent. Suleiman's favourite wife, Roxelana, was eager for her eldest son Selim to become the next sultan. But Selim was an alcoholic and Roxelana's third son, Bayezid, had shown far greater military ability. The two princes quarrelled and eventually Bayezid rebelled against his father. His letter of remorse never reached Suleiman and he was forced to flee abroad to avoid execution. In 1559, Bayezid arrived in Iran where Tahmasp gave him a warm welcome. Suleiman was eager to negotiate his son's return, but Tahmasp rejected his promises and threats until, in 1561, Suleiman offered him land and 400,000 gold pieces. In September of that year, Tahmasp and Bayezid were enjoying a banquet at Tabriz when Tahmasp suddenly pretended he had received news that the Ottoman prince was engaged in a plot against his life. An angry mob gathered and Tahmasp had Bayezid put into custody, alleging it was for his own safety. Tahmasp then handed the prince over to the Ottoman ambassador. Shortly afterwards, Bayezid was killed by agents sent by his own father.[12]

Shah Tahmasp I and the Mughal Emperor Humayun in Isfahan.

Final years

In 1574, Tahmasp fell ill and discord broke out among the Qizilbash once more, this time over which prince was to succeed him. The shah's Georgian and Circassian wives had also introduced a new faction into the court. Seven of Tahmasp's surviving sons were by Georgian or Circassian mothers and two by a Turcoman.[13] Of the latter, Mohammed Khodabanda was regarded as unfit to rule because he was almost blind, and his younger brother, Ismail, had been imprisoned by Tahmasp since 1555. Nevertheless, one court faction supported Ismail, while another backed Haydar Mirza, the son of a Georgian. Tahmasp himself was believed to favour Haydar but he prevented his supporters from killing Ismail.[14]

Tahmasp died as a result of poison, although it is unclear whether this was by accident or on purpose. On his death, as expected, fighting broke out between the different court factions. Haydar was killed and Ismail emerged triumphant as Shah Ismail II.[15]

In 1574, Tahmasp also had the 36th Nizari Ismaili Shia Imam Murād Mīrzā executed, due to the perceived political threat he posed.

Tahmasp and the arts

Tahmasp was an enthusiastic patron of the arts with a particular interest in the Persian miniature, especially book illustration. He had been trained in drawing himself, and had some talent. The most famous example of such work is the *Shāhnāma-yi Shāh Tahmāspī* (*King's Book of Kings*), commissioned for Tahmasp by his father and containing 250 miniatures by the leading court artists of the era.[16] However in the 1540s he is recorded as losing interest in the arts, and his imperial atelier largely dispersed.

Offspring

He married (first) his maternal first cousin, Kadamali Sultan Begum, née Sultanum Begum daughter of Musa Sultan bin 'Isa Beg Musullu, of the Aq Quyunlu, m. (second) Sultan Agha Khanum, a Circassian, sister of Shamkhal Kara-Musal Sultan,Governor of Sakki, m. (third), Sultanzada Khanum a Georgian, m. (fourth) Zahra Baji daughter of Prince Ot'ar Shalikashvili of Samtskhe from Shalikashvili family of Georgia, m. (fifth) Khan-Parwar Khanum a Georgian, m. (sixth) Huri-Khan Khanum, a Georgian, m. (seventh) daughter of the Governor of Daghestan, m. (eighth)Aisha Begum, daughter of Sufian Khan, Khan of Khiva, m. (ninth) Zainab Sultan Khanum, m. (tenth) Zahra Baji a Georgian.[17]

Sons

- Prince *Shahzadeh* Soltan 'Ali Quli Mirza (*b.*1528-*d.*1529)
- Mohammed Khodabanda
- Ismail II
- Prince *Shahzadeh* Soltan Murad Mirza (*b.* 1538-*d.* 5 September 1545) Governor of Kandahar 1545.
- Prince *Shahzadeh* Soltan Soleiman Mirza (*b.* at Nikhichivan, 1554-*k.* at Qazvin, 2 November 1576) Governor of Fars 1555–1557, and Mashhad 1576.
- Prince *Shahzadeh* Soltan Heidar Mirza (*b.* 1555-*k.* at Qazvin, 14 May 1576) having had issue, a daughter, married Hasan Khan Ustajalu. She had issue, 4 sons.
- Prince *Shahzadeh* Soltan Mostafa Mirza (*b.* 1557-*k.* at Qazvin, 2 November 1576) having had issue, 2 daughters, Princess Mahd-e-Olia married Abbas I and a daughter who married Zulfikhar Khan Karamanlu and had issue, 2 sons.
- Prince *Shahzadeh* Soltan Mahmud Mirza (*b.* 1559-*k.* at Qazvin, 24 February 1577) Governor Shirvan 1566–1567, and of Lahijan 1567–1571, having had issue, a son, Prince Mohammad Baqer Mirza (*b.*1575 -*k.* 24 February 1577)
- Prince *Shahzadeh* Imam Qoli Mirza (*b.* 1562- *k.* at Qazvin, 24 February 1577), without issue .
- Prince *Shahzadeh* Soltan 'Ali Mirza (*b.* at Qazvin, 1563- *d.* at Isfahan, 31 January 1642) Governor of Ganja 1570–1577. He was blinded and imprisoned by Shah Abbas I at Alamut. m. Kabuli Begum, having had issue, one son, Prince *Shahzada* Soltan Mustafa Mirza.
- Prince *Shahzadeh* Soltan Ahmad Mirza (*b.* at Qazvin, 1564-*k.* at Qazvin, 24 February 1577)
- Prince *Shahzadeh* Zeinal Abedin Mirza (*d.* before 1576)
- Prince *Shahzadeh* Musa Mirza (*d.* before 1576)

Daughters

- Princess *Shahzadeh Alamiyan* Gowhar Soltan Beygom (*b.* 1540- *d.* 19 May 1577) married Prince 'Abu'l Fath Soltan Ibrahim Mirza (*b.* 1543- *k.* 24 February 1577) son of paternal uncle, Shahzada 'Abu'l Fath Muiz ud-din Bahram Mirza son of Ismail I, having had issue, one daughter, Princess Gowhar Shad Begum (*b.* 1561-*d.* ?)
- Princess *Shahzadi Alamiyan* Pari-Khan Khanum (*b.*1548-*k.* 17 February 1578) unmarried and without issue.
- Princess *Shahzadi Alamiyan* Khadija Sultan Begum, m. (first) her cousin, Jamshid Khan Gilani (*b.* 1557- *k.* 1580) son of Soltan Mahmud Mirza Gilani, governor of Fuman. m. (second) as his third wife, Mir Nimatu'llah Yazdi Zu'l-Nurain, eldest son of Amir Ghiyas ed-din Muhammad Yazdi Mir-i-Miran, having had issue, two sons, by her first husband.
- Princess *Shahzadi Alamiyan* Zainab Beygom (*b.* 1550-*d.* at Isfahan, 14 May 1641) unmarried and without issue
- Princess *Shahzadi Alamiyan* Mariam Soltan Beygom (*b.* 1551-*d.* at Isfahan, 1608), m. (first) Khan Ahmad Khan governor of Gilan, m. (second) as his second wife, Mir Nematu'llah Yazdi Zu'l-Nurain, eldest son of Amir Ghiyas ed-din Muhammad Yazdi Mir-i-Miran, having had issue one daughter and one son, by her first husband.
- Princess *Shahzadi Alamiyan* Beygom Khanoum, m. Musib Beyg Khan, son of Muhammad Khan Takahi.
- Princess *Shahzadi Alamiyan* Khanish Begom. (*b.* after 1562-*d.* at Isfahan, 1590) m. as his first wife, Mir Nimatu'llah Yazdi Zu'l-Nurain, eldest son of Amir Ghiyas ed-din Muhammad Yazdi Mir-i-Miran, having had issue, two sons.
- Princess *Shahzadi Alamiyan* Fatemeh Soltan Beygom, m. Amir Khan Musallu (*k.* at Qahqaha, 1584) son of Muhsmmad Beyg Mosalla.
- Princess *Shahzadi Alamiyan* Shahbanu Khanum, m. his first wife, her cousin, Amir Salman Khan Ustajalu son of Shah 'Ali Quli Mirza Ustajalu.[18]

See also

- Safavid conversion of Iran from Sunnism to Shiism

References

[1] The Royal Ark (http://www.royalark.net/Persia/safawi3.htm)
[2] The Royal Ark (http://www.royalark.net/Persia/safawi3.htm)
[3] *Women in Iran: From the Rise of Islam to 1800* ed. Nashat and Beck (University of Illinois Press, 2003) p.145
[4] Andrew J. Newman *Safavid Iran* (I.B.Tauris) p.23
[5] This section: Savory pp.51–58
[6] Nahavandi and Bomati p.283
[7] Savory pp.62–63
[8] Savory p.63
[9] Savory p.64-65
[10] Nahavandi and Bomati pp.286–287
[11] Savory pp.66–67
[12] Nahavandi and Bomati pp.284–286
[13] Her name was Sultanum Bekum Mawsillu (Andrew J. Newman, *Safavid Iran*, I.B.Tauris, 2004, p.42)
[14] Savory p.69
[15] *Cambridge History of Iran* Vol.6, p.248
[16] Savory p.129
[17] The Royal Ark (http://www.royalark.net/Persia/safawi3.htm)
[18] The Royal Ark (http://www.royalark.net/Persia/safawi3.htm)

Sources

- Roger Savory *Iran under the Safavids* (Cambridge University Press, 2007 reissue) pp. 96–100
- H. Nahavandi, Y. Bomati, *Shah Abbas, empereur de Perse (1587–1629)* (Perrin, Paris, 1998) pp. 222–226

Abbas I of Persia

Shah 'Abbās I		
Shahanshah of Persia		
[[File:Shah Abbas.jpg		]]
Shah of Iran		
Reign	1 October 1587 – 19 January 1629	
Predecessor	Mohammad I	
Successor	Safi	
Dynasty	Safavid	
Father	Mohammed Khodabanda	
Mother	Khayr al-Nisa Begum	
Born	27 January 1571 Herat, Afghanistan	
Died	19 January 1629 (aged 57) Mazandaran, Iran	
Religion	Shia Islam	

Shāh 'Abbās the Great (or **Shāh 'Abbās I**) (Persian: شاه عبَاس بزرگ) (27 January 1571 – 19 January 1629) was Shah (king) of Iran, and generally considered the greatest ruler of the Safavid dynasty. He was the third son of Shah Mohammad.[1]

Abbas came to the throne during a troubled time for Iran. Under his weak-willed father, the country was riven with discord between the different factions of the Qizilbash army, who killed Abbas' mother and elder brother. Meanwhile, Iran's enemies, the Ottoman Empire and the Uzbeks, exploited this political chaos to seize territory for themselves. In 1587, one of the Qizilbash leaders, Murshid Qoli Khan, overthrew Shah Mohammed in a coup and placed the 16-year-old Abbas on the throne. But Abbas was no puppet and soon seized power for himself. He reduced the influence of the Qizilbash in the government and the military and reformed the army, enabling him to fight the Ottomans and Uzbeks and reconquer Iran's lost provinces. He also took back land from the Portuguese and the Mughals. Abbas was a great builder and moved his kingdom's capital from Qazvin to Isfahan. In his later years, the shah became suspicious of his own sons and had them killed or blinded.

Early years

Shah Abbas I and his court.

Abbas was born in Herat (now in Afghanistan, then one of the two chief cities of Khorasan) to the royal prince Mohammed Khodabanda and his wife Khayr al-Nisa Begum (known as "Mahd-i Ulya"), the daughter of the governor of Mazandaran province, who claimed descent from the fourth Shi'a Imam Zayn al-Abidin.[2] [3] At the time of his birth, Abbas' grandfather Shah Tahmasp I was ruler of Iran. Abbas' parents gave him to be nursed by Khani Khan Khanum, the mother of the governor of Herat, Ali Qoli Khan Shamlu. When Abbas was four, Tahmasp sent his father to stay in Shiraz where the climate was better for Mohammed's fragile health. Tradition dictated that at least one prince of the royal blood should reside in Khorasan, so Tahmasp made Abbas nominal governor of the province, despite his young age, and Abbas was left behind in Herat.[4]

In 1578, Abbas' father became Shah of Iran. Abbas' mother soon came to dominate the government, but she had little time for Abbas, preferring to promote the interests of his elder brother Hamza. The queen antagonised leaders of the powerful Qizilbash army, who plotted against her and strangled her in July 1579. Mohammed was a weak ruler who was incapable of preventing Iran's rivals, the Ottoman Empire and the Uzbeks, invading the country or stopping factional feuding among the Qizilbash. The young crown prince Hamza was more promising and led a campaign against the Ottomans, but he was murdered in mysterious circumstances in 1586. Attention now turned to Abbas.[5] [6]

At the age of 14, Abbas had come under the power of Murshid Qoli Khan, one of the leaders of the Qizilbash in Khorasan. When a large Uzbek army invaded Khorasan in 1587, Murshid decided the time was right to overthrow the ineffectual Shah Mohammed. He rode to the Safavid capital Qazvin with the young prince and proclaimed him king. Mohammed made no protest against his deposition and handed the royal insignia over to his son on 1 October 1587. Abbas was 16 years old.[7] [8]

Absolute monarch

Abbas takes control

Shah 'Abbās King of the Persians.
Copper engraving by Dominicus Custos, from his *Atrium heroicum Caesarum* pub. 1600-1602.

The kingdom Abbas inherited was in a desperate state. The Ottomans had seized vast territories in the west and the north-west (including the major city of Tabriz) and the Uzbeks had overrun half of Khorasan in the north-east. Iran itself was riven by fighting between the various factions of the Qizilbash, who had mocked royal authority by killing the queen in 1579 and the grand vizier in 1583.

First, Abbas settled his score with his mother's killers, executing four of the ringleaders of the plot and exiling three others.[9] His next task was to free himself from the power of the "kingmaker", Murshid Qoli Khan. Murshid made Abbas marry Hamza's widow and a Safavid cousin, and began distributing important government posts among his own friends, gradually confining Abbas to the palace. Meanwhile the Uzbeks continued their conquest of Khorasan. When Abbas heard they were besieging his old friend Ali Qoli Khan Shamlu in Herat he pleaded with Murshid to take action. Fearing a rival, Murshid did nothing until the news came that Herat had fallen and the Uzbeks had slaughtered the entire population. Only then did he set out on campaign to Khorasan. But Abbas planned to avenge the death of Ali Qoli Khan and he suborned four Qizilbash leaders to kill Murshid after a banquet on 23 July 1589. With Murshid gone, Abbas could now rule Iran in his own right.[10] [11]

Abbas decided he must re-establish order within Iran before he took on the foreign invaders. To this end he made a humiliating peace treaty – known as the Treaty of Istanbul – with the Ottomans in 1589/90, ceding them the provinces of Azerbaijan, Karabagh, Ganja and Qarajadagh as well as parts of Georgia, Luristan and Kurdistan.[12] [13]

Reducing the power of the Qizilbash

Anthony Shirley and Robert Shirley (*pictured in 1622*) helped modernize the Persian Army.

The Qizilbash had provided the backbone of the Iranian army from the very beginning of Safavid rule and they also occupied many posts in the government. To counterbalance their power, Abbas turned to another element in Iranian society, the *ghulams* (a word literally meaning "slaves"). These were Georgians, Armenians and Circassians who had converted to Islam and taken up service in the army or the administration. Abbas promoted such *ghulams* to the highest offices of the state. They included the Georgian Allahverdi Khan, who became leader of the *ghulam* regiments in the army as well as governor of the rich province of Fars. Abbas removed provincial governorships from some Qizilbash leaders and transferred Qizilbash groups to the lands of other Qizilbash tribes, thus weakening Qizilbash tribal unity.[14] Budgetary problems were resolved by restoring the shah's control of the provinces formerly governed by the Qizilbash chiefs, the revenues of which supplemented the royal treasury.

Reforming the army

Abbas needed to reform the army before he could hope to confront the Ottoman and Uzbek invaders. He also used military reorganisation as another way of sidelining the Qizilbash.[15] Instead, he created a standing army of 40,000 *ghulams* and Iranians to fight alongside the traditional, feudal force provided by the Qizilbash. The new army regiments had no loyalty but to the shah. They consisted of 10,000–15,000 cavalry armed with muskets and other weapons, a corps of musketeers (12,000 strong) and one of artillery (also 12,000 strong). In addition Abbas had a personal bodyguard of 3,000 *ghulams*.[16]

Abbas also greatly increased the amount of cannons at his disposal, permitting him to field 500 in a single battle. Ruthless discipline was enforced and looting was severely punished. Abbas was also able to draw on military advice from a number of European envoys, particularly from the English adventurers Sir Anthony, and his brother Robert Shirley, who arrived in 1598 as envoys from the Earl of Essex on an unofficial mission to induce Persia into anti-Ottoman alliance.[17]

Reconquest

War against the Uzbeks

Abbas' first campaign with his reformed army was against the Uzbeks who had seized Khorasan and were ravaging the province. In April 1598 he went on the attack. One of the two main cities of the province, Mashhad, was easily recaptured but the Uzbek leader Din Mohammed Khan was safely behind the walls of the other chief city, Herat. Abbas managed to lure the Uzbek army out of the town by feigning a retreat. A bloody battle ensued on 9 August 1598, in the course of which the Uzbek khan was wounded and his troops retreated (the khan was murdered by his own men on the way). Abbas' north-east frontier was now safe for the time being and he could turn his attention to the Ottomans in the west.[18]

War against the Ottomans

Since the treaty of 1589–90 Abbas had been regarded as almost an Ottoman vassal. The Safavids had never beaten their western neighbours in a straight fight. In 1602, Abbas decided he would no longer put up with Ottoman insults. After a particularly arrogant series of demands from the Turkish ambassador, the shah had him seized, had his beard shaved and sent it to his master, the sultan, in Constantinople. This was a declaration of war.[19] Abbas first recaptured Nahavand and destroyed the fortress in the city, which the Ottomans had planned to use as an advance base for attacks on Iran.[20] The next year, Abbas pretended he was setting off on a hunting expedition to Mazandaran with his men. This was merely a ruse to deceive the Ottoman spies in his court – his real target was Azerbaijan.[21] He changed course for Qazvin where he assembled a large army and set off to retake Tabriz, which had been in Ottoman hands for decades.

"Abbas King of Persia", as seen by Thomas Herbert in 1627.

For the first time, the Iranians made great use of their artillery and the town – which had been ruined by Ottoman occupation – soon fell.[22] Abbas set off to besiege Yerevan, the capital of Armenia, and one of the main Turkish strongholds in the Caucasus. It finally fell in June 1604 and with it the Ottomans lost the loyalty of most Armenians, Georgians and other Caucasians. But Abbas was unsure how the new sultan, Ahmed I, would respond and withdrew from the region using scorched earth tactics.[23] For a year, neither side made a move, but in 1605, Abbas sent his general Allahverdi Khan to meet Ottoman forces on the shores of Lake Van. On 6 November 1605 the Iranians led by Abbas scored a decisive victory over the Ottomans at Sufiyan, near Tabriz.[24] The Persian victory was recognized in the Treaty of Nasuh Pasha in 1612.

Several years of peace followed as the Ottomans carefully planned their response. But their secret training manoeuvres were observed by Iranian spies. Abbas learnt the Ottoman plan was to invade via Azerbaijan, take Tabriz then move on to Ardabil and Qazvin, which they could use as bargaining chips to exchange for other territories.[25] The shah decided to lay a trap. He would allow the Ottomans to enter the country, then destroy them. He had Tabriz evacuated of its inhabitants while he waited at Ardabil with his army. In 1618, an Ottoman army of 50,000 led by the grand vizier, invaded and easily seized Tabriz. The vizier sent an ambassador to the shah demanding he make peace and return the lands taken since 1602. Abbas refused and pretended he was ready to set fire to Ardabil and retreat further inland rather than face the Ottoman army. When the vizier heard the news, he decided to march on Ardabil right away. This was just what Abbas wanted. His army of 40,000 was hiding at a crossroads on the way and they ambushed the Ottoman army in a battle which ended in complete victory for the Iranians.[26]

In 1623, Abbas decided to take back Mesopotamia which had been lost by his grandfather Tahmasp. Profiting from the confusion surrounding the accession of the new sultan Murad IV, he pretended to be making a pilgrimage to the Shi'ite shrines of Kerbala and Najaf, but used his army to seize Baghdad. He was distracted by the rebellion in Georgia in 1624 which allowed an Ottoman force to besiege Baghdad, but the shah came to its relief the next year and crushed the Turkish army decisively. In 1638, however, after Abbas' death, the Ottomans retook Baghdad and the Iranian–Ottoman border became finalised.[27]

Kandahar and the Mughals

Jahangir's dream: Mughal picture showing Jahangir (right) embracing Abbas

Iran was traditionally allied with Mughal India against the Uzbeks, who coveted the province of Khorasan. The Mughal emperor Humayun had given Abbas' grandfather, Shah Tahmasp, the province of Kandahar as a reward for helping him back to his throne. In 1590, profiting from the confusion in Iran, Humayun's successor Akbar seized Kandahar. Abbas continued to maintain cordial relations with the Mughals, while always asking for the return of Kandahar. Finally, in 1620, a diplomatic incident in which the Iranian ambassador refused to bow down in front of the Emperor Jahangir led to war. India was embroiled in civil turmoil and Abbas found he only needed a lightning raid to take back Kandahar in 1622. After the conquest, he was very conciliatory to Jahangir, claiming he had only taken back what was rightly his and disavowing any further territorial ambitions. Jahangir was not appeased but he was unable to recapture the province.[28] [29]

Muhammad 'Ali Beg was the ambassador sent to the Mughal court by Shah Abbas of Iran, arriving in time for the New Year festival in March 1631. He remained there until October 1632, during which time his portrait was painted by the royal artist, Hashim.

War against the Portuguese

During the 16th century the Portuguese had established bases in the Persian Gulf. In 1602, the Iranian army under the command of Imam-Quli Khan Undiladze managed to expel the Portuguese from Bahrain.[30] In 1622, with the help of four English ships, Abbas retook Hormuz from the Portuguese in the Capture of Ormuz (1622). He replaced it as a trading centre with a new port, Bandar Abbas, nearby on the mainland, but it never became as successful.[31]

The island of Hormuz was captured by an Anglo-Persian force in the 1622 Capture of Ormuz.

The shah and his subjects

Isfahan: a new capital

Abbas moved his capital from Qazvin to the more central and more Persian Isfahan in 1598. Embellished by a magnificent series of new mosques, baths, colleges, and caravansarais, Isfahan became one of the most beautiful cities in the world. As Roger Savory writes, "Not since the development of Baghdad in the eighth century A.D. by the Caliph al-Mansur had there been such a comprehensive example of town-planning in the Islamic world, and the scope and layout of the city centre clear reflect its status as the capital of an empire."[32] Isfahan became the centre of Safavid architectural achievement, with the mosques Masjed-e Shah and the Masjed-e Sheykh Lotfollah and other monuments like the Ali Qapu, the Chehel Sotoun palace, and the Naghsh-i Jahan Square.

Arts

Abbas' painting ateliers (of the Isfahan school established under his patronage) created some of the finest art in modern Iranian history, by such illustrious painters as Reza Abbasi, Muhammad Qasim and others. Despite the ascetic roots of the Ṣafavid dynasty and the religious injunctions restricting the pleasures lawful to the faithful, the art of Abbas' time denotes a certain relaxation of the strictures. Historian James Saslow interprets the portrait by Muhammad Qasim as showing that the Muslim taboo against wine, as well as that against male intimacy, "were more honored in the breach than in the observance". Abbas brought 300 Chinese potters to Iran to enhance local production of Chinese-style ceramics.[33] From E. Sykes's "Persia and Its People": "Early in the seventeenth century, Shah Abbas imported Chinese workmen into his country to teach his subjects the art of making porcelain, and the Chinese influence is very strong in the designs on this ware. Chinese marks are also copied, so that to scratch an article is sometimes the only means of proving it to be of Persian manufacture, for the Chinese glaze, hard as iron, will take no mark."[34] [35]

Religious attitude and religious minorities

Like all other Safavid monarchs, Abbas was a Shi'ite Muslim. He had a particular veneration for Imam Hussein. In 1601, he made a pilgrimage on foot from Isfahan to Mashhad, site of the shrine of Imam Reza, which he restored (it had been despoiled by the Uzbeks).[36] Since Sunni Islam was the religion of Iran's main rival, the Ottoman Empire, Abbas often treated Sunnis living in western border provinces harshly.[37]

Kelisa-e Vank (the Armenian Vank Cathedral) in New Julfa

Abbas was generally tolerant of Christianity. The Italian traveller Pietro della Valle was astonished at the shah's knowledge of Christian history and theology and establishing diplomatic links with European Christian states was a vital part of the shah's foreign policy.[38] Christian Armenia was a key province on the border between Abbas' realm and the Ottoman Empire. From 1604 Abbas implemented a "scorched earth" policy in the region to protect his north-western frontier against any invading Ottoman forces, a policy which involved the forced resettlement of many Armenians from their homelands. Many were transferred to New Julfa, a town the shah had built for the Armenians near his capital Isfahan. Thousands of Armenians died on the journey.[39] Those who survived enjoyed considerable religious freedom in New Julfa, where the shah built them a new cathedral. Abbas' aim was to boost the Iranian economy by encouraging the Armenian merchants who had moved to New Julfa. As well as religious liberties, he also offered them interest-free loans and allowed the town to elect its own mayor (*kalantar*).[40] Other Armenians were transferred to the provinces of Gilan and Mazandaran. These were less lucky. Abbas wanted to establish a second capital in Mazandaran, Farahabad, but the climate was unhealthy and malarial. Many settlers died and others gradually abandoned the city.[41] [42] [43]

In 1614–15, Abbas suppressed a rebellion by the Christian Georgians of Kakheti, killing 60–70,000 and deporting over 100,000 Georgian peasants to Iran.[44] He later had the Georgian queen Ketevan tortured to death when she refused to renounce Christianity.[45] [46]

Contacts with Europe

Abbas' tolerance towards Christians was part of his policy of establishing diplomatic links with European powers to try to enlist their help in the fight against their common enemy, the Ottoman Empire. The idea of such an anti-Ottoman alliance was not a new one – over a century before, Uzun Hassan, then ruler of part of Iran, had asked the Venetians for military aid – but none of the Safavids had made diplomatic overtures to Europe and Abbas' attitude was in marked contrast to that of his grandfather, Tahmasp I, who had expelled the English traveller Anthony Jenkinson from his court on hearing he was a Christian.[47] For his part, Abbas declared that he "preferred the dust from the shoe soles of the lowest Christian to the highest Ottoman personage."[48]

The ambassador Husain Ali Beg led the first Persian embassy to Europe (1599–1602).

Fresco in the Doge's Palace in Venice depicting Doge Mariano Grimani receiving the Persian Ambassadors, 1599

In 1599, Abbas sent his first diplomatic mission to Europe. The group crossed the Caspian Sea and spent the winter in Moscow, before proceeding through Norway, Germany (where it was received by Emperor Rudolf II) to Rome where Pope Clement VIII gave the travellers a long audience. They finally arrived at the court of Philip III of Spain in 1602. Although the expedition never managed to return to Iran, being shipwrecked on the journey around Africa, it marked an important new step in contacts between Iran and Europe and Europeans began to be fascinated by the Iranians and their culture – Shakespeare's 1601–02 *Twelfth Night*, for example, makes two references (at II.5 and III.4) to 'the Sophy', then the English term for the Shahs of Iran.[49] [50] Henceforward, the number of diplomatic missions to and fro greatly increased.[51]

The shah had set great store on an alliance with Spain, the chief opponent of the Ottomans in Europe. Abbas offered trading rights and the chance to preach Christianity in Iran in return for help against the Ottomans. But the stumbling block of Hormuz remained, a port which had fallen into Spanish hands when the King of Spain inherited the throne of Portugal in 1580. The Spanish demanded Abbas break off relations with the English East India Company before they would consider relinquishing the town. Abbas was unable to comply. Eventually Abbas became frustrated with Spain, as he did with the Holy Roman Empire, which wanted him to make his 170,000 Armenian subjects swear allegiance to the Pope but did not trouble to inform the shah when the Emperor Rudolf signed a peace treaty with the Ottomans. Contacts with the Pope, Poland and Moscow were no more fruitful.[52]

More came of Abbas' contacts with the English, although England had little interest in fighting against the Ottomans. The Sherley brothers arrived in 1598 and helped reorganise the Iranian army. The English East India Company also began to take an interest in Iran and in 1622 four of its ships helped Abbas retake Hormuz from the Portuguese in the capture of Hormuz. It was the beginning of the East India Company's long-running interest in Iran.[53]

Family tragedies and death

Shah Abbas in later life with a page. By Muhammad Qasim (1627).[54]

Of Abbas' five sons, three had survived past childhood, so the Safavid succession seemed secure. He was on good terms with the crown prince, Mohammed Baqir Mirza (born 1587; better known in the West as Safi Mirza). In 1614, however, during a campaign in Georgia, the shah heard rumours that the prince was conspiring against his life with a leading Circassian, Fahrad Beg. Shortly after, Mohammed Baqir broke protocol during a hunt by killing a boar before the shah had chance to put his spear in. This seemed to confirm Abbas' suspicions and he sunk into melancholy; he no longer trusted any of his three sons. In 1615, he decided he had no choice but to have Mohammed killed. A Circassian named Behbud Beg executed the Shah's orders and the prince was murdered in a hammam in the city of Resht. The shah almost immediately regretted his action and was plunged into grief.[55]

In 1621, Abbas fell seriously ill. His heir, Mohammed Khodabanda, thought he was on his deathbed and began to celebrate his accession to the throne with his Qizilbash supporters. But the shah recovered and punished his son with blinding, which would disqualify him from ever taking the throne.[56] The blinding was only partially successful and the prince's followers planned to smuggle him out of the country to safety with the Great Mughal whose aid they would use to overthrow Abbas and install Mohammed on the throne. But the plot was betrayed, the prince's followers were executed and the prince himself imprisoned in the fortress of Alamut where he would later be murdered by Abbas' successor, Shah Safi.[57]

Imam Qoli Mirza, the third and last son, now became the crown prince. Abbas groomed him carefully for the throne but, for whatever reason, in 1627, he had him partially blinded and imprisoned in Alamut.[58]

Unexpectedly, Abbas now chose as heir the son of Mohammed Baqir Mirza, Sam Mirza, a cruel and introverted character who was said to loathe his grandfather because of his father's murder. It was he who in fact did succeed Shah Abbas at the age of seventeen in 1629, taking the name Shah Safi. Abbas's health was troubled from 1621 onwards. He died at his palace in Mazandaran in 1629 and was buried in Kashan.[59]

Character and legacy

According to Roger Savory: "Shah Abbas I possessed in abundance qualities which entitle him to be styled 'the Great'. He was a brilliant strategist and tactician whose chief characteristic was prudence. He preferred to obtain his ends by diplomacy rather than war, and showed immense patience in pursuing his objectives."[60] In Michael Axworthy's view, Abbas "was a talented administrator and military leader, and a ruthless autocrat. His reign was the outstanding creative period of the Safavid era. But the civil wars and troubles of his childhood (when many of his relatives were murdered) left him with a dark twist of suspicion and brutality at the centre of his personality."[61]

The Cambridge History of Iran rejects the view that the death of Abbas marked the beginning of the decline of the Safavid dynasty as Iran continued to prosper throughout the 17th century, but blames him for the poor statemanship of the later Safavid shahs: "The elimination of royal princes, whether by blinding or immuring them in the harem, their exclusion from the affairs of state and from contact with the leading aristocracy of the empire and the generals, all the abuses of the princes' education, which were nothing new but which became the normal practice with Abbas at the court of Isfahan, effectively put a stop to the training of competent successors, that is to say, efficient princes prepared to meet the demands of ruling as kings."[62]

Abbas gained strong support from the common people. Sources report him spending much of his time among them, personally visiting bazaars and other public places in Isfahan.[63] Short in stature but physically strong until his health declined in his final years, Abbas could go for long periods without needing to sleep or eat and could ride great distances. At the age of 19 Abbas shaved off his beard, keeping only his moustache, thus setting a fashion in Iran.[64]

Offspring

Sons

- Prince *Shahzadeh* Soltan Mohammad Baqer *Feyzi Mirza* (*b*. 15 September 1587, Mashhad, Khorasan-*k*. 25 January 1615, Rasht, Gilan), was Governor of Mashhad 1587–1588, and of Hamadan 1591–1592. Married (1st) at Esfahan, 1601, Princess Fakhri-Jahan, daughter of Ismail II. Married (2nd) Del Aram, a Georgian. Married (3rd) Marta daughter of Eskandar Mirza. He had issue, two sons:
  - (By Del Aram) Prince *Shahzadeh* Soltan Abul-Naser Sam Mirza, succeeded as Safi.
  - (By Fakhri-Jahan) Prince *Shahzadeh* Soltan Soleyman Mirza (*k*. August 1632 at Alamut, Qazvin). He had issue.
- Prince *Shahzadeh* Soltan Hasan Mirza (*b*. September 1588, Mazandaran – *d*. 18 August 1591, Qazvin)
- Prince *Shahzadeh* Soltan Hosein Mirza (*b*. 26 February 1591, Qazvin – *d*. before 1605)
- Prince *Shahzadeh* Tahmasph Mirza
- Prince *Shahzadeh* Soltan Mohammad Mirza (*b*. 18 March 1591, Qazvin – *k*. August 1632, Alamut, Qazvin) Blinded on the orders of his father, 1621.
- Prince *Shahzadeh* Soltan Ismail Mirza (*b*. 6 September 1601, Esfahan – *k*. 16 August 1613)
- Prince *Shahzadeh* Imam Qoli Amano'llah Mirza (*b*. 12 November 1602, Esfahan – *k*. August 1632, Alamut, Qazvin) Blinded on the orders of his father, 1627. He had issue, one son:
  - Prince *Shahzadeh* Najaf Qoli Mirza (*b*. 1625-*k*. August 1632, Alamut, Qazvin)

Daughters

- Princess *Shahzadeh 'Alamiyan* Shazdeh Beygom (*d*. before 1629), married Mirza Mohsen Razavi. She had issue, two sons.
- Princess *Shahzadeh 'Alamiyan* Zobeydeh Beygom (*b*. 4 December 1586 -*k*. 20 February 1632). She had issue, three sons and one daughter, including: Jahan-Banoo Begum, married in 1623, Simon II of Kartli son of Bagrat VII of Kartli by his wife, Queen Anna, daughter of Alexander II of Kakheti. She had issue, a daughter: Princess 'Izz-e-Sharif.

- Princess *Shahzadeh 'Alamiyan* Khan Agha Beygom, married Mirza Abu Talib 'Ala ud-din Muhammad al-Husaini al-Marashi, son of Mir Rafi ud-din Muhammad Khalifa Isfahani. She had issue, four sons and four daughters.
- Princess *Shahzadeh 'Alamiyan* Havva Beygom (*d.* 1617, Zanjan)
- Princess *Shahzadeh 'Alamiyan* Shahbanoo Beygom.
- Princess *Shahzadeh 'Alamiyan* Malek-Nesa Beygom (*d.* 1629)

See also

- García de Silva Figueroa
- History of Iran
- Persian embassy to Europe (1599–1602)
- Persian embassy to Europe (1609–1615)
- Safavid conversion of Iran from Sunnism to Shiism

References

[1] Chambers Biographical Dictionary, ISBN 0-550-18022-2, page 1
[2] Savory p.71
[3] Newman p.42
[4] Nahavandi and Bomati pp.27–28
[5] Nahavandi and Bomati 29–34
[6] Savory pp.73–75
[7] Nahavandi and Bomati 34–36
[8] Savory p.75
[9] Nahavandi and Bomati p.36
[10] Nahavandi and Bomati pp.36–39
[11] Newman p.50
[12] Savory pp.76–77
[13] Newman p.52
[14] Savory p.78
[15] Michael Axworthy *Iran: Empire of the Mind* pp.134–35
[16] Savory p.79
[17] Nahavandi and Bomati pp.141–144
[18] Savory p.83-4
[19] Nahavandi and Bomati pp.147–148
[20] Savory p.85
[21] Nahavandi and Bomati pp.148–149
[22] Nahavandi and Bomati pp.149–150
[23] Nahavandi and Bomati pp.150–151
[24] Savory p.87
[25] Nahavandi and Bomati p.153
[26] Nahavandi and Bomati pp.153–156
[27] Nahavandi and Bomati pp.158–159
[28] Nahavandi and Bomati pp.120–125
[29] Abraham Eraly *The Mughal Throne* (Phoenix, 2000) pp.263–265
[30] Juan R. I. Cole. "Rival Empires of Trade and Imami Shiism in Eastern Arabia, 1300–1800" (http://links.jstor.org/sici?sici=0020-7438(198705)19:2<177:REOTAI>2.0.CO;2-X). JSTOR. p. 186. . Retrieved 23 August 2010.
[31] Nahavandi and Bomati pp.159–162
[32] Roger Savory *Iran Under the Safavids* p.96
[33] Newman, Andrew J. (2006), "Monumental Challenge and Monumental Responses: the reign of Abbas I (1587–1629)", *Safavid Iran: rebirth of a Persian empire*, I. B. Tauris, p. 69, ISBN 978-1-86064-667-6
[34] Francis Barrow Pearce (1920). *Zanzibar: the island metropolis of eastern Africa* (http://books.google.com/books?id=XRRzAAAAMAAJ&pg=PA359#v=onepage&q&f=false). GREAT BRITAIN: Dutton. p. 359. . Retrieved March/12/2012. "interest to quote the following extract from E. Sykes's Persia and Its People: "Early in the seventeenth century, Shah Abbas imported Chinese workmen into his country to teach his subjects the art of making porcelain, and the Chinese influence is very strong in the designs on this ware. Chinese marks are also copied, so that to scratch an article is sometimes the only means of proving it to be of Persian manufacture, for

the Chinese glaze, hard as iron, will take no mark.""
[35] Francis Barrow Pearce (1920). *Zanzibar: the island metropolis of eastern Africa* (http://books.google.com/books?id=XRRzAAAAMAAJ&pg=PA430#v=onepage&q&f=false). GREAT BRITAIN: Dutton. p. 430. . Retrieved March/12/2012. "Shah Abbas, 202; his name men- tioned on bronze guns at Zanzibar, 200; imports Chinese artisans to teach the art of pottery-making, 350"
[36] Nahavandi and Bomati pp.96–99
[37] Nahavandi and Bomati pp.111–112
[38] Nahavandi and Bomati p.107
[39] According to Bomati and Nahavandi (p.103), of 56,000 who left Armenia, only 30,000 reached the new town.
[40] Bomati and Nahavandi p.209
[41] This paragraph:Nahavandi and Bomati 100–104
[42] This paragraph:*Cambridge History of Iran* Volume 6 p.454
[43] This paragraph: *The Armenian People from Ancient to Modern Times: Foreign Dominion to Statehood: the Fifteenth Century to the Twentieth Century* edited by Richard G. Hovannisian (Palgrave Macmillan,2004) pp.19–20
[44] R.G. Suny *The Making of the Georgian Nation* (Indiana University Press, 1994) p.50
[45] Suny p.50
[46] Assatiani and Bendianachvili *Histoire de la Géorgie* (L'Harmattan, 1997) p.188
[47] Laurence Lockhart in *The Legacy of Persia* ed. A. J. Arberry (Oxford University Press, 1953 p.347)
[48] Nahavandi and Bomati p.114
[49] *Twelfth Night* (http://books.google.com/books?id=DNKKsveMfhQC&pg=PA177&lpg=PA177&dq=the+sophy+twelfth+night&source=bl&ots=TjopMHWII_&sig=NvSFCpvIGyR_jVvlYeQ8Eu8A5z4&hl=en&ei=XCt2TPejPIvA4gbszqClBg&sa=X&oi=book_result&ct=result&resnum=4&ved=0CBoQ6AEwAw#v=onepage&q=the sophy twelfth night&f=false)
[50] Richard Wilson, "When Golden Time Convents": Twelfth Night and Shakespeare's Eastern Promise, *Shakespeare*, Volume 6, Issue 2 June 2010 , pages 209 – 226 (http://www.informaworld.com/smpp/content~content=a923903386~db=all~jumptype=rss)
[51] Nahavandi and Bomati pp.128–130
[52] Nahavandi and Bomati pp.130–137
[53] Nahavandi and Bomati pp.161–162
[54] Nahavandi and Bomati, illustration opposite p.162
[55] This paragraph: Nahavandi and Bomati p.235-7
[56] Savory p.95
[57] Nahavandi and Bomati pp.240–241
[58] Nahavandi and Bomati pp.241–242
[59] Bomati and Nahavandi pp.243–6
[60] Savory p.101
[61] Axworthy p.134
[62] H.R. Roemer in *Cambridge History of Iran*, Volume 6, p.278
[63] Savory p.103
[64] Nahavandi and Bomati pp.44–47, 57–58

Sources

- H. Nahavandi, Y. Bomati, *Shah Abbas, empereur de Perse (1587–1629)* (Perrin, Paris, 1998)
- Roger Savory *Iran under the Safavids* (Cambridge University Press, 2007 reissue)
- *The Cambridge History of Iran* Volume 6
- Andrew J. Newman *Safavid Iran* (I.B.Tauris, 2006)

External links

- *Shah Abbās: The Remaking of Iran*, The British Museum (http://www.britishmuseum.org/whats_on/future_exhibitions/shah_abbas.aspx), in association with Iran Heritage Foundation, 19 February – 14 June 2009,
- John Wilson, *Iranian treasures bound for Britain*, BBC Radio 4, 19 January 2009, BBC Radio 4's live magazine (http://news.bbc.co.uk/today/hi/today/newsid_7833000/7833651.stm), *Front Row* (http://www.bbc.co.uk/iplayer/console/b00gqy39) (audio report).
- "Shah 'Abbas: The Remaking of Iran" (http://en.qantara.de/webcom/show_article.php/_c-310/_nr-633/i.html)

Safi of Persia

Shah Safi

Mirza (royal title)
Shahanshah of Persia

Reign	28 January 1629 – 12 May 1642
Coronation	28 January 1629
Titles	Shahanshah, Sahib-i-Qiran, Sultan bar Salatin [1]
Born	1611
Died	12 May 1642 (aged 30/31)
Predecessor	Abbas I
Successor	Abbas II
Royal House	Safavi
Father	Shahzadeh Soltan Mohammad Baqer Mirza
Mother	Dilaram Khanum

Shah Safi (Persian: شاه صفی) was Shah of Iran from 1629 to 1642. He was the sixth ruler of the Safavid dynasty.

Safi was given the name Sam Mirza when he was born. He was the son of Mohammed Baqir Mirza, the eldest son of Shah Abbas I, and Dilaram Khanum, a Georgian wife.[2] In 1615, Abbas had Mohammed Baqir killed, fearing he was plotting against his life. Over the next few years, the suspicious Abbas killed or blinded his other sons, leaving his grandson Safi heir to the throne.

Safi was crowned on 28 January 1629 at the age of eighteen. He ruthlessly eliminated anyone he regarded as a threat to his power, executing almost all the Safavid royal princes as well as leading courtiers and generals. He paid little attention to the business of government and had no cultural or intellectual interests (he had never learned to read or write properly), preferring to spend his time drinking wine or indulging in his addiction to opium. Supposedly, however, he abhorred tobacco smoke as much as his grandfather did, going as far as to have those caught smoking tobacco in public killed by pouring molten lead in their mouths.

The dominant political figure of Safi's reign was Saru Taqi, appointed grand vizier in 1634. Saru Taqi was incorruptible and highly efficient at raising revenues for the state, but he could also be autocratic and arrogant.

Iran's foreign enemies took the opportunity to exploit Safi's perceived weakness. The Ottomans made incursions in the west in 1630 and 1634 and in 1638 they succeeded in capturing Baghdad, which would remain in their hands until World War I. Nevertheless, the Treaty of Zuhab which ensued in 1639 put an end to all further wars between the Safavids and the Ottomans. Apart from the Ottoman attacks, Iran was troubled by the Uzbeks and Turkmens in the east and lost Kandahar to the Mughals in 1638.

Safi died on 12 May 1642.

References

[1] The Royal Ark (http://www.royalark.net/Persia/safawi3.htm)
[2] Sussan Babaie and others: *Slaves of the Shah* (I.B. Tauris, 2003) p.104

Sources

• *Cambridge History of Iran* Volume 6, p. 278ff.

Mahmud Hotaki

Shah Mahmud Hotaki	
Emir of Afghanistan/Shah of Iran	

Sketch work of Mir Mahmud Shah	
Reign	Hotaki Empire: 1717–1725
Coronation	1717 and 1722
Full name	Mir Mahmud Shah Hotak
Titles	Shah, Padshah, Mir, Sultan
Born	1697
Died	April 22, 1725
Place of death	Isfahan
Predecessor	Abdul Aziz Hotak
Successor	Ashraf Hotaki
Dynasty	Hotaki dynasty
Father	Mirwais Khan Hotak
Religious beliefs	Sunni Islam

Shah **Mahmud Hotaki**, (Pashto, Persian, Urdu, Arabic: اش دومحم وتن ی), also known as **Mahmud Ghilzai** (1697? — April 22, 1725), was an Afghan ruler of the Hotaki dynasty who defeated and overthrew the Safavid dynasty to become the king of Persia from 1722 until his death in 1725.[1]

He was the eldest son of Mirwais Hotak, the chief of the Ghilzai-Pashtun tribe of Afghanistan, who had made the Kandahar region independent from Persian rule in 1709.[2] When Mirwais died in 1715, he was succeeded by his brother, Abdul Aziz, but the Ghilzai Afghans persuaded Mahmud to seize power for himself and in 1717 he overthrew and killed his uncle.[3]

Mahmud takes the throne of Persia

The Abdali Pashtuns inhabited the region of Khorasan while the Ghilzais controlled the Kandahar region (*Candahar*) to the southeast.

In 1720, Mahmud and the Ghilzais defeated the rival ethnic Afghan tribe of the Abdalis. However, Mahmud had designs on the Persian empire itself. He had already launched an expedition against Kerman in 1719 and in 1721 he besieged the city again. Failing in this attempt and in another siege on Yazd, in early 1722, Mahmud turned his attention to the shah's capital Isfahan, after first defeating the Persians at the Battle of Gulnabad. Rather than biding his time within the city and resisting a siege in which the small Afghan army was unlikely to succeed, Sultan Husayn marched out to meet Mahmud's force at Golnabad. Here, on March 8, the Persian royal army was thoroughly routed and fled back to Isfahan in disarray. The shah was urged to escape to the provinces to raise more troops but he decided to remain in the capital which was now encircled by the Afghans. Mahmud's siege of Isfahan lasted from March to October, 1722. Lacking artillery, he was forced to resort to a long blockade in the hope of starving the Persians into submission. Sultan Husayn's command during the siege displayed his customary lack of decisiveness and the loyalty of his provincial governors wavered in the face of such incompetence. Starvation and disease finally forced Isfahan into submission (it is estimated that 80,000 of its inhabitants died during the siege). On October 23, Sultan Husayn abdicated and acknowledged Mahmud as the new shah of Persia.[4]

Mahmud's reign as shah

In the early days of his rule, Mahmud displayed benevolence, treating the captured royal family well and bringing in food supplies to the starving capital. But he was confronted with a rival claimant to the throne when Hosein's son, Tahmasp declared himself shah in November. Mahmud sent an army against Tahmasp's base, Qazvin. Tahmasp escaped and the Afghans took the city but, shocked at the treatment they received at the hands of the conquering army, the population rose up against them in January 1723. The revolt was a success and Mahmud was worried about the reaction when the surviving Afghans returned to Isfahan to bring news of the defeat. Fearing a revolt by his subjects, Mahmud invited his Persian ministers and nobles to a meeting under false pretences and had them slaughtered. He also executed up to 3,000 of the Persian royal guards. At the same time, the Ottomans and the Russians took advantage of the chaos in Persia to seize land for themselves, limiting the amount of territory under Mahmud's control.[5]

His failure to impose his rule across Persia made Mahmud depressed and suspicious. He was also concerned about the loyalty of his own men, since many Afghans preferred his cousin Ashraf Khan. In February 1725, believing a rumour that one of Sultan Husayn's sons, Safi Mirza, had escaped, Mahmud ordered the execution of all the other Safavid princes who were in his hands, with the exception of Sultan Husayn himself. When Sultan Husayn tried to stop the massacre, he was wounded, but his action led to Mahmud sparing the lives of two of his young children.[6]

Death

Mahmud began to succumb to insanity as well as physical deterioration. On April 22, 1725, a group of Afghan officers freed Ashraf Khan from the prison where he had been confined by Mahmud and launched a palace revolution which placed Ashraf on the throne. Mahmud died three days later, either from his illness – at it was claimed at the time – or murder by suffocation.

> ...Thereafter his disorder rapidly increased, until he himself was murdered on April 22 by his cousin Ashraf, who was thereupon proclaimed king. Mír Maḥmúd was at the time of his death only twenty-seven years of age, and is described as "middle-sized and clumsy; his neck was so short that his head seemed to grow to his shoulders; he had a broad face and flat nose, and his beard was thin and of a red colour; his looks were wild and his countenance austere and disagreeable; his eyes, which were blue and a little squinting, were generally downcast, like a man absorbed in deep thought."[7]
>
> —Edward G. Browne, 1924

References

[1] "AN OUTLINE OF THE HISTORY OF PERSIA DURING THE LAST TWO CENTURIES (A.D. 1722–1922)" (http://persian.packhum.org/persian/pf?file=90001014&ct=29). *Edward Granville Browne*. London: Packard Humanities Institute. p. 29. . Retrieved 2010-09-24.
[2] Dupree, Mir Wais Hotak (1709–1715) (http://www.afghanan.net/afghanistan/mirwais.htm)
[3] Axworthy p.38
[4] Axworthy pp.39–55
[5] Axworthy pp.64–65
[6] Axworthy pp.65–67
[7] "AN OUTLINE OF THE HISTORY OF PERSIA DURING THE LAST TWO CENTURIES (A.D. 1722–1922)" (http://persian.packhum.org/persian/pf?file=90001014&ct=31). *Edward Granville Browne*. London: Packard Humanities Institute. p. 31. . Retrieved 2010-09-24.

Sources

- Michael Axworthy, *The Sword of Persia: Nader Shah, from Tribal Warrior to Conquering Tyrant* Hardcover 348 pages (26 July 2006) Publisher: I.B. Tauris (http://www.ibtauris.com/ibtauris/display.asp?K=510000000117668&cid=ibtauris&sf_01=CAUTHOR&st_01=Michael+Axworthy&sf_02=CTITLE&sf_03=KEYWORD&sf_04=VX_ISBN;=&m=1&dc=1) Language: English ISBN 1-85043-706-8

External links

- An outline of the History of Persia during the last two centuries (1722–1922), The Afghan Invasion (1722–1730) (http://persian.packhum.org/persian/pf?file=90001014&ct=29)
- Encyclopædia Britannica Online – Last Afghan empire (http://www.britannica.com/EBchecked/topic/7798/Afghanistan/21392/Last-Afghan-empire)

Ashraf Hotaki

Ashraf Khan Hotaki	
Emir of Afghanistan / Shah of Persia	
 Shah Ashraf Hotaki	
Reign	Hotaki Empire: 1725–1729
Coronation	1715 and 1725
Full name	Ashraf Khan Hotaki[1]
Titles	Khan, Shah, Mir
Birthplace	Kandahar Province
Died	1730
Place of death	Balochistan
Predecessor	Mahmud Hotaki
Successor	Hussain Hotaki
Dynasty	Hotaki dynasty
Father	Abdul Aziz Hotak[2]
Religious beliefs	Sunni Islam

Shah Ashraf Hotaki, (Pashto, Persian, Urdu, Arabic: ى تو فرشأ اش), also known as **Ashraf Ghilzai** (died 1730), son of Abdul Aziz Hotak, was the fourth ruler of the Hotaki dynasty. An Afghan from the Ghilzai Pashtuns, he served as a commander in the army of Shah Mahmud during their conquest of the Persia Empire. Ashraf participated in the Battle of Gulnabad against the Persians and became victorious. In 1725, he succeeded to the throne (Shah of Persia) after the death of his cousin Mahmud.

The nephew of Mirwais Hotak, his reign was noted for the sudden decline in the Hotaki Afghan Empire under increasing pressure from Turkish, Russian, and Persian forces.[3]

Ashraf Khan halted both the Russian and Turkish onslaughts. He defeated the Ottoman Empire in a battle near Kermanshah, after the enemy had come close to Isfahan. This led to peace negotiations with the Sublime Porte, which were briefly disrupted after Ashraf's ambassador insisted his master should be Caliph of the East and the Ottoman Sultan Caliph of the West. This caused great umbrage to the Ottomans, but a peace agreement was finally signed at Hamadan in the autumn of 1727.[4]

Ultimately, though it was a little-known Afsharid Turkmen rebel, Nader Shah, who defeated Ashraf's Ghilzai forces at the Battle of Damghan in October 1729, driving them back to what is now Afghanistan.[3]

Death

During the retreat, Ashraf is believed to have been captured and murdered by Baloch bandits in 1730.[5] This was probably a retaliation for killing Mahmud, and was ordered by Hussain Hotaki who was ruling from Kandahar at the time.

> *Ashraf, having taken Yazd and Kirmán, marched into Khurásán with an army of thirty thousand men to give battle to Ṭahmásp, but he was completely defeated by Nádir on October 2 at Dámghán. Another decisive battle was fought in the following year at Múrchakhúr near Iṣfahán. The Afgháns were again defeated and evacuated Iṣfahán to the number of twelve thousand men, but, before quitting the city he had ruined, Ashraf murdered the unfortunate ex-Shah Husayn, and carried off most of the ladies of the royal family and the King's treasure. When Ṭahmásp II entered Iṣfahán on December 9 he found only his old mother, who had escaped deportation by disguising herself as a servant, and was moved to tears at the desolation and desecration which met his eyes at every turn. Nádir, having finally induced Ṭahmásp to empower him to levy taxes on his own authority, marched southwards in pursuit of the retiring Afgháns, whom he overtook and again defeated near Persepolis. Ashraf fled from Shíráz towards his own country, but cold, hunger and the unrelenting hostility of the inhabitants of the regions which he had to traverse dissipated his forces and compelled him to abandon his captives and his treasure, and he was finally killed by a party of Balúch tribesmen.*[3]
>
> —Edward G. Browne, 1924

Ashraf Khan's death marked the end of Hotaki rule in Persia, but the country of Afghanistan was still under Shah Hussain Hotaki's control until Nader Shah's 1738 conquest of Kandahar where the young Ahmad Shah Durrani was held prisoner. It was only a short pause before the establishment of the last Afghan Empire[6] (modern state of Afghanistan) by Ahmad Shah Durrani in 1747.[7] [8]

See also

- Hotaki dynasty
- History of Iran
- History of Afghanistan

References

[1] Mujtaba, Bahaudin Ghulam; Sayed Tayeb Jawad (2006). *Afghanistan: Realities of War and Rebuilding* (http://books.google.com/books?id=Yd8vAQAAIAAJ&source=gbs_navlinks_s). Ilead Academy. p. 10. ISBN 0-9774211-1-2, 9780977421114. . Retrieved 2010-08-22.

[2] Vogelsang, Willem (2002). *The Afghans* (http://books.google.com/books?id=9kfJ6MlMsJQC&lpg=PP1&pg=PA224#v=onepage&q&f=false). Wiley Blackwell. p. 224. ISBN 0-631-19841-5. . Retrieved 2010-08-22.

[3] "AN OUTLINE OF THE HISTORY OF PERSIA DURING THE LAST TWO CENTURIES (A.D. 1722–1922)" (http://persian.packhum.org/persian/pf?file=90001014&ct=31). *Edward Granville Browne*. London: Packard Humanities Institute. p. 31. . Retrieved 2010-09-24.

[4] Jonas Hanway, *The Revolutions of Persia* (1753), p.254.

[5] "AN OUTLINE OF THE HISTORY OF PERSIA DURING THE LAST TWO CENTURIES (A.D. 1722–1922)" (http://persian.packhum.org/persian/pf?file=90001014&ct=30). *Edward G. Browne*. London: Packard Humanities Institute. p. 30. . Retrieved 2010-09-24.

[6] "Last Afghan empire" (http://www.britannica.com/EBchecked/topic/7798/Afghanistan/21392/Last-Afghan-empire). *Louis Dupree, Nancy Hatch Dupree and others*. Encyclopædia Britannica Online. . Retrieved 2009-10-17.

[7] "AFGHANISTAN x. Political History" (http://iranicaonline.org/articles/afghanistan-x-political-history). *D. Balland*. Encyclopædia Iranica. . Retrieved 2010-09-25.

[8] Houtsma, Martijn Theodoor (1987). *E.J. Brill's first encyclopaedia of Islam 1913–1936* (http://books.google.com/books?id=GEl6N2tQeawC&lpg=PP1&pg=PA146#v=onepage&q&f=false). **2**. BRILL. p. 146. ISBN 90-04-09796-1, 9789004097964. . Retrieved 2010-09-25.

External links

- An Outline Of The History Of Persia During The Last Two Centuries (A.D. 1722–1922) (http://persian.packhum.org/persian/pf?file=90001014&ct=29)
- Hotak Rule (http://www.afghansite.com/afghanistan/afghanistan.asp?File=28)

Abbas III

Shah Abbas III	
Shahanshah of Persia	
Reign	1732–1736
Coronation	7 September 1732
Titles	Shahanshah, Sahib-i-Qiran, Sultan bar Salatin [1]
Died	1739 or 1740
Place of death	Sabzevar
Predecessor	Tahmasp II
Successor	Nader Shah
Royal House	Safavi
Father	Tahmasp II

Abbas III (Persian: شاه عباس سوم) (died 1739?; reigned 1732–1736) was a son of Shah Tahmasp II of the Safavid dynasty. After the deposition of his father by Nader Khan (the future Nader Shah) the infant Abbas was appointed nominal ruler of Iran on 7 September 1732.[2] Nader Khan, who was the real ruler of the country, assumed the positions of deputy of state and viceroy. Abbas III was deposed on March 1736, when Nader Khan had himself crowned as Nader Shah. This marked the official end of the Safavid dynasty. Abbas was sent to join his father in prison in Sabzevar, Khorasan.[3]

In 1738, Nader Shah set out on campaign to Afghanistan and India, leaving his son Reza Qoli Mirza to rule his realm in his absence. Hearing rumours that his father had died, Reza made preparations for assuming the crown. According to the most "authoritative account",[4] Mohammed Hosein Khan Qajar, who had been entrusted with supervising Abbas and his father in captivity, warned Reza that the townspeople of Sabzevar would rise up in revolt, free Tahmasp II and place him on the throne again on hearing the news of Nader's death. Reza gave Mohammed Hosein orders to execute Tahmasp and his sons to forestall this. Mohammed Hosein strangled Tahmasp, cut the young Abbas down with his sword and had his brother Esmail killed too. According to Michael Axworthy, the dating of these events is speculative, but they probably took place in May or June 1739.[5] Other sources (*Encyclopaedia Iranica*, Lockhart) prefer 1740.

References

[1] The Royal Ark (http://www.royalark.net/Persia/safawi3.htm)

[2] Axworthy p.123

[3] Axworthy p.165

[4] Axworthy p.220

[5] Axworthy p.221

Bibliography

- Michael Axworthy *Sword of Persia: Nader Shah: From Tribal Warlord to Conquering Tyrant* (IB Tauris, 2006)
- L. Lockhart, *Nadir Shah*, London, 1938
- R.M. Savory, *Encyclopaedia Iranica*

Nader Shah

Nader Shah	
Shahanshah of Persia	

Nader Shah's portrait from the collection of the Victoria and Albert Museum.	
Reign	1736–1747
Born	1688 or 1698
Birthplace	Dastgerd, (Khorasan, Iran)
Died	June 19, 1747
Place of death	Quchan (Khorasan, Iran)
Predecessor	Abbas III
Successor	Adil Shah

Nāder Shāh Afshār (Persian: نادر شاه افشار; also known as *Nāder Qoli Beg* - نادر قلی بیگ or *Tahmāsp Qoli Khān* - تهماسپ قلی خان) (November, 1688[1] or August 6, 1698[2] – June 19, 1747) ruled as Shah of Iran (1736–47) and was the founder of the Afsharid dynasty. Because of his military genius, some historians have described him as the *Napoleon of Persia*[3] or the *Second Alexander*.[4] Nader Shah was a member of the Turkic Afshar tribe of northern Persia,[5] which had supplied military power to the Safavid state since the time of Shah Ismail I.[6]

Nader rose to power during a period of anarchy in Iran after a rebellion by the Hotaki Afghans had overthrown the weak Persian Shah Sultan Husayn, and both the Ottomans and the Russians had seized Persian territory for themselves. Nader reunited the Persian realm and removed the invaders. He became so powerful that he decided to depose the last members of the Safavid dynasty, which had ruled Iran for over 200 years, and become shah himself in 1736. His numerous campaigns created a great empire that briefly encompassed what is now part of or including Iran, Iraq, Afghanistan, Pakistan, Tajikistan, Turkmenistan, Uzbekistan, India, Georgia, Armenia, Azerbaijan, Russia, Oman and the Persian Gulf but his military spending had a ruinous effect on the Persian economy.[1]

Nader idolized Genghis Khan and Timur, the previous conquerors from Central Asia. He imitated their military prowess and—especially later in his reign—their cruelty. His victories briefly made him the Middle East's most powerful sovereign, but his empire quickly disintegrated after he was assassinated in 1747. Nader Shah has been described as "the last great Asian military conqueror".[7] He is credited for restoring Iranian power as an eminence between the Ottomans and the Mughals.[8] [9] [10] [11]

Early life

Nader Shah was born in the fortress of Dastgerd [12] into the Qereqlu (in Persian قرقلو) clan of the Afshars, a semi-nomadic Qizilbash tribe settled in the northern valleys of Khorasan, a province in the north-east of the Persian Empire.[13] His father, Emam Qoli, was a herdsman who may also have been a camel driver and coatmaker.[14] He died while Nader was still young.[15] According to legends, Nader and his mother were carried off as slaves by marauding Uzbek or Turkmen tribesmen, but Nader managed to escape. He joined a band of brigands while still a boy and eventually became their leader. Under the patronage of Afshar chieftains, he rose through the ranks to become a powerful military figure. Nader married the two daughters of Baba Ali Beg, a local chief.[1]

Fall of the Safavid dynasty

Nader grew up during the final years of the Safavid dynasty which had ruled Persia since 1502. At its peak, under such figures as Abbas the Great, Safavid Persia had been a powerful empire, but by the early 18th century the state was in serious decline and the reigning shah, Sultan Husayn, was a weak ruler. When Sultan Husayn attempted to quell a rebellion by the Ghilzai Afghans in Kandahar, the governor he sent was killed. Under their leader Mahmud Hotaki, the rebellious Afghans moved westwards against the shah himself and in 1722 they defeated a vastly superior force at the Battle of Gulnabad and then besieged the capital, Isfahan.[16] After the shah failed to escape to rally a relief force elsewhere, the city was starved into submission and Sultan Husayn abdicated, handing power to Mahmud. In Khorasan, Nader at first submitted to the local Afghan governor of Mashhad, Malek Mahmud, but then rebelled and built up his own small army. Sultan Husayn's son had declared himself Shah Tahmasp II, but found little support and fled to the Qajar tribe, who offered to back him. Meanwhile, Persia's imperial rivals, the Ottomans and the Russians, took advantage of the chaos in the country to seize territory for themselves.[17]

Fall of the Hotaki dynasty

Tahmasp and the Qajar leader Fath Ali Khan (the ancestor of Agha Mohammad Khan Qajar) contacted Nader and asked him to join their cause and drive the Ghilzai Afghans out of Khorasan. He agreed and thus became a figure of national importance. When Nader discovered that Fath Ali Khan was in treacherous correspondence with Malek Mahmud and revealed this to the shah, Tahmasp executed him and made Nader the chief of his army instead. Nader subsequently took on the title Tahmasp Qoli (Servant of Tahmasp). In late 1726, Nader recaptured Mashhad.[18]

Tomb of Nader Shah, a tourist attraction in Mashhad

Nader chose not to march directly on Isfahan. First, in May 1729, he defeated the Abdali Afghans near Herat. Many of the Abdali Afghans subsequently joined his army. The new shah of the Ghilzai Afghans, Ashraf, decided to move against Nader but in September 1729, Nader defeated him at the Battle of Damghan and again, decisively, in November at Murchakhort. Ashraf fled and Nader finally entered Isfahan, handing it over to Tahmasp in December. The citizens' rejoicing was cut short when Nader plundered them to pay his army. Tahmasp made Nader governor over many eastern provinces, including his native Khorasan, and married him to his sister. Nader pursued and defeated Ashraf, who was murdered by his own followers.[19] In 1738 Nader Shah besieged and destroyed the last Hotaki seat of power at Kandahar. He built a new city near Kandahar, which he named "Naderabad".[1]

Ottoman campaign

Painting of Nader Shah

In the spring of 1730, Nader attacked the Ottomans and regained most of the territory lost during the recent chaos. At the same time, the Abdali Afghans rebelled and besieged Mashhad, forcing Nader to suspend his campaign and save his brother, Ebrahim. It took Nader fourteen months to defeat the Abdali Afghans.

Relations between Nader and the Shah had declined as the latter grew jealous of his general's military successes. While Nader was absent in the east, Tahmasp tried to assert himself by launching a foolhardy campaign to recapture Yerevan. He ended up losing all of Nader's recent gains to the Ottomans, and signed a treaty ceding Georgia and Armenia in exchange for Tabriz. Nader saw that the moment had come to ease Tahmasp from power. He denounced the treaty, seeking popular support for a war against the Ottomans. In Isfahan, Nader got Tahmasp drunk then showed him to the courtiers asking if a man in such a state was fit to rule. In 1732 he forced Tahmasp to abdicate in favor of the Shah's baby son, Abbas III, to whom Nader became regent.

Nader decided he could win back the territory in Armenia and Georgia by seizing Ottoman Baghdad and then offering it in exchange for the lost provinces, but his plan went badly amiss when his army was routed by the Ottoman general Topal Osman Pasha near the city in 1733. Nader decided he needed to regain the initiative as soon as possible to save his position because revolts were already breaking out in Persia. He faced Topal again with a larger force and defeated and killed him. He then besieged Baghdad, as well as Ganja in the northern provinces, earning a Russian alliance against the Ottomans. Nader scored a great victory over a superior Ottoman force at Baghavard and by the summer of 1735, Persian Armenia and Georgia were his again. In March 1735, he signed a treaty with the Russians in Ganja by which the latter agreed to withdraw all of their troops from Persian territory.[20][21]

Nader becomes shah

In January 1736, Nader held a *qoroltai* (a grand meeting in the tradition of Genghis Khan and Timur) on the Moghan plain in Azerbaijan. The leading political and religious figures attended. Nader suggested he should be proclaimed the new shah in place of the young Abbas III. Everyone agreed, many—if not most—enthusiastically, the rest fearing Nader's anger if they showed support for the deposed Safavids. Nader was crowned Shah of Iran on March 8, 1736, a date his astrologers had chosen as being especially propitious.[22]

Religious policy

Nader Shah and two of his sons on the reverse side of Nader Shah's Sword

The Safavids had introduced Shi'a Islam as the state religion of Iran. Nader was probably brought up as a Shi'a [23] but later espoused the Sunni[24] faith as he gained power and began to push into the Ottoman and Mughal Empires. He believed that Safavid Shi'ism had intensified the conflict with the Sunni Ottoman Empire. His army was a mix of Shi'a and Sunni and included his own Qizilbash as well as Uzbeks, Afghans and others. He wanted Persia to adopt a form of religion that would be more acceptable to Sunnis and suggested that Persia adopt a form of Shi'ism he called "Ja'fari", in honour of the sixth Shi'a imam Ja'far al-Sadiq. He banned certain Shi'a practices which were particularly offensive to Sunnis, such as the cursing of the first three caliphs. Personally, Nader is said to have been indifferent towards religion and the French Jesuit who served as his personal physician reported that it was difficult to know which religion he followed and that many who knew him best said that he had none.[25] Nader hoped that "Ja'farism" would be accepted as a fifth school (*mazhab*) of Sunni Islam and that the Ottomans would allow its adherents to go on the hajj, or pilgrimage, to Mecca, which was within their territory. In the subsequent peace negotiations, the Ottomans refused to acknowledge Ja'farism as a fifth *mazhab* but they did allow Persian pilgrims to go on the *hajj*. Nader was interested in gaining rights for Persians to go on the *hajj* in part because of revenues from the pilgrimage trade.[1] Nader's other primary aim in his religious reforms was to weaken the Safavids further since Shi'a Islam had always been a major element in support for the dynasty. He had the chief mullah of Persia strangled after he was heard expressing support for the Safavids. Among his reforms was the introduction of what came to be known as the *kolah-e Naderi*. This was a hat with four peaks which symbolised the first four caliphs.[1] [22]

Invasion of India

Afsharid forces negotiate with a Mughal Nawab.

In 1738, Nader Shah conquered Kandahar, the last outpost of the Hotaki dynasty. His thoughts now turned to the Mughul Empire of India. This once powerful Muslim state was falling apart as the nobles became increasingly disobedient and the Hindu Marathas made inroads on its territory from the south-west. Its ruler Muhammad Shah was powerless to reverse this disintegration. Nader asked for Afghan rebels to be handed over, but the Mughal emperor refused. Nader used the pretext of his Afghan enemies taking refuge in India to cross the border and capture Ghazni, Kabul, Peshawar, Sindh and Lahore. He then advanced deeper into India crossing the river Indus before the end of year.

He defeated the Mughal army at the huge Battle of Karnal on 13 February 1739. After this victory, Nader captured Mohammad Shah and entered with him into Delhi.[10] When a rumour broke out that Nader had been assassinated, some of the Indians attacked and killed Persian troops. Nader reacted by ordering his soldiers to plunder the city. During the course of one day (March 22) 20,000 to 30,000 Indians were killed by the Persian troops, forcing Mohammad Shah to beg for mercy.[26]

"Battle of Karnal" painting by adel adili

In response, Nader Shah agreed to withdraw, but Mohammad Shah paid the consequence in handing over the keys of his royal treasury, and losing even the Peacock Throne to the Persian emperor. The Peacock Throne thereafter served as a symbol of Persian imperial might. Among a trove of other fabulous jewels, Nader also gained the Koh-i-Noor and Darya-ye Noor diamonds (Koh-i-Noor means "Mountain of Light" in Persian, Darya-ye Noor means "Sea of Light"). The Persian troops left Delhi at the beginning of May 1739. Nader's soldiers also took with them thousands of elephants, horses and camels, loaded with the booty they had collected. The plunder seized from India was so rich that Nader stopped taxation in Iran for a period of three years following his return.[27]

After India

The Indian campaign was the zenith of Nader's career. Afterwards he became increasingly despotic as his health declined markedly. Nader had left his son Reza Qoli Mirza to rule Persia in his absence. Reza had behaved highhandedly and somewhat cruelly but he had kept the peace in Persia. Having heard rumours that his father had died, he had made preparations for assuming the crown. These included the murder of the former shah Tahmasp and his family, including the nine-year old Abbas III. On hearing the news, Reza's wife, who was Tahmasp's sister, committed suicide. Nader was not impressed with his son's waywardness and reprimanded him, but he took him on his expedition to conquer territory in Transoxiana. In 1740 he conquered Khanate of Khiva. After the Persians had forced the Uzbek khanate of Bokhara to submit, Nader wanted Reza to marry the khan's elder daughter because she was a descendant of his hero Genghis Khan, but Reza flatly refused and Nader married the girl himself. Nader also conquered Khwarezm on this expedition into Central Asia.

Nader now decided to punish Daghestan for the death of his brother Ebrahim Qoli on a campaign a few years earlier. In 1741, while Nader was passing through the forest of Mazanderan on his way to fight the Daghestanis, an assassin took a shot at him but Nader was only lightly wounded. He began to suspect his son was behind the attempt and confined him to Tehran. Nader's increasing ill health made his temper ever worse. Perhaps it was his illness that made Nader lose the initiative in his war against the Lezgin tribes of Daghestan. Frustratingly for him, they resorted to guerrilla warfare and the Persians could make little headway against them. Nader accused his son of being behind the assassination attempt in Mazanderan. Reza angrily protested his innocence, but Nader had him blinded as punishment, although he immediately regretted it. Soon afterwards, Nader started executing the nobles who had witnessed his son's blinding. In his last years, Nader became increasingly paranoid, ordering the assassination of large numbers of suspected enemies.

With the wealth he gained, Nader started to build a Persian navy. With lumber from Mazandaran, he built ships in Bushehr. He also purchased thirty ships in India.[1] He recaptured the island of Bahrain from the Arabs. In 1743 he conquered Oman and its main capital the city of Muscat. In 1743 Nader started another war against the Ottoman Empire. Despite having a huge army at his disposal, in this campaign Nader showed little of his former military brilliance. It ended in 1746 with the signing of a peace treaty, in which the Ottomans agreed to let Nader occupy Najaf.[28]

Domestic policies

Nader changed the Iranian coinage system. He minted silver coins, called *Naderi*, that were equal to the Mughal rupee.[1] Nader discontinued the policy of paying soldiers based on land tenure.[1] Like the late Safavids he resettled tribes. Nader Shah transformed the *Shahsevan*, a nomadic group living around Azerbaijan whose name literally means "shah lover", into a tribal confederacy which defended Iran against the Ottomans and Russians.[29] [30] In addition, he increased the number of soldiers under his command and reduced the number of soldiers under tribal and provincial control.[1] His reforms may have strengthened the country, but they did little to improve Iran's suffering economy.[1]

Death and legacy

A Western view of Nader in his later years from a book by Jonas Hanway (1753). The background shows a tower of skulls.[31]

Nader became crueller and crueller as a result of his illness and his desire to extort more and more tax money to pay for his military campaigns. More and more revolts broke out and Nader crushed them ruthlessly, building towers from his victims' skulls in imitation of his hero Timur. In 1747, Nader set off for Khorasan where he intended to punish Kurdish rebels. Some of his officers feared he was about to execute them and plotted against him. Nader Shah was assassinated on 19 June 1747, at Fathabad in Khorasan. He was surprised in his sleep by Salah Bey, captain of the guards, and stabbed with a sword. Nader was able to kill two of the assassins before he died.[32] [33]

After his death, he was succeeded by his nephew Ali Qoli, who renamed himself Adil Shah ("righteous king"). Adil Shah was probably involved in the assassination plot.[20] Adil Shah was deposed within a year. During the struggle between Adil Shah, his brother Ibrahim Khan and Nader's grandson Shah Rukh almost all provincial governors declared independence, established their own states, and the entire Empire of Nader Shah fell into anarchy. Finally, Karim Khan founded the Zand dynasty and became ruler of Iran by 1760, while Ahmad Shah Durrani had already proclaimed independence in the east, marking the foundation of modern Afghanistan.

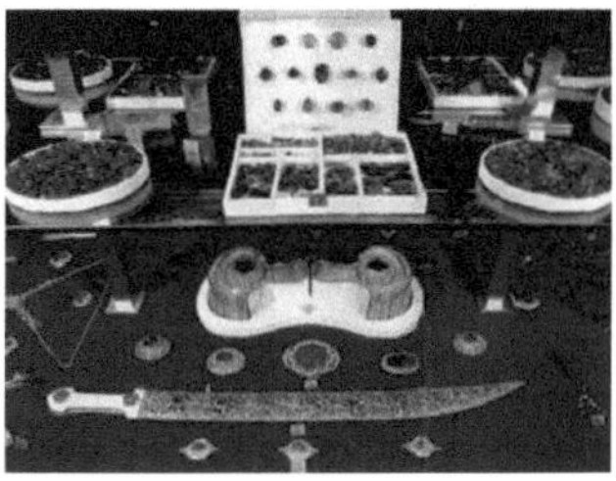

Nader Shah's dagger with a small portion of his jewelry. Now part of the Iranian Crown Jewels.

Nader Shah was well known to the European public of the time. In 1768, Christian VII of Denmark commissioned Sir William Jones to translate a Persian language biography of Nader Shah written by his Minister Mirza Mehdi Khan Astarabadi into French.[34] It was published in 1770 as *Histoire de Nadir Chah*.[35] Nader's Indian campaign alerted the British East India Company to the extreme weakness of the Mughal Empire and the possibility of expanding to fill the power vacuum. Without Nader, "eventual British [in India] would have come later and in a different form, perhaps never at all - with important global effects".[36]

The descendants of Nader Shah through his grandson Shah Rukh (Shahrokh) are under the Afshar Naderi surname.

See also

- Muslim conquest in the Indian subcontinent
- Safavid conversion of Iran from Sunnism to Shiism
- Treaty of Kerden
- Talpur

Nader Shah's tomb was designed by Hooshang Seyhoon.

References

[1] Ernest Tucker (March 29, 2006). "Nāder Shāh 1736-47" (http://www.iranicaonline.org/articles/nader-shah). *Encyclopædia Iranica*. .

[2] Nader's exact date of birth is unknown but August 6 is the "likeliest" according to Axworthy p.17 (and note) and *The Cambridge History of Iran* (Vol. 7 p.3); other biographers favour 1688.

[3] Axworthy p. xvii

[4] Biography of Nadir Shah Afshar "The Persian Napoleon" (1688-1747) (http://www.xs4all.nl/~kvenjb/madmonarchs/nadir/nadir_bio.htm)

[5] Michael Axworthy's biography of Nader, *The Sword of Persia* (I.B. Tauris, 2006), pp. 17-19: "His father was of lowly but respectable status, a herdsman of the Afshar tribe ... The Qereqlu Afshars to whom Nader's father belonged were a semi-nomadic Turcoman tribe settled in Khorasan in north-eastern Iran ... The tribes of Khorasan were for the most part ethnically distinct from the Persian-speaking population, speaking Turkic or Kurdish languages. Nader's mother tongue was a dialect of the language group spoken by the Turkic tribes of Iran and Central Asia, and he would have quickly learned Persian, the language of high culture and the cities as he grew older. But the Turkic language was always his preferred everyday speech, unless he was dealing with someone who knew only Persian."

[6] Stephen Erdely and Valentin A. Riasanovski. *The Uralic and Altaic Series*, Routledge, 1997, ISBN 0-7007-0380-2, p. 102

[7] *Cambridge History of Iran* Vol.7, p.59

[8] Vali Nasr, "The Shia Revival: How Conflicts within Islam Will Shape the Future" (New York 2006)

[9] "Nadir Shah" (http://www.britannica.com/EBchecked/topic/401451/Nadir-Shah). Encyclopædia Britannica Online. . Retrieved 2010-09-24.

[10] "AN OUTLINE OF THE HISTORY OF PERSIA DURING THE LAST TWO CENTURIES (A.D. 1722-1922)" (http://persian.packhum.org/persian/pf?file=90001014&ct=33). *Edward G. Browne*. London: Packard Humanities Institute. p. 33. . Retrieved 2010-09-24.

[11] Durand, Sir Henry Mortimer (1908). *Nadir Shah* (http://books.google.com.au/books?id=-qMsAAAAYAAJ&source=gbs_navlinks_s). A. Constable and co. ltd. pp. 352. . Retrieved 2010-09-24.

[12] http://histories.cambridge.org/extract?id=chol9780521200950_CHOL9780521200950A002

[13] Axworthy p.17-18

[14] Axworthy p.17

[15] Axworthy p.20

[16] "AN OUTLINE OF THE HISTORY OF PERSIA DURING THE LAST TWO CENTURIES (A.D. 1722-1922)" (http://persian.packhum.org/persian/pf?file=90001014&ct=30). *Edward G. Browne*. London: Packard Humanities Institute. p. 30. . Retrieved 2010-09-24.

[17] This section: Axworthy pp.17-56

[18] Axworthy pp.57-74

[19] Axworthy pp.75-116

[20] Elton L. Daniel, "The History of Iran" (Greenwood Press 2000) p.94

[21] Lawrence Lockhart *Nadir Shah* (London, 1938)

[22] This section: Axworthy pp.137-174

[23] Axworthy p.34

[24] Mattair, Thomas R. (2008). *Global security watch--Iran: a reference handbook* (http://books.google.com/books?id=Qtmp_2GjVA8C&lpg=PP1&pg=PA3#v=onepage&q&f=false). ABC-CLIO. p. 3. ISBN 0-275-99483-X, 9780275994839. . Retrieved 2010-09-24.

[25] Axworthy p.168

[26] Axworthy p.8

[27] This section: Axworthy pp.1-16, 175-210

[28] This section: Axworthy pp.175-274

[29] Floor, Willem. Journal of the American Oriental Society, Vol. 119, No. 3. (Jul. - Sep., 1999), pp. 543. Book review of Richard Tapper's *Frontier Nomads of Iran: A Political and Social History of the Shahsevan.*

[30] Daniel, Elton L. *The History of Iran.* Greenwood Publishing Group: 2000. p. 90.

[31] Axworthy p,273

[32] History of Iran: Afsharid Dynasty (Nader Shah) (http://www.iranchamber.com/history/afsharids/afsharids.php)

[33] Axworthy pp.243-286

[34] Sir William Jones (174... - Online Information article about Sir William jones (174 (http://encyclopedia.jrank.org/JEE_JUN/JONES_SIR_WILLIAM_1746_1794_.html)

[35] Axworthy p.330

[36] Axworthy p.xvi

Sources

- Michael Axworthy, *The Sword of Persia: Nader Shah, from Tribal Warrior to Conquering Tyrant* Hardcover 348 pages (26 July 2006) Publisher: I.B. Tauris Language: English ISBN 1-85043-706-8
- *The Cambridge History of Iran*, Volume 7 (Cambridge University Press, 1968)

Additional reading

- Lawrence Lockhart *Nadir Shah* (London, 1938)
- Ernest Tucker, *Nadir Shah's Quest for Legitimacy in Post-Safavid Iran* Hardcover 150 pages (4 October 2006) Publisher: University Press of Florida Language: English ISBN 0-8130-2964-3
- Michael Axworthy, *Iran: Empire of the Mind: A History from Zoroaster to the Present Day* (Paperback) ISBN 0-14-103629-X Publisher Penguin 6 November 2008

External links

- Nader Shah's portrait (http://www.qajarpages.org/nadershah.html)
- Nader Shah Mausoleum and Museum (http://www.farsinet.com/mashhad/nader_shah.html)

Peace of Amasya

The **Peace of Amasya** (Persian:پیمان آماسیه) was a treaty agreed to on May 29, 1555 between Shah Tahmasp of Safavid Iran and Sultan Suleiman the Magnificent of the Ottoman Empire at the city of Amasya, following the Ottoman–Safavid War of 1532–1555.

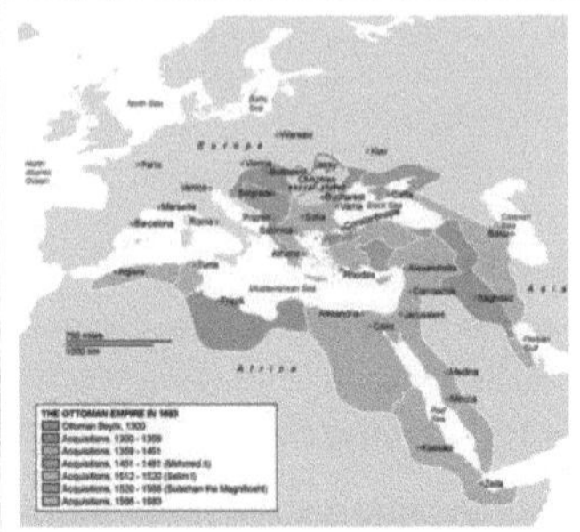

Suleiman's conquests in the Ottoman-Safavid War (1532–1555) gave him access to the Persian Gulf.

The treaty defined the border between Iran and the Ottoman empire and was followed by twenty years of peace. By this treaty Armenia and Georgia were divided equally between the two, the Ottoman Empire obtained most of Iraq, including Baghdad, which gave them access to the Persian Gulf, while the Persians retained their former capital Tabriz and the north-western borders as they were prior to the wars. The frontier thus established ran across the mountains dividing eastern and western Georgia (under native vassal princes), through Armenia, and via the western slopes of the Zagros down to the Persian Gulf. The Ottomans, further, gave permission for Persian pilgrims to go to the holy places of Mecca and Medina as well as to the Shia sites of pilgrimages in Iraq.[1]

References

[1] Shaw, Stanford J. (1976), *History of the Ottoman Empire and modern Turkey*, Volume 1, p. 109. Cambridge University Press, ISBN 0-521-29163-1

Fat'h-Ali Shah Qajar

Fat'h-Ali Shah Qajar	
Shahanshah of Persia	

Reign	17 June 1797 – 23 October 1834
Full name	Fat'h Ali Shah
Born	5 September 1772
Birthplace	Damghan, Iran
Died	23 October 1834 (aged 62)
Place of death	Isfahan, Iran
Buried	Qom
Predecessor	Mohammad Khan Qajar
Successor	Mohammad Shah Qajar
Dynasty	Qajar dynasty
Father	Hussein Qoli Khan Qajar
Religious beliefs	Shia Islam

Fat'h Ali Shah Qajar (Persian: فتح علی شاه قاجار; var. Fathalishah, Fathali Shah, Fath Ali Shah; 5 September 1772 – 23 October 1834) was the second Qajar Emperor/Shah of Persia. He reigned from 17 June 1797 until his death.

Early life

Fat'h Ali Shah was the son of Hossein Qoli Khan Qajar, brother of Agha Mohammad Khan. He was governor of Fars when his uncle was assassinated in 1797. Fat'h Ali shah's real name was "Bābā Khān" but he was crowned as Fat'h Ali Shah. He became suspicious of his chancellor Hajji Ebrahim Khan Kalantar and ordered his execution. Hajji Ebrahim Khan had been chancellor to Zand and Qajar rulers for some fifteen years.

Much of his reign was marked by the resurgence of Persian arts and painting, as well as a deeply elaborate court culture with extremely rigid etiquette. In particular during his reign, portraiture and large-scale oil painting reached a height previously unknown under any other Islamic dynasty, largely due to his personal patronage.

Fat'h Ali also ordered the creation of much royal regalia, including coronations chairs, "Takht-e-Tâvoos" (Persian: تخت طاووس) or Peacock throne and "Takht-e-Nāderī" (Persian: تخت نادری) or Naderi throne, which was also used by later kings, and the "Tāj-i-Kīyānī" (Persian: تاج کیانی), or Kiani Crown, a modification of the crown of the same name created by his uncle Agha Mohammad Khan. This, like most of his regalia, was studded with a large number of pearls and gems.

In 1797, he was given a complete set of the *Britannica's* 3rd edition, which he read completely; after this feat, he extended his royal title to include "Most Formidable Lord and Master of the *Encyclopædia Britannica*."[1]

Russo-Persia wars

Russo-Persian War (1804–1813)

During the early reign of Fat'h Ali Shah, Imperial Russia took control of Georgia claimed by the Persians. The war broke between Persia and Russia when Fat'h Ali Shah ordered the invasion of Georgia in 1804, under pressure from the Shia clergy, who were urging a war against Russia. The war began with notable victories for the Persians, but Russia shipped in advanced weaponry and cannons that disadvantaged the technologically inferior Qajar forces, who did not have artillery to match. Russia continued with a major campaign against Persia; Persia asked for help from Britain on the grounds of a military agreement with that country (the military agreement was signed after the rise of Napoleon in France). However, Britain refused to help Persia claiming that the military agreement concerned a French attack not Russian.

Portraet

General Gardane, with colleagues Jaubert and Joanin, at the Persian court of Fat'h Ali Shah in 1808.

Persia had to ask for help from France, sending an ambassador to Napoleon and concluding a Franco-Persian alliance with the signature of the Treaty of Finkenstein. However, just when the French were ready to help Persia, Napoleon made peace with Russia. At this time, John Malcolm arrived in Persia and promised support but Britain later changed its mind and asked Persia to retreat. Russian troops invaded Tabriz in 1813 and Persia was forced to sign the Treaty of Gulistan with Russia.

Treaty of Gulistan

On account of consecutive defeats of Persia and after the fall of Lankaran on 1 January 1813, Shah was forced to sign the Treaty of Gulistan. The text of treaty was prepared by a British diplomat; Sir Gore Ouseley; and was signed by Nikolai Fyodorovich Rtischev from the Russian side"[2] and Hajji Mirza Abol Hasan Khan from the Iranian side on 24 October 1813 in the village of Gulistan.[2]

By this treaty all of the cities, towns, and villages of Georgia, villages and towns on the coast of the Black Sea, all of the cities, towns and villages of the Khanates in South Caucasus and part of the Talysh Khanate, including Megrelia, Abkhazia, Imeretia, Guria, Baku khanate, Shirvan Khanate, Derbent, Karabakh khanate, Ganja khanate, Shaki Khanate and Quba Khanate became part of Russia.[3] In return Russia pledged to support Abbas Mirza as heir to the Persian throne after the death of Fat'h Ali Shah.

Wars with the Afghans

In 1816 Qajar forces conquered the city of Herat which had been ruled by the Afghans since the death of Nader Shah in 1747. Fierce Afghan guerrilla attacks forced the Persians to abandon the city shortly after. Two years later the Shah sent his son Mohammad Vali Mirza to recapture the city but was defeated at the Battle of Kafir Qala.

Russo-Persian War (1826–1828)

In 1826, 13 years after the Treaty of Gulistan Shah on the advice of British agents, decided to occupy the lost territories. Abbas Mirza invaded the Talysh Khanate and Karabakh khanate with an army of 35000 on 16 July 1828. The Khans quickly switched sides and surrendered their principal cities, Lenkoran, Quba and Baku to the Persians. In May 1827, Ivan Paskevich, Governor of Caucasus, invaded Echmiadzin, Nakhichevan, Abbasabad and on 1 October Erivan. Fourteen days later, General Eristov entered Tabriz. On January 1828 when Russians reached the shores of Lake Urmia, Abbas Mirza urgently signed the Treaty of Turkmenchay on 2 February 1828.

Persia in 1808 according to a British map, before defeats to Russia in the north on 1813 and the Treaty of Gulistan, and the loss of Herat to Great Britain in 1857 after the Treaty of Paris

Treaty of Turkmenchay

Turkmenchay Treaty was signed on 21 February 1828 by Hajji Mirza Abol Hasan Khan and General Ivan Paskievich . By this treaty Erivan khanate, Nakhchivan khanate, Talysh Khanate, Ordubad and Mughan became under the rule of Imperial Russia.[4] Iran pledged to pay Russia 10 Million in Gold and in return Russia pledged to support Abbas Mirza as heir to the Persian throne after the death of Fat'h Ali Shah . The treaty also stipulated the resettlement of

Armenians from Persia to the Caucasus, which also included an outright liberation of Armenian captives who were brought and had lived in Iran since 1804 or as far back as 1795.

Later life

Fat'h Ali later employed writers and painters to make a book about his wars with Russia, inspired by the Shahnameh of Ferdowsi. This book, considered by many to be the most important Persian book written in the Qajar period, is called the Shahanshahnama.

In 1829, Alexandr Griboyedov, the Russian diplomat and playwright was killed in the encirclement of the Russia embassy in Tehran. To apologize, the Shah sent Tsar Nicholas I one of the biggest diamonds of his crown Jewelry, Shah Diamond.

Fat'h Ali Shah Qajar *firman* in Shikasta Nastaʿlīq script, January 1831.

When his beloved son and crown prince Abbas Mirza died on 25 October 1833, Fat'h Ali named his grandson Mohammed Mirza as his crown prince. Fat'h Ali died a year later, on 23 October 1834.

He is instantly recognizable in all 25 known portraits – mainly due to his immense, deeply black beard, which reached well beneath his narrow waist.

Besides eulogistic chronicles, the only real sources that allow us to judge his personality are those of British, French and Russian diplomats. These vary greatly: earlier in his reign they tend to portray him as vigorous, manly and highly intelligent. Later they begin to point out his extreme indolence and avarice. Some Victorian sources refer to his baths in the Nigaristan palace which featured a *sursura* or slide, an inclined marble plane down which "naked harem beauties swooped [...] into the arms of their lord and master before being playfully dunked in a pool."[5] [6]

Offspring

Fath Ali Shah Qajar with his Kiani crown, and the Royal Mace of Iran at his knees, painted by Mir Ali

His first son, Mohammad Ali Mirza Dowlatsh, was seven months older than his brother Abbas Mirza (Fat'h Ali Shah's Crown Prince), but on account of his mother, Ziba Chehreh Khanoum, non-Qajar origin he was unable to claim the title "Valiahd." (Crown prince).

Sons

- Mohammad Ali Mirza (1788–1821)
- Mohammad Qoli Mirza (1788–1874)
- Mohammad Vali Mirza (1789–1869)
- Abbas Mirza (1789–1833)
- Kiomarth Mirza "Il-Khani"
- Hossein Ali Mirza (1789–1835)
- Hassan Ali Mirza
- Mohammad Taqi Mirza
- Ali Naqi Mirza
- Sheikh Ali Mirza
- Ali Shah Mirza
- Abdollah Mirza
- Imam Verdi Mirza
- Mohammad Reza Mirza
- Mahmud Mirza
- Heydar Qoli Mirza
- Homayoun Mirza
- Allah Verdi Mirza
- Esma'il Mirza
- Ahmad Ali Mirza

See also

- Sarah Shahi
- Tangeh Savashi near Tehran, where Fat'h Ali Shah had a relief carved into the side of a mountain pass.
- Imperial Crown Jewels of Persia
- Naderi throne
- Darya-ye Noor Diamond
- Peacock throne
- Shah Diamond

References

[1] *Banquet at Guildhall in the City of London, Tuesday 15 October 1968: Celebrating the 200th Anniversary of the Encyclopædia Britannica and the 25th Anniversary of the Honorable William Benton as its Chairman and Publisher.* United Kingdom: Encyclopædia Britannica International, Ltd.. 1968.

[2] Treaty of Gulistan (http://www.hrono.ru/dokum/ruper1813.html)

[3] John F. Baddeley, "The Russian Conquest of the Caucasus", Longman, Green and Co., London: 1908, p. 90

[4] Treaty of Turkmenchay (http://www.hist.msu.ru/ER/Etext/FOREIGN/turkman.htm)

[5] John H. Waller, *Beyond the Khyber Pass: the road to British disaster in the First Afghan War*, Random House, 1990, p. 59.

[6] *The Literary World* (http://books.google.co.uk/books?id=VDkZAAAAYAAJ&q=sursura#search_anchor). 1882. p. 85. . Retrieved 1 December 2012. Wording also available here (http://fultonhistory.com/newspaper 2/Auburn NY News And Bulletin/Auburn NY News And Bulletin 1883 pdf/Newspaper Auburn NY News &Bulletin 1883 - 0070.PDF) under "The Shah's Palaces"

External links

- Portrait of Fat'h Ali Shah by *Mihr Ali*, *Qajar Pages* (http://www.qajarpages.org/fathalishah.html)
- Complete list of Fath Ali Shah's Descendants, Manoutchehr M. Eskandari-Qajar (Kadjar), *Qajar Pages* (http://www.qajarpages.org/fathalishahchildren.html)
- Fath Ali Shah's Standing Portrait by *Mihr Ali*, 1809–1810, Hermitage Museum, *Hermitage Museum* (http://www.hermitagemuseum.org/fcgi-bin/db2www/descrPage.mac/descrPage?selLang=English&indexClass=PICTURE_EN&PID=\LVR-1107&numView=1&ID_NUM=3&thumbFile=/tmplobs/C_404Y585AWIJ0OI466.jpg&embViewVer=last&comeFrom=quick&sorting=no&thumbId=6&numResults=5&tmCond=fath+ali&searchIndex=TAGFILEN&author=Mihr Ali)
- Seated Portrait of Fath Ali Shah, by *Mihr Ali*, 1813–1814, presented by the Shah in 1817 to Aleksey Petrovich Yermolov, Hermitage Museum, *Hermitage Museum* (http://www.hermitagemuseum.org/fcgibin/db2www/descrPage.mac/descrPageselLang=English&indexClass=PICTURE_EN&PID=5CLVR1108&numView=1&ID_NUM=1&thumbFile=2Ftmplobs2FPBQ6D2240RDP6UO24F6.jpg&embViewVer=last&comeFrom=quick&sorting=no&thumbId=6&numResults=5&tmCond=fath+ali&searchIndex=TAGFILEN&author=Mihr Ali)
- Genealogy of Fath Ali (Royalark.net) (http://www.royalark.net/Persia/qajar3.htm)
- Fath 'Ali Shah seated on the Peacock Throne, miniature from Shāhanshāhnāmeh of sābā British Library (http://www.imagesonline.bl.uk/results.asp?image=074205&imagex=5&searchnum=2)
- *Divān-i-Khāqān*, Fat'h Ali Shah's own Poems, presented by the Shah to the Prince Regent in 1812, now in Royal Library at Windsor Castle, *royal collection* (http://www.royalcollection.org.uk/eGallery/object.asp?object=1005020&row=0&detail=magnify)
- Louvre (http://cartelen.louvre.fr/cartelen/visite?srv=rs_display_res&critere=fath+ali&operator=AND&nbToDisplay=5&langue=fr)
- Crown (http://www.artres.com/c/htm/CSearchZ.aspx?o=&Total=1&FP=33529218&E=22SIJMFUEGBD&SID=JMGEJNUJP9CE&Pic=1&SubE=2UNTWAGHGCSE)
- Photos of qajar kings (http://www.yadegaran.com)

Treaty of Istanbul (1590)

The **Treaty of Istanbul of 1590** or **Treaty of Constantinople** also known as **Treaty of Ferhat Pasha**[1] (Turkish: *Ferhat Paşa Antlaşması*) was a treaty between Ottoman Empire and Safavid Persia after a long war between 1577 and 1589. (signed on the 21. of May 1590)

The war

The war began in Southern Caucasus. At the Battle of Çıldır, the Ottoman army defeated the Persian army and invaded most of South Caucasus . After Ottoman army returned to its main base in Constantinople, however, Persians began to regain some of their former territories. During the next phase, another Ottoman army (mostly composed of Crimean Tatars) arrived and Ottomans were able to attack once more. In 1583 after the battle named Battle of Torches (Ottoman Turkish:مشللر محاربسی), so called because the battle continued during night, Ottomans were able to reconquer South Caucasus. During the next six years the only important event was the conquest of Tabriz by the Ottomans. Confronted by other problems (i.e., civil wars and the war against Uzbeks), Abbas I, the shah of Persia agreed to sign a treaty with unfavorable terms.[2]

The treaty

The treaty is known as Treaty of Istanbul or treaty of Constantinople. (Ferhat Pasha was the commander (Turkish: *serdar*) of Ottoman army). According to the treaty, Ottoman Empire kept its gains in the war. These included most of Caucasus (present day Georgia, Armenia and Azerbaijan) as well as Tabriz and north-west Iran.[3] Persians also agreed to pay obeisance to religious leaders of Sunni faith.

Aftermath

This treaty was a big success for Ottoman empire, for vast areas had been annexed. However, it was not long lasting. Abbas I after solving problems at home, waited for a suitable moment to regain his former possessions. When Ottoman Empire was engaged in Jelali revolts he was able to regain most of his loses which Ottoman Empire had to accept in the treaty of Nasuh Pasha, 22 years after the treaty of Constantinople.

See also

- Timeline of Turkish history
- Timeline of Iranian history
- Ottoman–Safavid War (1578–1590)

References

[1] Alexander Mikaberidze, *Conflict and Conquest in the Islamic World: A Historical Encyclopedia*, ABC-CLIO, 2011, p. 698. (http://books.google.com/books?id=WjQfo3a1eVMC&pg=PA698&dq="also+known+as+Treaty+of+Ferhat+Pasha,+after+the+commander+of+the+Ottoman+forces"&hl=en&ei=JapZTq3LM8GImQWUst2tDA&sa=X&oi=book_result&ct=result&resnum=1&ved=0CCkQ6AEwAA#v=onepage&q="also known as Treaty of Ferhat Pasha, after the commander of the Ottoman forces"&f=false)

[2] Prof.Yaşar Yücel-Prof Ali Sevim:*Türkiye tarihi III*, AKDTYKTTK Yayınları, 1991, pp 21-23, 43-44

[3] Nicolae Jorga: *Geschiste des Osmanichen vol III*, (trans: Nilüfer Epçeli) Yeditepe Yayınları, 2009, ISBN 975-6480-20-3 ,p.213

Treaty of Nasuh Pasha

Treaty of Nasuh Pasha (Persian: عهدنامه حوصن پاشا, Turkish: *Nasuh Paşa Antlaşması*) was a treaty between the Ottoman Empire and Safavid Persia after the war of 1603–1612, signed on 20 November, 1612.

Background

Shah Abbas the Great of Persia had to concede wast areas in Northwest Iran and Caucasus to the Ottoman Empire by the Treaty of Istanbul in 1590. After solving problems at home Shah Abbas was planning to regain the losses. He waited for a suitable moment to attack. He saw his chance in 1603 when 14 years old sultan Ahmet I ascended to Ottoman throne. Ottoman Empire was engaged in a long and costly war against the Holy Roman Empire, so called Long war (1593–1606) and there were a series of revolts in Anatolia named Jelali revolts.

The war

Shah Abbas and his general Allahverdi Khan in a surprise attack began to regain the territories lost in 1590. Although Ottoman Empire was able to raise an army against Persia, the two commanders (Ottoman language: *serdar*, سردار) Cigalazade Yusuf Sinan Pasha (1605) and Kuyucu Murat Pasha (1611) both died in Diyarbakır, the winter camp of the army (natural cause) and Ottoman army suffered from the lack of able commandership. Finally, Grand vizier Nasuh Pasha (also called Damat Nasuh Pasha) agreed to sign a treaty.[1]

The terms

1. The Ottoman Empire agreed to return all territory gained by the treaty of Istanbul of 1590.
2. The border line became the line drawn in the treaty of Amasya in 1555.
3. In turn Persia agreed to pay an annual tribute of 200 loads of silk.
4. The route of Persian pligrams to Haj was changed (over Syria instead of Iraq).

Aftermath

This treaty is the first treaty in Ottoman history by which the empire agreed to lose territory.[2] (The first treaty in European theater in which the empire agreed to lose treaty would be the Treaty of Karlowitz in 1699.) On the other hand, this treaty was a big success for Abbas. By the treaty he both increased his realm and restored Safavid prestige. However, Abbas refused to pay the tribute. So the war newed in 1615.

References

[1] Prof.Yaşar Yücel-Prof Ali Sevim:*Türkiye tarihi vol.III*, AKDTYKTTK Yayınları, 1991, 43-44
[2] On line education document (in Turkish) (http://www.egitimvadisi.net/forum/index.php?topic=142.0 -)

Treaty of Serav

Treaty of Serav (Persian: عهدنامه سراب, Turkish: *Serav Antlaşması*) was a treaty between Ottoman Empire and Safavid Persia after the war of 1615 - 1618. (signed on the 26th of September, 1618)

Background

By the treaty of Nasuh Pasha in 1612 Ottoman Empire had agreed to turn back Caucasus and Northwest Iran to Safavid Persia. Safavid Empire on the other hand agreed to pay an annual tribute of 200 loads of silk as a part of reparations . However, Shah Abbas I (so called "the Great") of Persia refused to pay the tribute. The war renewed in 1615.

The war

The Ottoman commander in chief Grand Vizier Öküz Kara Mehmed Pasha tried to capture Yerevan (modern Armenia) which was recently abandoned by the treaty of Nasuh Pasha. But he lifted the siege after 44 days. The target of the next commander in chief Damat Halil Pasha was Ardabil. This time Abbas sued for peace.[1]

The terms

The terms of the treaty was similar to those of treaty of Nasuh Pasha with several minor rectifications of the border line. Also, the annual tribute of the Persian side was reduced from 200 loads to 100 loads.[2]

Aftermath

This treaty proved that a stalemate between Ottoman empire and Safavid Persia had been reached and neither side might gain substantial territories in the long run. In the following decades there were times when Ottomans succeeded to storm Tabriz and there were times when the Persians successfully captured Baghdad. But these victories were all temporary and the balance of power between the two states continued up to 20th century.

See also

- Treaty of Zuhab
- Treaty of Kerden

References

[1] Academic essay (Turkish)[[Category:Articles with Turkish language external links (http://host.nigde.edu.tr/remzikilic/makale/index.php?entry=entry091112-135422)]]
[2] Prof.Yaşar Yücel-Prof Ali Sevim:*Türkiye tarihi Vol. III*, AKDTYKTTK Yayınları, 1991, 44-45

Treaty of Zuhab

The **Treaty of Zuhab** (Persian: قصرشیرین, also called **Treaty of Qasr-e Shirin** (**Treaty of Kasr-ı Şirin**)) was an accord signed between Safavid Empire and the Ottoman Empire on May 17, 1639.[1] The accord ended the war that had begun in 1623 and was the last conflict in almost 150 years of intermittent wars between the two states over territorial disputes. The treaty divided territories in the Middle East by granting Yerevan in the southern Caucasus to Iran and all of Mesopotamia (including Baghdad) to the Ottomans. Nevertheless, border disputes between Persia and the Ottoman Empire did not end. Between 1555 and 1918, Persia and the Ottomans signed no less than 18 treaties that would re-address their disputed borders. The exact demarcation according to this treaty would not begin until the 19th century, essentially laying out the rough outline for the frontier between modern day Iran and the states of Turkey and Iraq (the former Ottoman-Persian border until 1918, when the Ottoman empire lost its territories in the Middle East following their defeat in World War I.)

Notes

[1] Somel, Selçuk Akşin, *Historical Dictionary of the Ottoman Empire*, (Scarecrow Press Inc., 2003), 306.

References

- Somel, Selçuk Akşin, *Historical Dictionary of the Ottoman Empire*, Scarecrow Press Inc., 2003.

See also

- Safavid dynasty
- Ottoman Empire
- History of Iran
- Iran–Iraq War
- List of treaties
- Treaty of Akhal

Treaty of Ahmet Pasha

The **Treaty of Ahmet Pasha** (Persian: , Turkish: *Ahmet Paşa Antlaşması*) was a treaty signed on 10 January 1732 between the Ottoman Empire and the Safavid Persia.

Background

In the 17th century, a stalemate between the Ottoman and Safavid empires had been reached by the treaties of Serav and Zuhab. However, during the short rule of Afghanistan based Hotaki dynasty, chaos in Iran resulted in clashes along frontiers, especially in Caucasus. Meanwhile, Peter I of Russia began to occupy North Caucasus and Azerbaijan. Fearing a Russian controlled Caucasus, the Ottomans decided to capture Tblisi to balance the Russian advance. But this operation resulted in a long Ottoman Safavid war.

War

Between 1723 and 1730, the Ottomans were able to control South Caucasus by capturing Yerevan and Gence in addition to Tblisi. In the southern fronts (i.e., Western Iran), Ottomans captured Tabriz, Urmia, Kermanshah and Hamedan. But after Tahmasp II of Safavids began controlling Iran, the Ottoman advance was checked. Tired of war, both sides decided to end the war. Ahmet Pasha (Ottomon) and Mehmet Rıza Kulu (Persian) signed the treaty.[1] [2]

Terms of treaty

The terms of the treaty were:

1. Ottoman Empire kept its gains in Caucasus,[3]
2. Ottoman gains in West Iran were conceded to Persia, and
3. Aras River became the new border line in South Caucasus.

Aftermath

The treaty proved to be an armistice rather than a permanent treaty. Because, neither Ottoman sultan Mahmut I approved the loss of Tabriz nor Nader Shah, then the commander in chief of the Persian army, the losses in Caucasus. During Nader Shah's reign, Afsharid Persia was able to regain its losses.

See also

- Treaty of Kerden
- Treaty of Constantinople (1724)

References

[1] Prof.Yaşar Yücel-Prof Ali Sevim:*Türkiye tarihi vol.III*, AKDTYKTTK Yayınları, 1991, pp. 282-290
[2] Prof.Yaşar Yücel-Prof Ali Sevim:*Türkiye tarihi vol.IV*, AKDTYKTTK Yayınları, 1991, pp. 3-4
[3] On line document (in Turkish) (http://www.egitimvadisi.net/forum/index.php/topic,66.0/prev_next,next.html#new)

Treaty of Istanbul (1736)

Treaty of İstanbul was a treaty between Ottoman Empire and Safavid Persia signed on 24 September 1736.

Background

Treaty of Ahmet Pasha (1732) between the two countries had been found unsatisfactory in both countries. Ottoman Sultan Mahmut I didn't approve the abandonment of Tabriz and Nader, the future Shah of Iran dethroned Tahmasp II for accepting Ottoman control of the Caucasus.

War

Shortly after the treaty of Ahmet Pasha, Nader declared war and attacked Iraq and Caucasus. In Iraq, after some initial victories such as capturing Kirkuk, he was repelled. But in Caucasus he was more successful. In less than two years Ottomans had to abandon both Tblisi and Yerevan.[1] Meanwhile Russian Empire was about to attack Ottoman Empire in Crimea and Ukraina [2] (see Russo-Turkish War, 1735-1739).[3] . So the porte was compelled to sign a peace treaty.

Terms of the treaty

Initial discussions between **Ali Pasha** (Ottoman side) and **Mirza Muhammed** (Persian side) in Iran about the territorial changes presented no big problem. But the competition between the two sects of Islam resulted heated discussions in the following sessions in İstanbul.[4] Finally the terms were;

1. Ottoman Empire recognized Nader as the new shah of Persia.[5]
2. Ottoman Empire conceded Caucasus to Persia
3. The Ottomans also agreed to allow the Iranian hajis (pligrims) to Mecca (then under Ottoman control)

Aftermath

By the treaty of Ahmet Pasha Persia had gained West Iran and by the treaty of İstanbul Persia had further gained Caucasus. But Nader had other plans. He was also planning to annex Iraq and possibly Eastern Anatolia. That's why he renewed war in 1743. But this time, he was not successful. (see Treaty of Kerden)

References and notes

[1] Prof. Yaşar Yüce-Prof. Ali Sevim: *Türkiye tarihi Cilt IV*, AKDTYKTTK Yayınları, İstanbul, 1991, p 4-8

[2] Nicolae Jorga: *Geschichte des Osmanischen* Reiches IV, (trans by Nilüfer Epçeli),İstanbul 2009, ISBN 975-6480-21-1 p.345

[3] Moreover Austrian Empire, the ally of Russia was also preparing for the war. see Austro-Turkish War, 1737-1739,

[4] The dominant sect in Ottoman Empire was Sunni and in Persia Shia

[5] (http://www.dataci.net/istanbul-antlasmasi-1736.html)

Treaty of Kerden

Treaty of Kerden (Turkish: *Kerden Antlaşması*, Persian: (نادرگ هماندهع) was signed between Ottoman Empire and Afsharid Iran on the 4th of September, 1746.

Background

During the last years of Safavid dynasty in Iran, Ottomans were able to annex most of Caucasus and west Iran. Meanwhile, Afghans were able to annex a part of Khorasan. The shah had to appoint Nadir, a Afshar Turkmen[1] warlord, as his commander in chief. Under Nadir's commandship, Iran was able to regain most of her losses. After the victories, it was an easy matter for Nadir to seize the throne. In 1736, Nadir Shah founded the Afsharid dynasty[2] (to be continued up to 1796.) Nadir Shah was planning to found an empire. After reconquering former territories of Iran, he further tried to annex the eastern territories of the Ottoman empire (East Anatolia and Iraq). But his operations in Ottoman lands (1743–1746) were mostly fruitless and he decided to end hostilities in Ottoman lands. He also proposed to reconcile the two sects (mazhab) of Islam. (Ottomans were of Sunni faith and most of Iranians were of Shia faith.) He planned to force Ottomans, then the champion of Sunnis, to accept Shia as a fifth legal sect of Islam.[3]

The terms of the treaty

The treaty was signed in Kerden (a location[4] near Qazwin,Iran). The representatives were Hasan Ali Haji (Safavid side) and Mustafa Nazif (Ottoman side) [5]

1. The boundary line between the two countries was the same boundary line drawn roughly a century ago by the treaty of Zuhab of 1639. (i.e., modern Turkey-Iran and Iraq-Iran border lines.)
2. The Ottomans agreed to legitimise the Afsaharid dynasty as the rulers of Iran.[6]
3. The Ottomans also agreed to allow the Iranian hajis (pilgrims) to Mecca (then under Ottoman control)
4. Exchange of consulates (Turkish: *şehbender*) were permitted in both countries.
5. Both sides agreed to liberate the prisoners of war.
6. Iran abandoned to force the Ottomans to decleare Shia as the fifth legal sect of Islam

References and notes

[1] Encyclopaedia Britannica (http://www.history.co.uk/encyclopedia/ndir-shh.html)

[2] Afshar is a name of a Turkmen tribe

[3] Nicolae Jorga: *Geschiste des Osmanichen vol IV*, (trans: Nilüfer Epçeli) Yeditepe Yayınları, 2009, ISBN 975-6840-19-X , p. 371

[4] The military camp of Shah Nadir

[5] Prof. Yaşar Yüce-Prof. Ali Sevim: *Türkiye tarihi Cilt IV*, AKDTYKTTK Yayınları, İstanbul, 1991 p 26

[6] During the war Ottomans were backing Safavid prince Safi Mirza as the legal shah of Iran.

Treaty of Erzurum

Treaty of Erzurum refers to either of two treaties, in 1823 and 1847, that settled boundary disputes between the Ottoman Empire and Iran.

First Treaty

Although the Treaty of Zuhab in 1639 had established the boundary between Ottoman Empire and Iran, the border in the mountainous Zuhab region remained a site of intermittent conflict in the subsequent two centuries. Ottoman incursions in north west Iran in pursuit of rebel tribesmen made Persians to retaliated. this attack prompted Sultan Mahmud II in 1821 to declare war on Iran. Fath Ali Shah Qajar's army barely managed to defeat the superior Ottoman army. The first Treaty of Erzurum was signed in July 1823, but it essentially confirmed the 1639 border and thus failed to resolve the disputes that had led to conflict.

Second Treaty

A series of border incidents in the 1830s again brought Iran and the Ottoman Empire to the brink of war. Britain and Russia offered to mediate, and a second Treaty of Erzurum was signed on May 31, 1847.[1] This treaty divided the disputed region between Iran and the Ottoman Empire and provided for a boundary commission to delimit the entire border. The boundary commission's work encountered several political setbacks but finally completed its task in 1914.

References

[1] Victor Prescott and Gillian D. Triggs, *International Frontiers and Boundaries: Law, Politics and Geography* (Martinus Nijhoff Publishers, 2008: ISBN 90-04-16785-4), p. 6.

Bibliography

- Lambton, Ann K. S. "The Qajar Dynasty." In *Qājār Persia: Eleven Studies*, edited by Ann K. S. Lambton. Austin: University of Texas Press, 1987.

External links

Persian text of Erzurum treaty (http://ariarman.org/Amirkabir-Diplomatic.htm)

Article Sources and Contributors

knowledge, Korosh-kabir, Korsmeyer, Kotniski, Kourosh ziabari, Kouroush, Kowey, Kozuch, Kraf001, KrakatoaKatie, Kralizec!, Krawiec, Krizpo, Krstajic, Krukowski, Ksanexx, Kubigula, Kukini, Kungfuadam, Kurdo777, Kurtis, Kuru, Kusma, Kusunose, Kwamikagami, Kwh, Kyle Barbour, Kzzl, Lachaume, Lacrimosus, LadnaV, Lakshmix, Lanny Budd, Lapskingwiki, Larno Man, Lars Washington, Laserpointer35, Laur, Laurentleap, Ld. Ata, Leandrod, Lectonar, Leftdown, Leilan, Lelkesa, Lemmey, Lenardonio, Lennonist, Lenny2894, Leo71538, Leroy65X, Levineps, Lexington1, Licalor, Lightdarkness, Lightmouse, Likeitsmyjob, Lilac Soul, Ling.Nut, Llauren, Lochaber, Lockesdonkey, Locos epraix, Lofty, LonelyMarble, Longlivepalestine, Loodog, Look2See1, Looslion, Looxix, Lord Richard, LordFate, Lordflamel, Lotje, Lottamiata, Lowerarchy, Lradrama, Luban, Lucyin, Luna Santin, Lupin, Lupo, M karzarj, M samadi, M0rphzone, M2545, MER-C, MJCdetroit, ML, MONGO, MPerel, MRC37, MShefa, MTN, MZMcBride, Maahmaah, Macaddct1984, Macmanui, MadameArsenic, Madden, Made of nguyen, Madisonlax1, Maelnuneb, Magaga, Magician 60, Magioladitis, Magister Mathematicae, Mahanga, Mahmood khan, Mahmoud Mahmoud, MahmoudFalam, Mahmudmasri, Mahshidnadimi, Maikel, Majalinno, Majorly, Maks99, Malco23, Male1979, Malleus Fatuorum, Malo, Malpertuis, Mani1, ManiF, Mannerheim, Mansurfattahi, Manu kian maheri, Manu kian maheri93, Maor X, Marashie, Mardavich, Mardetanha, Marek69, Margheritapizza, Mark24387, Mark83, MarkBuckles, MarkJezrylle, Marmoulak, MarshallKe, Marshmallow00, Martarius, Martianboy74, MartinSpacek, Martinmuse, Martinwilke1980, MassRefuge, Massagetae, Master Jay, Masterpiece2000, Materialscientist, Mathemagician57721, MattieTK, MaxD, Mazdakabedi, Maziy300, McDogm, Mcattell, Mdozturk, Meanie, Megastrike14, Mehdi138, Mehijoon, Mehran, MehrdadNY, Mehrdadd, Mehrnia, Mehrshad123, MehrtashG, Melaen, Melca, Melih Yavuz, Memarzadeh, Menchi, Merendoglu, Merhaba1, Merrymanhappy, Mesgul82, Meters, Mewulwe, Mezigue, Mezmerizer, Mfa fariz, Mfn0426, Miaow Miaow, Mic, Michael Devore, Michael Hardy, MichaelBillington, Michaelas10, MichalS, Michigan10, Mickey gfss2007, Micronie, Microtony, Midgrid, Midme, Midnightcomm, Mihai, Mike Payne, Mike Theodore, Mike92591, Mikko Paananen, Millard73, Million Little Gods, Milowent, MinnesotanConfederacy, Minorities observer, Miraculousrandomness, Miranda, Miranda Meagan Keefe, Mirv, Miskin, MisterSheik, Misza13, Mitch1981, Mithraeum, Mitrebox, Mitso Bel, Mjb, Mjroots, Mjroots2, Mjw65, Mkidson, Mlenoirh, Mm40, Mn, Mnamiri, Mnasiri7, Mnmazur, Modster, Modulatum, Mogism, Moh.Hosseini, Mohonu, Mohseng, Molerat, Moogsi, Moonraker, Morwen, Moscow1677, Moslehi, Mosmof, Mostafameraji, Moswento, Mowens35, Mpt, Mr Adequate, Mr link, Mr. Yooper, MrFish, MrSmitherSpank, Mranderson56, Mrjahan, Mrostam, Ms2ger, Mschel, Mtheory1, Mttll, Mukadderat, Mukhtar1, Munci, Muslim10, Mwalkmi, Mwanner, Myland, Mysid, NYC2TLV, NYCJosh, Nabla, Nakon, NameIsRon, Naravorgaara, Naryathegreat, Nate1481, NathanBeach, Nathanheeter, Naur, Navardesh, Nave.notnilc, Navid991, Navidkhalili19, NawlinWiki, Ndyguy, Neddyseagoon, Negrulio, Neilbeach, Neilc, Nepaheshgar, Nestorius, NetProfit, Nethency, Netkinetic, Netoholic, NeutralWriter, Never Cry Wolf, Nevermind1, Nevermind123, NewLionDragon, Newguineafan, Newslikeariver, Newyork1501, Nexus359, Nhgaudreau, Niazewik, Nicetomeetyou, Nichalp, NicholasJB, Nick, Night eule, Nightmarenirvana, Nightryder84, Nightstallion, Nikkimaria, Nima1024, Nimanas, Nimavojdani, Nine Tail Fox, Ninjarrr, NisarKand, Nishkid64, Niz1vir, Nmpenguin, Nnayebi, Nnmmnnmm, No Guru, Nobs, Noiseidea123, Noobyguy, Noodlenut, Notedgrant, Notheruser, Nquotes, Nrbelex, Ntb, NuclearWarfare, Nuhnuhnuh nuuuhhh, Numbo3, Numen, Nunh-huh, Nurg, Nv8200p, ONUnicorn, Obradovic Goran, Ogress, Ohconfucius, Ohnoitsjamie, Oji The Unseen, Okedem, Olegwiki, Olessi, Olivier, Olorin28, Omicronpersei8, Omnipaedista, Omphaloscope, One, One Salient Oversight, OneGuy, Onef9day, Oneforlogic, Onorem, Opader, Opelio, Optimist on the run, Orartu, OrbitOne, OreL.D, Orenburg1, Oreo Priest, Orijentolog, Oroso, Orthologist, Otolemur crassicaudatus, OttawaAC, Otto Von Haasns, Oudysseos, Outerlimits, OwenX, Owenb1, P30Carl, PCHS-NJROTC, PDXblazers, PEAR, PFHLai, PK2, Pacific ocean62, Paddles, Padishah5000, Pahari Sahib, Pahlavannariman, Pajoohesh, Pakteenat, Pam55, Papiffleate, Paradoxic, Pare Mo, Parshan, Parthava, Parthian Scribe, ParthianShot, Pass a Method, Patchouli, Patrick-br, Paul August, Paul Siebert, Pax:Vobiscum, Pbehnam, Pbjb14, Pce3@ij.net, Pcu123456789, Peaky beaky, Peanut4, Pearle, PedEye1, Pedram-e, Peeteeree, Pejman, Pejman47, Pejman7, Pensionero, Peoples Red Revolution, Pepsidrinka, PericlesofAthens, Persia2099, Persiafshin, Persian Poet Gal, Persian Punk, PersianPride91, Persiang, Persiano, Peruvianllama, Peter Horn, PeterDoubt, PeterPredator, Peterange, PeterisP, Petrux, Peymanr, Pgan002, Pgk, Phasechange, Phead128, Phgao, Phildav76, Philip Trueman, Philippe, Philpill, Picapica, Piccadilly Sirkus, Piddle620, Pigman, Pink!Teen, Pinnerup, Pioelad, Pioneer-12, Piperh, Piruz Pars, Pkpat2011, PlovG, Pointlessness, Poka5532, Polaron, PoliticalJunkie, Politician, Pontificateus, Populus, Poring, Porteghal.com, Pournick, Pouya, Pownerus, Poyani, Prester John, PrestonH, Prosfilaes, Pruneau, Pryde 01, PseudoSudo, Pseudomonas, Psm, Psubrat2000, Publicus, Puerto.rico, Pug pug, Pylambert, QAZ, QSTA, QajarCoffee, Qatarihistorian, Qmwne235, Quebec99, Queenmomcat, Quigley, Quindraco, Quintote, Quite vivid blur, Qutezuce, Qwyrxian, Qxz, R'n'B, R1000R1000, R9tgokunks, RG2, RS1900, Raamin, Raayaar, Raayen, RadioKirk, Rajpoot91, Rama, Ramallite, Ran, RanEagle, Randor1980, Rangi, Rarelibra, Raven in Orbit, RaviC, RavinBill, Rayis, Razimpatel, Rdsmith4, Reader77, Realismadder, Recognizance, Red sphere, RedWolf, Redefining form, Redgolpe, Redthoreau, Redvers, Reedy, Refdoc, Reg porter, Regress, Rei, Relyem, Remember the dot, Renegade5005, Res Gestæ Divi Augusti, ResurrectionOfABeast, Retired username, Retrospector87, Rettetast, Revas, Reverse Gear, Revolutionary92, RexNL, Rezamalekpur, Rezashah4, Rfariborz, Rgoodermote, Rhodeislandblue, Riana, Rich Farmbrough, RichiH, Rick Block, RickK, Ridwan Gazi, Rjensen, Rjm656s, Rjwilmsi, Rkerver, Rklawton, Rkt2312, Rlandmann, Rmhermen, Robartin, Robbie Davis-Floyd, Robbiemuffin, Robertvan1, Robth, RockMFR, Rockee Balboa, Rockin the burb, Romanm, RonBeeCNC, Ronthewizard1994, Roomsinto, Roozbeh, Roshanak, Rossami, Rosywounds, Royalguard11, Rpdant767, Rrburke, Rsexton404, Rsking727, Runehelmet, Russavia, RustamDastani, RustyTheMailman, Ruud Koot, Rwendland, Ryanheitner, Ryanjunk, Ryulong, S rafiee, S.K., SBKT, SD5, SFC9394, SG, SHTON, SJP, SLFAW, SSCPUSA, SSZ, Sa.vakilian, Saeed, Sagecedar, Sahar83, Sahim, Saippuakauppias, Salom Khalitun, Salt Yeung, Salvio giuliano, Salvo46, Sam Vimes, Sam esfid, Samrolken, Samuelsen, Sandstein, Sandwich Eater, Sangak, Sango123, Sanya3, Saranghae honey, Sarastro777, Sarayuparin, Sardanaphalus, Sasanjan, Sassan, Savie Kumara, Sbn1984, Sceptre, Scipius, Sciurinæ, Scorpiona, Scott Burley, Scott777, Scythian77, Sduty, Sealpiano, Sean William, Sean.hoyland, Search4Lancer, Seb az86556, Secfan, Selda1982, Sentinel, Separesh, Seresin, Serte, Sertrel, Sevilledade, Sfaiku, Sfan00 IMG, Sfwm86, Shahab pb2002, Shahin.shn, Shahram9, Shaliya waya, Shambles, Shanel, Shanes, Shangrilaista, Sharnak, Shayan7, ShelfSkewed, Shervin moloudi, Shervink, Shia'ithna Islam, Shiham K, Shnyusasl, Shogun, Shredder2012, Siamax, Siavash1989, Siba, Siddiqui, Sidkul 2000, Sieb, Signsolid, Siksok, Silence, Simesa, Simfan34, SimonP, SimonX, Simplerelationship, Sina Kardar, Sina111, Sindbad behzad, Sinooher, Sintaku, Sir blue, Sirlearnsalot, Sirwanaa, Sisson, Sitenl, Sitsat, Sitush, Sjakkalle, Sjharten, Sk8a h8a, SkerHawx, Skomorokh, Skywalker, Skywayman, Slackermonkey, Slakr, Slicky, SlightlyMad, Sloane, SmartGuy Old, Smarty cady, Smaug, Smb1001, Smeash, Smith2006, SmthManly, Snigbrook, Snowmanradio, Soakleif, Sobreira, Socom49, Sohailstyle, Solhjoo, Solipsist, Solitaryxchild, Soman, Someone65, Sophie means wisdom, Sopranosmob781, SorenShadow, Soroosh.a, Soroushsabzi, SouthernComfort, Soxwon, Space.lion, Spahbod, Sparkandspice, Spectheintro, Spesaeterna, Spiff, Splash, SpookyMulder, Sppopuri, SpuriousQ, Ssbohio, Ssidpersian, Starghost, Starnestommy, Station1, Statuens, Steel, Steinbach, Stephan Schulz, Stephen HaasIII, Stephen MUFC, Stephenb, Stephenginns, Stephreef, SteveStrummer, Stevedegrace, Stevertigo, Stevey7788, Stickee, Storkian, Strait, Stromdal, Strykerz, Stuarta, Student7, Stumink, Stvfetterly, Stw, Sumanthk, Sun Creator, Superbeecat, Superdude99, Superm401, Supertask, Supertouch, Surepseh, Sverdrup, Swaskow, Swiftcheeze, Switchercat, Swooningdisaster, Sylvain1972, Syvanen, TBSchemer, THEunique, TJ Spyke, TMLutas, TUF-KAT, Tabletop, Taco325i, Taear, Tahashoeb, Tahmasp, Tajik, Takabeg, Takwish, Tangotango, Tannin, Targeman, Tariqabjotu, Tartessos75, Tashtastic, Tasksaway, Tatrgel, Tbhotch, Tdchbru, TeamZissou, TedE, Template namespace initialisation script, Tempshill, Tenntexan2, Terence, Texture, Tgeairn, Thaimoss, Thanatos666, Thatspeed, The Behnam, The Fear, The Human Trumpet Solo, The Letter J, The Mad Bomber, The Man in Question, The Right Honourable, The Singing Badger, The Son of Oink, The Transhumanist, The wub, TheEditrix2, TheFarix, TheLeopard, TheMcManusBro, ThePrznKonection, TheSeez, Thecunninglinguists, Thedrew77, Thegreyanomaly, Themalau, Thermaland, Thewallowmaker, Thingg, Thiseye, Thistheman, ThoHug, Thricecube, Thuresson, Tickle me, Tidal Air, Tidying Up, TigerShark, Tigeroo, TimBentley, Timir2, Timneu22, Timrem, Tired time, Tirerim, Titoxd, Tjbird9675, Tjdw, Tjmayerinsf, Tkeu, Tmopkisn, Tobby72, Tobias Conradi, Toddgee, Tombseye, TommyBoy, Tone, Tony1, Tony24644, Tophtucker, Totalbeatlesfreak, TotoBaggins, Tpbradbury, Trachys, TrafficBenBoy, Tragic romance, Traininvain80, Trekphiler, TreveX, Trevor MacInnis, Trimp, Tronno, TruthBeethoven, Tsuji, Tufftoon, TurkYusuf1, Twillisjr, Twinxor, Twocs, Uber nemo, Ucanlookitup, Uirauna, UkPaolo, Ulgca, Uncle Scrooge, Unforgiven1975, Ungomma, Unknown Master, Unknownable, Unomano, Urmazd, Use Your Illusion, User2004, User27091, Usergreatpower, Utcursch, VBGFscJUn3, VMajander, Vancouverguy, Vanished user 5zariu3jisj0j4irj, Vardion, Vegaswikian, Vekoler, Ventur, Vergil22, Viartis, Vice regent, Viewfinder, Violetriga, Virgule82, Viridae, Vissequ, Vitalperspective, Vittorio, Vitund, VivaWikipedia, Vividuppers, Vmrgrsergr, VolatileChemical, Volksgeist, Vonones, Vortexion, Vpendse, Vrastic, Vzbs34, WATP, WAvegetarian, WIMYV, WLRoss, WOSlinker, WTucker, Wafulz, Walkerma, Wavelength, Waxinggibbous999, Wayiran, Wayward, Wclark, Webclient101, Webfan29, Wellfauqu2, Welsh, WereSpielChequers, Wes!, Weyes, Wgin4444, WhatamIdoing, WhisperToMe, Whispering, Whitethunder79, Wik, Wiki Wikardo, Wiki alf, WikiDao, WikiInvestor, WikiLaurent, WikiMan225, Wikiacc, Wikidea, Wikidog783, Wikieditor06, Wikien2009, Wikilo12, Wikimol, Wikinick, WikipediaIRAN, Wikipediarules2221, Wikiwo123, WildlifeAnalysis, William Quill, William Saturn, WilliamThweatt, Williamb, Williamborg, Wimt, Winhunter, Wkboonec, Wknight94, Wmahan, Woer$, Wolfdaddydiso, Wolley3, Woohookitty, Writtenright, WrongDude, Wtmitchell, Xaosflux, Xashaiar, XavierGreen, Xdenizen, Xenophon777, Xerxes855, Xerxesisis, Xezbeth, Xgmx2, Xhard7986, Xnuala, Xod, Xooon, Xp54321, Xsoundx, Xullius, Xythianos, Xyzzyva, Yandman, Yangilbert, Yegane, Yelizandpaul, YellowMonkey, Yemal, YeshuaDavid, Yman88, Yonatan, Yono, You3f, Youknowthatoneguy, Yummifruitbat, Yungjohnson, Yuval a, Zach27, Zaharous, Zaius1000, Zamani nima2006, Zandweb, Zanimum, Zaparojdik, Zaphraud, Zazaban, Zburh, Zeamays, Zee18, Zereshk, Zimbabweed, Zinonymous, Zmmz, Zoe, Zonoor, Zora, Zoroastrian, Zscout370, Zuirama, Zundark, Zurkhardo, ZxxZxxZ, Zzampano, Zzorse, Zzuuzz, Zzyzx11, ^demon, Δ, Ζεύς, درابیه‌شت, علی وی ی، مانفی، مانی، هامن، پارسا آملی، 4539 anonymous edits

Ismail I *Source*: http://en.wikipedia.org/w/index.php?title=Ismail_I *Contributors*: 3210, Abdulnr, AdultSwim, Afghana, Aghvank, Ahmad2099, AladdinSE, Alai, AldirmaGonul, Alefbe, Amirreza.h, An Siarach, Andres rojas22, Arash the Bowman, Ardabil baki, ArenM, Artaxiad, Ashrf1979, Atabəy, AtticusX, Attilios, Aytakin, Babakexorramdin, Bahramm 2, Bakuemil, Behi83, Betacommand, BlueEyedCat, Bm79, Bobrayner, Bokan995, Boris Živ, Boxreca, Brothejr, CalumH93, Carlossuarez46, Chae jung, Closedmouth, Cmh, CommonsDelinker, Cplakidas, Darkred, Dimadick, Dougweller, Download, DragonTiger23, Drew R. Smith, EdJohnston, Editor2020, Elemesh, Emperor 2345, Emperor of world, Epbr123, Eupator, Falcon8765, Fishal, Folantin, Furius, GMIAMIR.2010, Ghazne, Glatisant, Glorius son of Azerbaijan, Gomada, Grandmaster, Griot-de, Hajji Piruz, HappyInGeneral, Hibernian, Hinio, Houshyar, Hvakshahtrah, II MusLiM HyBRiD II, Ipaat, Irishpunktom, IsaKazimi, Isse swe, Itai, Jake Nelson, Jandalhandler, Japanese Searobin, Jaraalbe, Jim1138, John Vandenberg, John of Reading, Johnbod, Johnstevens5, Joy, Kamuran otukenli, Kansas Bear, Kapitop, Kashk, Kettabb, Khatru2, Khoikhoi, Khosrow II, Kigali1, Kiumars, Koavf, Kober, Ktsquare, Kurdo777, Kwamikagami, Languagehat, Larno Man, Laurinavicius, Logan, Lysozym, Madhero88, Magioladitis, ManiF, Manocihr, Mardavich, Mehrdadd, Melikov Memmed, Michael Hardy, Mississippienne, Moink, Mursel, Navayli, Neda Reza Azad, Nepaheshgar, Neurolysis, Nicatazeri, Nick Number, Njamali, Nk, NovaSkola, Ocaasi, Ogress, Orartu, Pantherarosa, ParthianShot, Pejman47, Persia2099, PohranicniStraze, Polylerus, Proberts2003, Proger, R'n'B, Ratemonth, Refdoc, ResurrectionOfABeast, Rich Farmbrough, Ricky81682, Rjwilmsi, RogDel, Roozbeh, Rsrikanth05, Sa.vakilian, Sadads, Salvatore of Nina, SamuelTheGhost, Saposcat, Sarefo, SchreiberBike, Scythian1, Shadegan(goru), Shafei, Sixtytwohundred, SouthernComfort, Spasage, Spider 2200, Sthenel, Sun Creator, Tajik, Tajiki0013, Takabeg, The Mysterious El Willstro, TimBentley, TimBits, Timbits, Tmhm, Tommy2010, Twohashdy, Unflavoured, Vacation9, Vanisaac, Vox Magna, Waacstats, Wahabijaz, WereSpielChequers, Woohookitty, Xanadu, Xashaiar, Xythianos, Yurik, Zolaktaf, Zoroastren, Æthelwold, علی وی ی، نوگن‌ابن‌وگنی، 392 anonymous edits

Safavid dynasty *Source*: http://en.wikipedia.org/w/index.php?title=Safavid_dynasty *Contributors*: 1ForTheMoney, 79spirit, A.Abdullayev, AVand, Aaron north, Abdulnr, AdilBaguirov, Adler.fa, Adul, Afghana, AgadaUrbanit, Aghajanpour, Akbar Khan89, Al-Mujahid Fi Sabil Allah, Al-Zaidi, Al1encas1no, Alamukht, Alansohn, Alborz Fallah, Aldis90, Alecxo, Alefbe, Alex Middleton, AlexanderPar, Alimerzairan, AllyUnion, Altai Khan, Altenmann, Alucard (Dr.), AlysTarr, Amatulic, Amaury, Americophile, An Siarach, Anariko, Andrewpmk, AnnaFrance, Anonymous from the 21st century, AnsarParacha, Arab Hafez, Arad, Arash the Bowman, Archaeogenetics, Ardeshire Babakan, ArenM, Armundo bokan, Artaxiad, Arwel Parry, Asenine, Ashrf1979, Atabəy, Attilios, Auntof6, Aynabend, Azerbaboon, Azturkk, BALawrence, Babakexorramdin, Babakgh, Bahar67, Bahramm 2, Balthazarduju, Basilicofresco, BeachHome, Bebejairan, Beh-nam, Blanchette, BlueEyedCat, Blueberrybuttermilkpancakes, Bm79, Bonás, BoogaLouie, Brad Eleven, Brandmeister (old), BrendelSignature, Brenont, Cabolitae, Calm, Camembert, Can't sleep, clown will eat me, CanadianLinuxUser, Canglesea, Capricorn42, Captain-n00dle, Carlosguitar, Cattus, Ceyockey, Chae jung, Charles Matthews, Charliecow7, Chris the

Nader Shah *Source*: http://en.wikipedia.org/w/index.php?title=Nader_Shah *Contributors*: A8UDI, Abbyahmadi, Adel-adili, Adistarp, Agha Nader, Aksi great, Alansohn, Alborz Fallah, Alefbe, Alex Bakharev, AlexanderPar, Algebraist, Aligudarz1, Allens, Amatulic, Amir Sahameddin Ghiassi, Amir85, Amirreza.h, Andres rojas22, Anooshahpour, Arad, Arash the Bowman, Ardeshire Babakan, ArglebargleIV, Aryobarzan, Asidemes, Atabəy, Ave Caesar, Babakexorramdin, Bahramm 2, Balster neb, Beetstra, Beh-nam, Beypeople, Bhadani, Bobblehead, Boris Živ, Bozorg, Brandon, C.Fred, Chaharlang, Chrism, Chubbles, Clown in black and yellow, Colonies Chris, CommonsDelinker, Conversion script, Cpoteratchi, CreazySuit, D-Rock, D6, DBigXray, DarklyCute, Darkred, Dbachmann, Ddd0dd, Deeptrivia, DerDoc, DesiCarlyle, Deuterium 1, Deville, Dmanning, Douglasfrankfort, Dougweller, Downwards, DragonTiger23, Drbreznjev, Ducky59, Epbr123, Eregli bob, Fishal, Folantin, Fratrep, Futuro13, GZUS KRYST, Galoubet, Gh, Ghazne, Gmotamedi, Goethean, Gol, Graham87, GreatWhiteNortherner, Grutter, Hagedis, Hamidrafi23, Harryboyles, Hillel, Hohum, HotWinters, Houshyar, Hu12, Iranart, Itai, Izady, J.delanoy, J04n, JaGa, JamesBWatson, John Hill, John Vandenberg, Jwanders, Kansas Bear, Karent82, Kasir, Kaveh94, Kigali1, Koavf, Leotolstoy, Liftarn, M.Imran, MAx, Magicmonster, Mahanchian, Mardavich, Marmoulak, Mattis, Mav, Meenz, Metropolitan90, Mick Knapton, Mike.lifeguard, Mimihitam, Mirrori1, Mmehdi.g, Moreschi, Mrwojo, Neda Reza Azad, Nedim Ardoğa, Nepaheshgar, Netpilot43556, NielsenGW, NisarKand, Nokhodi, Nostradamus1, ObserverToSee, Oghuz turkish, Omarova, OnlySpeak, PanjshirPashtun, Pantherarosa, Paratov, Parsa1993, ParthianShot, Pearle, PhnomPencil, Pjahan, Pol098, Pouya, Prester John, Private Butcher, Profitoftruth85, Qnkhawar, R'n'B, Ramin1981, RandomTool2, Rbraunwa, Rich Farmbrough, Ricky81682, Rjwilmsi, Rohit.m.gutte, Roozbeh, Rostami55, Rune X2, Sa.vakilian, Sannse, Serein (renamed because of SUL), Sesel, Shauni, Shervin1982, Siddiqui, Silence, SimonP, SiobhanHansa, Sixtytwohundred, SkerHawx, Sodacan, SouthernComfort, SpacemanSpiff, Spasage, Sperling, Spider 2200, Spock44, Superslum, THEunique, Tajik, Tajiki0013, Takabeg, TarTirNazok, Tassedethe, Tbhotch, Teddybearcan, The Behnam, TheEgyptian, TheNeon, Thomf, Thorsen, TimBits, Tovojolo, Tqaisrani, Vargavandnick, Vssun, Vzbs34, Wachoviadeal, Wertuose, What makes a man turn neutral?, WingedEarth, WinnerWinner2468, Winter Gaze, Woohookitty, Xooon, Zakksez, ZandyehPrince, Zanganeh, Zereshk, Zlerman, ینوگنابنوگن, 274 anonymous edits

Image Sources, Licenses and Contributors

File:Ottoman flag alternative 2.svg *Source*: http://en.wikipedia.org/w/index.php?title=File:Ottoman_flag_alternative_2.svg *License*: unknown *Contributors*: Kerem Özcan. Original uploader was Kerem Ozcan at en.wikipedia. Later version(s) were uploaded by Nightmare X at en.wikipedia.

File:Osmanli-nisani.svg *Source*: http://en.wikipedia.org/w/index.php?title=File:Osmanli-nisani.svg *License*: unknown *Contributors*: Juris Tiltins

File:Ottoman Empire 16-17th century.jpg *Source*: http://en.wikipedia.org/w/index.php?title=File:Ottoman_Empire_16-17th_century.jpg *License*: unknown *Contributors*: lynxxx

File:Flag of Palaeologus Emperor.svg *Source*: http://en.wikipedia.org/w/index.php?title=File:Flag_of_Palaeologus_Emperor.svg *License*: unknown *Contributors*: Alkari, Cplakidas, Electionworld, Fry1989, Himasaram, Homo lupus, Mhmrodrigues, Peeperman, TFCforever, Wouterhagens, Ωριγένης, 5 anonymous edits

File:Blank.png *Source*: http://en.wikipedia.org/w/index.php?title=File:Blank.png *License*: unknown *Contributors*: Bastique, Chlewey, ChrisDHDR, It Is Me Here, Jed, Paradoctor, Patrick, Penubag, Rocket000, Roomba, Timeroot, Tintazul

File:Karamanid Dynasty flag.svg *Source*: http://en.wikipedia.org/w/index.php?title=File:Karamanid_Dynasty_flag.svg *License*: unknown *Contributors*: User:Jecowa

File:Coat of Arms of the House of Kotromanić.svg *Source*: http://en.wikipedia.org/w/index.php?title=File:Coat_of_Arms_of_the_House_of_Kotromanić.svg *License*: unknown *Contributors*: User:Sodacan

File:Coat of arms of the Second Bulgarian Empire.svg *Source*: http://en.wikipedia.org/w/index.php?title=File:Coat_of_arms_of_the_Second_Bulgarian_Empire.svg *License*: unknown *Contributors*: User:Пакко

File:Supposed Serbian Empire flag.svg *Source*: http://en.wikipedia.org/w/index.php?title=File:Supposed_Serbian_Empire_flag.svg *License*: unknown *Contributors*: User:B1mbo

File:Coa Kastrioti Family.svg *Source*: http://en.wikipedia.org/w/index.php?title=File:Coa_Kastrioti_Family.svg *License*: unknown *Contributors*: User:Jpb1301, User:Madboy74, User:MissMJ, User:Syryatsu

File:Mameluke Flag.svg *Source*: http://en.wikipedia.org/w/index.php?title=File:Mameluke_Flag.svg *License*: unknown *Contributors*: User:Ryucloud

File:Coat of Arms of Hungary Historic Design.png *Source*: http://en.wikipedia.org/w/index.php?title=File:Coat_of_Arms_of_Hungary_Historic_Design.png *License*: unknown *Contributors*: User:Buho07

File:Flag of Greece (1828-1978).svg *Source*: http://en.wikipedia.org/w/index.php?title=File:Flag_of_Greece_(1828-1978).svg *License*: unknown *Contributors*: -

File:Egypt_flag_1882.svg *Source*: http://en.wikipedia.org/w/index.php?title=File:Egypt_flag_1882.svg *License*: unknown *Contributors*: -

File:Flag_of_Bosnia_(1878-1908).svg *Source*: http://en.wikipedia.org/w/index.php?title=File:Flag_of_Bosnia_(1878-1908).svg *License*: unknown *Contributors*: User:David Liuzzo, User:Themightyquill

File:Flag of Serbia (1830-1882).svg *Source*: http://en.wikipedia.org/w/index.php?title=File:Flag_of_Serbia_(1830-1882).svg *License*: unknown *Contributors*: -

File:Flag of Albanian Provisional Government 1912-1914.gif *Source*: http://en.wikipedia.org/w/index.php?title=File:Flag_of_Albanian_Provisional_Government_1912-1914.gif *License*: unknown *Contributors*: Jaume Ollé

File:Flag of Romania.svg *Source*: http://en.wikipedia.org/w/index.php?title=File:Flag_of_Romania.svg *License*: unknown *Contributors*: User:AdiJapan

File:Flag of Bulgaria.svg *Source*: http://en.wikipedia.org/w/index.php?title=File:Flag_of_Bulgaria.svg *License*: unknown *Contributors*: User:SKopp

File:Cyprus Blue Ensign-1922.svg *Source*: http://en.wikipedia.org/w/index.php?title=File:Cyprus_Blue_Ensign-1922.svg *License*: unknown *Contributors*: -

File:Flag of Iraq 1924.svg *Source*: http://en.wikipedia.org/w/index.php?title=File:Flag_of_Iraq_1924.svg *License*: unknown *Contributors*: -

File:Palestine-Mandate-Ensign-1927-1948.svg *Source*: http://en.wikipedia.org/w/index.php?title=File:Palestine-Mandate-Ensign-1927-1948.svg *License*: unknown *Contributors*: AnonMoos, Ar-ras, Chesdovi, Fry1989, Greentubing, Homo lupus, Mattes, Obersachse, Orrling, OwenBlacker, Saudijp, Siebrand, Молох

File:Flag_of_France.svg *Source*: http://en.wikipedia.org/w/index.php?title=File:Flag_of_France.svg *License*: unknown *Contributors*: Anomie

File:Flag of French Tunisia.svg *Source*: http://en.wikipedia.org/w/index.php?title=File:Flag_of_French_Tunisia.svg *License*: unknown *Contributors*: Original uploader was Zscout370 at en.wikipedia

File:Flag of Kuwait.svg *Source*: http://en.wikipedia.org/w/index.php?title=File:Flag_of_Kuwait.svg *License*: unknown *Contributors*: User:SKopp

File:Flag of Italy (1861-1946).svg *Source*: http://en.wikipedia.org/w/index.php?title=File:Flag_of_Italy_(1861-1946).svg *License*: unknown *Contributors*: User:F l a n k e r

File:Battle of Nicopolis, 1396, Facsimile of a Miniature Conserved in the Topkapi Museum in Istanbul-.jpg *Source*: http://en.wikipedia.org/w/index.php?title=File:Battle_of_Nicopolis,_1396,_Facsimile_of_a_Miniature_Conserved_in_the_Topkapi_Museum_in_Istanbul-.jpg *License*: unknown *Contributors*: AnonMoos, Rapsar, Underlying lk

File:Moldovita murals 2010 41.jpg *Source*: http://en.wikipedia.org/w/index.php?title=File:Moldovita_murals_2010_41.jpg *License*: unknown *Contributors*: Man vyi

File:1526 - Battle of Mohács.jpg *Source*: http://en.wikipedia.org/w/index.php?title=File:1526_-_Battle_of_Mohács.jpg *License*: unknown *Contributors*: Rapsar, Underlying lk

File:Szigetvar 1566.jpg *Source*: http://en.wikipedia.org/w/index.php?title=File:Szigetvar_1566.jpg *License*: unknown *Contributors*: Lokman

File:Ottoman empire.svg *Source*: http://en.wikipedia.org/w/index.php?title=File:Ottoman_empire.svg *License*: unknown *Contributors*: user:André Koehne

Image:Vienna Battle 1683.jpg *Source*: http://en.wikipedia.org/w/index.php?title=File:Vienna_Battle_1683.jpg *License*: unknown *Contributors*: Bukk, Gryffindor, Interpretix, Jagatai, Pe-Jo, Thib Phil, Vincent Steenberg, Zolo

File:Ottoman Sultan selim III 1789.jpg *Source*: http://en.wikipedia.org/w/index.php?title=File:Ottoman_Sultan_selim_III_1789.jpg *License*: unknown *Contributors*: User:Isl@m

File:January Suchodolski - Ochakiv siege.jpg *Source*: http://en.wikipedia.org/w/index.php?title=File:January_Suchodolski_-_Ochakiv_siege.jpg *License*: unknown *Contributors*: Dmitry Rozhkov, Kaganer, Labattblueboy, Man vyi, Mathiasrex, Schekinov Alexey Victorovich

File:Jean Baptiste Vanmour Prozession des Sultans in Istanbul.jpg *Source*: http://en.wikipedia.org/w/index.php?title=File:Jean_Baptiste_Vanmour_Prozession_des_Sultans_in_Istanbul.jpg *License*: unknown *Contributors*: FA2010

File:Turkish troops storming Fort Shefketil (cropped).jpg *Source*: http://en.wikipedia.org/w/index.php?title=File:Turkish_troops_storming_Fort_Shefketil_(cropped).jpg *License*: unknown *Contributors*: User:Underlying lk

File:Opening ceremony of the First Ottoman Parliament at the Dolmabahce Palace in 1876.jpg *Source*: http://en.wikipedia.org/w/index.php?title=File:Opening_ceremony_of_the_First_Ottoman_Parliament_at_the_Dolmabahce_Palace_in_1876.jpg *License*: unknown *Contributors*: Rapsar, Underlying lk

File:Young Turk Revolution - Decleration - Armenian Greek Muslim Leaders.png *Source*: http://en.wikipedia.org/w/index.php?title=File:Young_Turk_Revolution_-_Decleration_-_Armenian_Greek_Muslim_Leaders.png *License*: unknown *Contributors*: Charles Roden Buxton

Image:Sultanvahideddin.jpg *Source*: http://en.wikipedia.org/w/index.php?title=File:Sultanvahideddin.jpg *License*: unknown *Contributors*: FSII, FunkMonk, Gryffindor, Infrogmation, Kürschner, Patrickneil, Te5, Молох, 5 anonymous edits

Image:Pera Museum 2.jpg *Source*: http://en.wikipedia.org/w/index.php?title=File:Pera_Museum_2.jpg *License*: unknown *Contributors*: FA2010, Funfood, G.dallorto, Gryffindor, JMCC1, Pera Müzesi, Qizilbash, Sreejithk2000 AWB, Takabeg

Image:Thomas allom, c1840, The Enterance to Divan.png *Source*: http://en.wikipedia.org/w/index.php?title=File:Thomas_allom,_c1840,_The_Enterance_to_Divan.png *License*: unknown *Contributors*: Original uploader was Tommiks at en.wikipedia

Image:Tughra Suleiman.jpg *Source*: http://en.wikipedia.org/w/index.php?title=File:Tughra_Suleiman.jpg *License*: unknown *Contributors*: User:Pagan

Image:1879-Ottoman Court-from-NYL.png *Source*: http://en.wikipedia.org/w/index.php?title=File:1879-Ottoman_Court-from-NYL.png *License*: unknown *Contributors*: Original uploader was Karabekir at en.wikipedia

File:Tott-33-Cadi leskier.jpg *Source*: http://en.wikipedia.org/w/index.php?title=File:Tott-33-Cadi_leskier.jpg *License*: unknown *Contributors*: Herm Baron von Tott

Image:Battle of Vienna.SultanMurads with janissaries.jpg *Source*: http://en.wikipedia.org/w/index.php?title=File:Battle_of_Vienna.SultanMurads_with_janissaries.jpg *License*: unknown *Contributors*: G. Jansoone

Image:Ottoman Navy at the Golden Horn.jpg *Source*: http://en.wikipedia.org/w/index.php?title=File:Ottoman_Navy_at_the_Golden_Horn.jpg *License*: unknown *Contributors*: Original uploader was Res Gestæ Divi Augusti at en.wikipedia

File:Battle of Vienna.Sipahis.jpg *Source*: http://en.wikipedia.org/w/index.php?title=File:Battle_of_Vienna.Sipahis.jpg *License*: unknown *Contributors*: G. Jansoone

File:Moderniranianlanguagesmap.jpg *Source*: http://en.wikipedia.org/w/index.php?title=File:Moderniranianlanguagesmap.jpg *License*: unknown *Contributors*: Americophile, David Eppstein, Davius, Firespeaker, FrancisTyers, KureCewlik81, Le Behnam, Rehman, Ruy Pugliesi, Takabeg, WhisperToMe, 4 anonymous edits

File:Kelileh va Demneh.jpg *Source*: http://en.wikipedia.org/w/index.php?title=File:Kelileh_va_Demneh.jpg *License*: unknown *Contributors*: User Zereshk on en.wikipedia

File:Hafeziyeh 0386.JPG *Source*: http://en.wikipedia.org/w/index.php?title=File:Hafeziyeh_0386.JPG *License*: unknown *Contributors*: User:Mardetanha

File:Mehmooni2.jpg *Source*: http://en.wikipedia.org/w/index.php?title=File:Mehmooni2.jpg *License*: unknown *Contributors*: Guitarpop, Kotoito, Marjan-shivA, ZxxZxxZ, 5 anonymous edits

File:Esfahan-shah-sq.jpg *Source*: http://en.wikipedia.org/w/index.php?title=File:Esfahan-shah-sq.jpg *License*: unknown *Contributors*: Bontenbal, Diligent, Kaveh, Siasard, Wiki-uk, 3 anonymous edits

File:Magnify-clip.png *Source*: http://en.wikipedia.org/w/index.php?title=File:Magnify-clip.png *License*: unknown *Contributors*: User:Erasoft24

File: .JPG *Source*: http://en.wikipedia.org/w/index.php?title=File:ورزشگاه_آزادی_در_سایه.JPG *License*: unknown *Contributors*: خودم

File:Iran ethnoreligious distribution 2004.jpg *Source*: http://en.wikipedia.org/w/index.php?title=File:Iran_ethnoreligious_distribution_2004.jpg *License*: unknown *Contributors*: ALE!, DanTD, Diyako, Fabienkhan, Feydey, Kotoito, Ninja irani, Ojanfar, Persia2099, WhisperToMe, Wikiacc, Wknight94, ZxxZxxZ, بی‌نام, 2 anonymous edits

File:Iran Population (1880-2005).JPG *Source*: http://en.wikipedia.org/w/index.php?title=File:Iran_Population_(1880-2005).JPG *License*: unknown *Contributors*: User:M samadi

File:Сефи_1й_1629-42.jpg *Source*: http://en.wikipedia.org/w/index.php?title=File:Сефи_1й_1629-42.jpg *License*: unknown *Contributors*: Original uploader was Евгений Ардаев at ru.wikipedia

File:1541-Battle in the war between Shah Isma'il and the King of Shirvan-Shahnama-i-Isma'il.jpg *Source*: http://en.wikipedia.org/w/index.php?title=File:1541-Battle_in_the_war_between_Shah_Isma'il_and_the_King_of_Shirvan-Shahnama-i-Isma'il.jpg *License*: unknown *Contributors*: Andres rojas22, Melikov Memmed

File: .jpg *Source*: http://en.wikipedia.org/w/index.php?title=File:جنگ_مرو.jpg *License*: unknown *Contributors*: Americophile, Andres rojas22, Ashrf1979, Nevit, Olivier2, Qizilbash, Takabeg, ینوگنابنوگن

File:The Battle between Shah Ismail and Shaybani Khan.jpg *Source*: http://en.wikipedia.org/w/index.php?title=File:The_Battle_between_Shah_Ismail_and_Shaybani_Khan.jpg *License*: unknown *Contributors*: Mu'in Musavvir , (Iranian, Safavid dynasty

File:Shah esmaeil01.jpg *Source*: http://en.wikipedia.org/w/index.php?title=File:Shah_esmaeil01.jpg *License*: unknown *Contributors*: Ashena iran

File:Sekumname1525 Chaldiran battle.jpg *Source*: http://en.wikipedia.org/w/index.php?title=File:Sekumname1525_Chaldiran_battle.jpg *License*: unknown *Contributors*: anonym. Original uploader was Евгений Ардаев at ru.wikipedia

File:Timurid.svg *Source*: http://en.wikipedia.org/w/index.php?title=File:Timurid.svg *License*: unknown *Contributors*: User:Stannered

File:Safavid_Flag.png *Source*: http://en.wikipedia.org/w/index.php?title=File:Safavid_Flag.png *License*: unknown *Contributors*: Orange Tuesday (talk) Original uploader was Orange Tuesday at en.wikipedia

File:Lion and Sun Emblem of Persia.svg *Source*: http://en.wikipedia.org/w/index.php?title=File:Lion_and_Sun_Emblem_of_Persia.svg *License*: unknown *Contributors*: User:Sodacan

File:Jomann_Imperium_Periscum.jpg *Source*: http://en.wikipedia.org/w/index.php?title=File:Jomann_Imperium_Periscum.jpg *License*: unknown *Contributors*: AndreasPraefcke, Ervaude, Flamarande, Geagea, Matthead, ینوگنابنوگن, 2 anonymous edits

File:Ottoman_flag.svg *Source*: http://en.wikipedia.org/w/index.php?title=File:Ottoman_flag.svg *License*: unknown *Contributors*: User:Kerem Ozcan

File:Сефи 1й 1629-42.jpg *Source*: http://en.wikipedia.org/w/index.php?title=File:Сефи_1й_1629-42.jpg *License*: unknown *Contributors*: Original uploader was Евгений Ардаев at ru.wikipedia

File:Map Safavid persia.png *Source*: http://en.wikipedia.org/w/index.php?title=File:Map_Safavid_persia.png *License*: unknown *Contributors*: User:Fabienkhan

File:Tahmasb-1.jpg *Source*: http://en.wikipedia.org/w/index.php?title=File:Tahmasb-1.jpg *License*: unknown *Contributors*: User:Nasser-sadeghi

File:Matthaus 1598.jpg *Source*: http://en.wikipedia.org/w/index.php?title=File:Matthaus_1598.jpg *License*: unknown *Contributors*: Chaos, Chipmunkdavis, FSII, Meno25, PM

File:Hondius 1610.jpg *Source*: http://en.wikipedia.org/w/index.php?title=File:Hondius_1610.jpg *License*: unknown *Contributors*: Chaos, Chipmunkdavis, FSII, Geagea, Kilom691, Meno25, PM

File:Shah Tahmasp and Humayun.jpeg *Source*: http://en.wikipedia.org/w/index.php?title=File:Shah_Tahmasp_and_Humayun.jpeg *License*: unknown *Contributors*: Original uploader was Irishpunktom at en.wikipedia

Image:Shah Abbas I engraving by Dominicus Custos - Antwerp artist printer and engraver.jpg *Source*: http://en.wikipedia.org/w/index.php?title=File:Shah_Abbas_I_engraving_by_Dominicus_Custos_-_Antwerp_artist_printer_and_engraver.jpg *License*: unknown *Contributors*: AndreasPraefcke, Ashrf1979, Haiduc, Martin H., Qizilbash, Takabeg, 2 anonymous edits

File:Moll 1720 Persian Empire.JPG *Source*: http://en.wikipedia.org/w/index.php?title=File:Moll_1720_Persian_Empire.JPG *License*: unknown *Contributors*: MOLL, Herman

File:Safavid Persian Empire.jpg *Source*: http://en.wikipedia.org/w/index.php?title=File:Safavid_Persian_Empire.jpg *License*: unknown *Contributors*: Ervaude, Geagea, Papatt, Takabeg, Väsk, Wayiran, Zykasaa

File:Husain Ali Beg.jpg *Source*: http://en.wikipedia.org/w/index.php?title=File:Husain_Ali_Beg.jpg *License*: unknown *Contributors*: 17th century book, Ritter

Image:Embassy to Europe.jpg *Source*: http://en.wikipedia.org/w/index.php?title=File:Embassy_to_Europe.jpg *License*: unknown *Contributors*: Carlo Caliari and Gabriele Calliari

File:Shah Abbas II.jpg *Source*: http://en.wikipedia.org/w/index.php?title=File:Shah_Abbas_II.jpg *License*: unknown *Contributors*: Sorosh Tavakoli from Stockholm, Sverige

File:Jomann Imperium Periscum.jpg *Source*: http://en.wikipedia.org/w/index.php?title=File:Jomann_Imperium_Periscum.jpg *License*: unknown *Contributors*: AndreasPraefcke, Ervaude, Flamarande, Geagea, Matthead, ینوگنابنوگن, 2 anonymous edits

File:Shah soleiman safavi.jpg *Source*: http://en.wikipedia.org/w/index.php?title=File:Shah_soleiman_safavi.jpg *License*: unknown *Contributors*: Ali Qoli Jabbador (ალი-ყული ჯაბადარი) Uploaded by: Zereshk

File:Safavid-helmet.jpg *Source*: http://en.wikipedia.org/w/index.php?title=File:Safavid-helmet.jpg *License*: unknown *Contributors*: User:Nasser-sadeghi

File:Qezelbash.JPG *Source*: http://en.wikipedia.org/w/index.php?title=File:Qezelbash.JPG *License*: unknown *Contributors*: User:E235

File:Safavid Persia - customs.jpg *Source*: http://en.wikipedia.org/w/index.php?title=File:Safavid_Persia_-_customs.jpg *License*: unknown *Contributors*: Americophile, Mghotbi 85, Takabeg, ZxxZxxZ

File:Sir Robert Shirley by Anthony Van Dyck 1622 Rome.jpg *Source*: http://en.wikipedia.org/w/index.php?title=File:Sir_Robert_Shirley_by_Anthony_Van_Dyck_1622_Rome.jpg *License*: unknown *Contributors*: Americophile, Andres rojas22, Paris 16, Shakko, World Imaging, Zolo

File:Polo game from poem Guy u Chawgan.jpg *Source*: http://en.wikipedia.org/w/index.php?title=File:Polo_game_from_poem_Guy_u_Chawgan.jpg *License*: unknown *Contributors*: Americophile, Amir85, Baba66, Grenavitar, Jastrow, Meisam, Persia2099, Qizilbash, Takabeg

File:Safavid dynastiet - clothes .jpg *Source*: http://en.wikipedia.org/w/index.php?title=File:Safavid_dynastiet_-_clothes_.jpg *License*: unknown *Contributors*: Unknown artist.

File:Jean Chardin 1739.jpg *Source*: http://en.wikipedia.org/w/index.php?title=File:Jean_Chardin_1739.jpg *License*: unknown *Contributors*: Ecummenic, Geagea, Sir Gawain, World Imaging, ZxxZxxZ

File:Safavid Legal system Karkan.jpg *Source*: http://en.wikipedia.org/w/index.php?title=File:Safavid_Legal_system_Karkan.jpg *License*: unknown *Contributors*: Mghotbi 85, ZxxZxxZ, 1 anonymous edits

File:19th century drawing of isfahan.jpg *Source*: http://en.wikipedia.org/w/index.php?title=File:19th_century_drawing_of_isfahan.jpg *License*: unknown *Contributors*: Marmoulak, ZxxZxxZ

File:Hotel Shah Abbas Sahn.jpg *Source*: http://en.wikipedia.org/w/index.php?title=File:Hotel_Shah_Abbas_Sahn.jpg *License*: unknown *Contributors*: Bontenbal, Iamunknown, Nevit, Olivier2, Woudloper, Überraschungsbilder

File:Chehel Sotoon.jpg *Source*: http://en.wikipedia.org/w/index.php?title=File:Chehel_Sotoon.jpg *License*: unknown *Contributors*: Arad Mojtahedi

File:Silk route.jpg *Source*: http://en.wikipedia.org/w/index.php?title=File:Silk_route.jpg *License*: unknown *Contributors*: User:Splette

File:Vank Cathedral interior.jpg *Source*: http://en.wikipedia.org/w/index.php?title=File:Vank_Cathedral_interior.jpg *License*: unknown *Contributors*: Thomas

File:Chahar bagh school drawing.jpg *Source*: http://en.wikipedia.org/w/index.php?title=File:Chahar_bagh_school_drawing.jpg *License*: unknown *Contributors*: Marmoulak, Wiki-uk, ZxxZxxZ

Image:Canons of medicine.JPG *Source*: http://en.wikipedia.org/w/index.php?title=File:Canons_of_medicine.JPG *License*: unknown *Contributors*: BajanZindy, Bontenbal, Fabienkhan, VIGNERON, Zereshk, 2 anonymous edits

image:Masjede Shah painting.jpg *Source*: http://en.wikipedia.org/w/index.php?title=File:Masjede_Shah_painting.jpg *License*: unknown *Contributors*: Marcel 84, Marmoulak, Nevit, ZH2010, ZxxZxxZ

File:Safavid-star.png *Source*: http://en.wikipedia.org/w/index.php?title=File:Safavid-star.png *License*: unknown *Contributors*: myself

File:Qazvin - Chehel Sotun.jpg *Source*: http://en.wikipedia.org/w/index.php?title=File:Qazvin_-_Chehel_Sotun.jpg *License*: unknown *Contributors*: User:Simorg

File:The Mantiq al-tair.jpg *Source*: http://en.wikipedia.org/w/index.php?title=File:The_Mantiq_al-tair.jpg *License*: unknown *Contributors*: Habibulla Meshedi Original uploader was Евгений Ардаев at ru.wikipedia

Image:Georgian prince by Reza Abbasi.jpg *Source*: http://en.wikipedia.org/w/index.php?title=File:Georgian_prince_by_Reza_Abbasi.jpg *License*: unknown *Contributors*: -

File:Safavid dynasty timeline-en.svg *Source*: http://en.wikipedia.org/w/index.php?title=File:Safavid_dynasty_timeline-en.svg *License*: unknown *Contributors*: User:Nasser-sadeghi, User:P30Carl

File:Cairo - Islamic district - Al Azhar Mosque and University.JPG *Source*: http://en.wikipedia.org/w/index.php?title=File:Cairo_-_Islamic_district_-_Al_Azhar_Mosque_and_University.JPG *License*: unknown *Contributors*: user:Maveric149

File:Umayyad750ADloc.png *Source*: http://en.wikipedia.org/w/index.php?title=File:Umayyad750ADloc.png *License*: unknown *Contributors*: User:Gabagool

File:Madhhab Map2.png *Source*: http://en.wikipedia.org/w/index.php?title=File:Madhhab_Map2.png *License*: unknown *Contributors*: User:Ghibar

File:Kairouan Mosque Stitched Panorama.jpg *Source*: http://en.wikipedia.org/w/index.php?title=File:Kairouan_Mosque_Stitched_Panorama.jpg *License*: unknown *Contributors*: MAREK SZAREJKO from CLONMEL, IRELAND - POLAND

File:Safavid Flag.png *Source*: http://en.wikipedia.org/w/index.php?title=File:Safavid_Flag.png *License*: unknown *Contributors*: Orange Tuesday (talk) Original uploader was Orange Tuesday at en.wikipedia

File:Flag of the Mughal Empire.svg *Source*: http://en.wikipedia.org/w/index.php?title=File:Flag_of_the_Mughal_Empire.svg *License*: unknown *Contributors*: Orange Tuesday (talk)

File:Afsharid Imperial Standard (3 Stripes).svg *Source*: http://en.wikipedia.org/w/index.php?title=File:Afsharid_Imperial_Standard_(3_Stripes).svg *License*: unknown *Contributors*: User:Orange Tuesday

File:Black flag.svg *Source*: http://en.wikipedia.org/w/index.php?title=File:Black_flag.svg *License*: unknown *Contributors*: User:PavelD

File:Map of The Hotakids.png *Source*: http://en.wikipedia.org/w/index.php?title=File:Map_of_The_Hotakids.png *License*: unknown *Contributors*: Arab Hafez at en.wikipedia

File:Flag of Afghanistan.svg *Source*: http://en.wikipedia.org/w/index.php?title=File:Flag_of_Afghanistan.svg *License*: unknown *Contributors*: User:Zscout370

File:Flag of Pakistan.svg *Source*: http://en.wikipedia.org/w/index.php?title=File:Flag_of_Pakistan.svg *License*: unknown *Contributors*: User:Zscout370

File:Flag of Turkmenistan.svg *Source*: http://en.wikipedia.org/w/index.php?title=File:Flag_of_Turkmenistan.svg *License*: unknown *Contributors*: User:Vzb83

File:The Surrender of Kandahar.jpg *Source*: http://en.wikipedia.org/w/index.php?title=File:The_Surrender_of_Kandahar.jpg *License*: unknown *Contributors*: unknown

File:Map of Afghanistan during the Safavid and Moghul Empire.jpg *Source*: http://en.wikipedia.org/w/index.php?title=File:Map_of_Afghanistan_during_the_Safavid_and_Moghul_Empire.jpg *License*: unknown *Contributors*: Bowen, Emanuel, d. 1767.

File:Mirwais-Hotak.jpeg *Source*: http://en.wikipedia.org/w/index.php?title=File:Mirwais-Hotak.jpeg *License*: unknown *Contributors*: Asadullah Danish

File:Noimage.png *Source*: http://en.wikipedia.org/w/index.php?title=File:Noimage.png *License*: unknown *Contributors*: User:Mike bzh

File:SHAH-MAHMUD-HOTAK.jpg *Source*: http://en.wikipedia.org/w/index.php?title=File:SHAH-MAHMUD-HOTAK.jpg *License*: unknown *Contributors*: Maulvi Abdurab Ahadi

File:Ashraf Shah Hotaki 1725-1729.jpg *Source*: http://en.wikipedia.org/w/index.php?title=File:Ashraf_Shah_Hotaki_1725-1729.jpg *License*: unknown *Contributors*: User:Officer

File:Shah-Husain-Hotak.jpg *Source*: http://en.wikipedia.org/w/index.php?title=File:Shah-Husain-Hotak.jpg *License*: unknown *Contributors*: Maulvi Abdurab Ahadi

File:Flag_of_the_Durrani_Empire.svg *Source*: http://en.wikipedia.org/w/index.php?title=File:Flag_of_the_Durrani_Empire.svg *License*: unknown *Contributors*: -

File:Zand Dynasty flag.svg *Source*: http://en.wikipedia.org/w/index.php?title=File:Zand_Dynasty_flag.svg *License*: unknown *Contributors*: User:Orange Tuesday

File:Afsharid_Imperial_Standard_(3_Stripes).svg *Source*: http://en.wikipedia.org/w/index.php?title=File:Afsharid_Imperial_Standard_(3_Stripes).svg *License*: unknown *Contributors*: User:Orange Tuesday

File:Afsharid1736-1802.svg *Source*: http://en.wikipedia.org/w/index.php?title=File:Afsharid1736-1802.svg *License*: unknown *Contributors*: User:Tintazul

File:Sind silver rupeeunder the Afsharid Shahs of Iran 18th century.jpg *Source*: http://en.wikipedia.org/w/index.php?title=File:Sind_silver_rupeeunder_the_Afsharid_Shahs_of_Iran_18th_century.jpg *License*: unknown *Contributors*: User:Uploadalt

File:ImamHusaynMosqueKarbalaIraqPre2006.JPG *Source*: http://en.wikipedia.org/w/index.php?title=File:ImamHusaynMosqueKarbalaIraqPre2006.JPG *License*: unknown *Contributors*: User:Toushiro

File:Investiture of Ali Edinburgh codex.jpg *Source*: http://en.wikipedia.org/w/index.php?title=File:Investiture_of_Ali_Edinburgh_codex.jpg *License*: unknown *Contributors*: Ibn al-Kutbi

File:ImamAliMosqueNajafIraq.JPG *Source*: http://en.wikipedia.org/w/index.php?title=File:ImamAliMosqueNajafIraq.JPG *License*: unknown *Contributors*: User:Toushiro

File:Dhulfiqar.svg *Source*: http://en.wikipedia.org/w/index.php?title=File:Dhulfiqar.svg *License*: unknown *Contributors*: Myself

File:Kalema-tut-shahadat.jpg *Source*: http://en.wikipedia.org/w/index.php?title=File:Kalema-tut-shahadat.jpg *License*: unknown *Contributors*: Md iet (talk). Original uploader was Md iet at en.wikipedia

File:Inscription on Aqmar main gate 1.jpg *Source*: http://en.wikipedia.org/w/index.php?title=File:Inscription_on_Aqmar_main_gate_1.jpg *License*: unknown *Contributors*: Md iet (talk). Original uploader was Md iet at en.wikipedia

File:Inscription on Aqmar Mosque.jpg *Source*: http://en.wikipedia.org/w/index.php?title=File:Inscription_on_Aqmar_Mosque.jpg *License*: unknown *Contributors*: User:Md iet

File:Aqmar corner.jpg *Source*: http://en.wikipedia.org/w/index.php?title=File:Aqmar_corner.jpg *License*: unknown *Contributors*: User:Md iet

File:Orde van Ali Perzie 1880 Gritzner 001.jpg *Source*: http://en.wikipedia.org/w/index.php?title=File:Orde_van_Ali_Perzie_1880_Gritzner_001.jpg *License*: unknown *Contributors*: Maximilian Gritzner, died 1903

File:Islam by country.png *Source*: http://en.wikipedia.org/w/index.php?title=File:Islam_by_country.png *License*: unknown *Contributors*: Baba66, Bjarki S, Cwbm (commons), Dbachmann, Denisutku, Denniss, Hibernian, Koavf, MacedonianBoy, Maphobbyist, Mediatus, Nicolay Sidorov, Sercan.ergün, Timeshifter, Túrelio, 3 anonymous edits

File:Muharram procession 2, Manama, Bahrain (Feb 2005).jpg *Source*: http://en.wikipedia.org/w/index.php?title=File:Muharram_procession_2,_Manama,_Bahrain_(Feb_2005).jpg *License*: unknown *Contributors*: Ashrf1979, Chanad, Denniss, Gire 3pich2005, Saforrest, 3 anonymous edits

File:Kerbela Hussein Moschee.jpg *Source*: http://en.wikipedia.org/w/index.php?title=File:Kerbela_Hussein_Moschee.jpg *License*: unknown *Contributors*: SFC Larry E. Johns, USA

File:Tree shia islam n3 .pdf *Source*: http://en.wikipedia.org/w/index.php?title=File:Tree_shia_islam_n3_.pdf *License*: unknown *Contributors*: User:Md iet

File:Flag of Persia (1910).svg *Source*: http://en.wikipedia.org/w/index.php?title=File:Flag_of_Persia_(1910).svg *License*: unknown *Contributors*: SeNeKa

Image:Monument+karim khan.jpg *Source*: http://en.wikipedia.org/w/index.php?title=File:Monument+karim_khan.jpg *License*: unknown *Contributors*: Evgenia Kononova Original uploader was Ipaat at en.wikipedia

File:Zand-majles.jpg *Source*: http://en.wikipedia.org/w/index.php?title=File:Zand-majles.jpg *License*: unknown *Contributors*:

File:Vakeel-mosque.jpg *Source*: http://en.wikipedia.org/w/index.php?title=File:Vakeel-mosque.jpg *License*: unknown *Contributors*: Bontenbal, Fabienkhan, Romanm, Shauni, Zereshk

File:State Flag of Iran (1964).svg *Source*: http://en.wikipedia.org/w/index.php?title=File:State_Flag_of_Iran_(1964).svg *License*: unknown *Contributors*: Orange Tuesday (talk). Original uploader was Orange Tuesday at en.wikipedia

File:Flag of Russia.svg *Source*: http://en.wikipedia.org/w/index.php?title=File:Flag_of_Russia.svg *License*: unknown *Contributors*: Anomie, Zscout370

File:Imperial Emblem of the Qajar Dynasty (Lion and Sun).svg *Source*: http://en.wikipedia.org/w/index.php?title=File:Imperial_Emblem_of_the_Qajar_Dynasty_(Lion_and_Sun).svg *License*: unknown *Contributors*: User:Sodacan

File:Map Iran 1900-en.png *Source*: http://en.wikipedia.org/w/index.php?title=File:Map_Iran_1900-en.png *License*: unknown *Contributors*: User:Fabienkhan

File:Persia 1814.jpg *Source*: http://en.wikipedia.org/w/index.php?title=File:Persia_1814.jpg *License*: unknown *Contributors*: Begoon, DieBuche, Ervaude, Kaveh, Kintetsubuffalo, Zykasaa, 2 anonymous edits

File:Molla2.jpg *Source*: http://en.wikipedia.org/w/index.php?title=File:Molla2.jpg *License*: unknown *Contributors*: Americophile, Bontenbal, Fabienkhan, Shakko, Shauni, ینوگنابنوگن

File:A Zoroastrian Family Teheran 1910.JPG *Source*: http://en.wikipedia.org/w/index.php?title=File:A_Zoroastrian_Family_Teheran_1910.JPG *License*: unknown *Contributors*: Lx 121, Pymwiki, Ravichandar84, Tim!

File:1800 19th Street NW.JPG *Source*: http://en.wikipedia.org/w/index.php?title=File:1800_19th_Street_NW.JPG *License*: unknown *Contributors*: User:AgnosticPreachersKid

File:Yek toman qajar.jpg *Source*: http://en.wikipedia.org/w/index.php?title=File:Yek_toman_qajar.jpg *License*: unknown *Contributors*: Zereshk

Image:Mohammad Khan Qajar.jpg *Source*: http://en.wikipedia.org/w/index.php?title=File:Mohammad_Khan_Qajar.jpg *License*: unknown *Contributors*: User EvKnight13 on en.wikipedia

File:Fath Ali Shah(hermitage2).jpg *Source*: http://en.wikipedia.org/w/index.php?title=File:Fath_Ali_Shah(hermitage2).jpg *License*: unknown *Contributors*: Mihr Ali

Image:Mohammadshah.jpg *Source*: http://en.wikipedia.org/w/index.php?title=File:Mohammadshah.jpg *License*: unknown *Contributors*: Deadstar, Meisam, Shakko, TobiasVetter, ی‌ن‌وگ‌ن‌اب‌ن‌وگ‌ن

File:Nāser al-Dīn Schah.jpg *Source*: http://en.wikipedia.org/w/index.php?title=File:Nāser_al-Dīn_Schah.jpg *License*: unknown *Contributors*: , 51 Rue d Anjou St. Honore Paris

File:Mozaffar-ed-Din Shah Qajar - 1.jpg *Source*: http://en.wikipedia.org/w/index.php?title=File:Mozaffar-ed-Din_Shah_Qajar_-_1.jpg *License*: unknown *Contributors*: BehnamFarid

Image:Mohammad Ali Shah.jpg *Source*: http://en.wikipedia.org/w/index.php?title=File:Mohammad_Ali_Shah.jpg *License*: unknown *Contributors*: Original uploader was Fabienkhan at fr.wikipedia

Image:AhmadShahQajar2.jpg *Source*: http://en.wikipedia.org/w/index.php?title=File:AhmadShahQajar2.jpg *License*: unknown *Contributors*: Bain News Service

File:Mossadegh US04.jpg *Source*: http://en.wikipedia.org/w/index.php?title=File:Mossadegh_US04.jpg *License*: unknown *Contributors*: AgnosticPreachersKid, R. Engelhardt, Ugo14

File:Iraj mirza picture.jpg *Source*: http://en.wikipedia.org/w/index.php?title=File:Iraj_mirza_picture.jpg *License*: unknown *Contributors*: Original uploader was Zereshk at en.wikipedia

File:Gholam-Hossein-Banan.jpg *Source*: http://en.wikipedia.org/w/index.php?title=File:Gholam-Hossein-Banan.jpg *License*: unknown *Contributors*: Original uploader was BehnamFarid at en.wikipedia

File:Fereydoon Moshiri.jpg *Source*: http://en.wikipedia.org/w/index.php?title=File:Fereydoon_Moshiri.jpg *License*: unknown *Contributors*: BehnamFarid, Meisam

File: .jpg *Source*: http://en.wikipedia.org/w/index.php?title=File:ن‌اردل‌اچ_گ‌ن‌ج_ردـل‌ی‌عام‌س‌ا_ه‌اش.jpg *License*: unknown *Contributors*: مول‌عم‌ان

File:Ritratto dell ambasciatore Gabriel de Luetz d Aramont Tiziano Vecellio 1541 1542 oil on canvas 76 x 74 cm.jpg *Source*: http://en.wikipedia.org/w/index.php?title=File:Ritratto_dell_ambasciatore_Gabriel_de_Luetz_d_Aramont_Tiziano_Vecellio_1541_1542_oil_on_canvas_76_x_74_cm.jpg *License*: unknown *Contributors*: Bukk, Vikiyazar, World Imaging

File:The Rock and Walled City of Van (1893).jpg *Source*: http://en.wikipedia.org/w/index.php?title=File:The_Rock_and_Walled_City_of_Van_(1893).jpg *License*: unknown *Contributors*: H. F. B. Lynch

File:Yavuz Sultan I. Selim Han.jpg *Source*: http://en.wikipedia.org/w/index.php?title=File:Yavuz_Sultan_I._Selim_Han.jpg *License*: unknown *Contributors*: Belli değil

File:Tughra of Selim I.JPG *Source*: http://en.wikipedia.org/w/index.php?title=File:Tughra_of_Selim_I.JPG *License*: unknown *Contributors*: Unknown court calligrapher.

File:Nakkaş Selim.jpg *Source*: http://en.wikipedia.org/w/index.php?title=File:Nakkaş_Selim.jpg *License*: unknown *Contributors*: Seyyīd Loḳmān Çelebī, Nakkaş ʿOs̱mān

File:Abraham Ortelius - Tvrcici imperii descriptio.jpg *Source*: http://en.wikipedia.org/w/index.php?title=File:Abraham_Ortelius_-_Tvrcici_imperii_descriptio.jpg *License*: unknown *Contributors*: Cirt, Geagea, Rlevse, Strainu, Takabeg

File:Selim I - deathbed.jpg *Source*: http://en.wikipedia.org/w/index.php?title=File:Selim_I_-_deathbed.jpg *License*: unknown *Contributors*: unknown

File:EmperorSuleiman.jpg *Source*: http://en.wikipedia.org/w/index.php?title=File:EmperorSuleiman.jpg *License*: unknown *Contributors*: AlperenGezer, Anetode, Bravkov1990, Hekerui, IvanLanin, Magog the Ogre, Pieter Kuiper, Rapsar

File:Tughra of Suleiman I the Magnificent.svg *Source*: http://en.wikipedia.org/w/index.php?title=File:Tughra_of_Suleiman_I_the_Magnificent.svg *License*: unknown *Contributors*: User:Afernand74

File:OsmanenDeutscheKavallerie-1-.jpg *Source*: http://en.wikipedia.org/w/index.php?title=File:OsmanenDeutscheKavallerie-1-.jpg *License*: unknown *Contributors*: AnRo0002, Doncsecz, Rapsar

Image:Semailname 47b.jpg *Source*: http://en.wikipedia.org/w/index.php?title=File:Semailname_47b.jpg *License*: unknown *Contributors*: BomBom, Geni, GrawpSock, Gryffindor, Johnbod, Lynxxx, Omegastar19, Rapsar, 1 anonymous edits

File:John Sigismund of Hungary with Suleiman the Magnificient in 1556.jpg *Source*: http://en.wikipedia.org/w/index.php?title=File:John_Sigismund_of_Hungary_with_Suleiman_the_Magnificient_in_1556.jpg *License*: unknown *Contributors*: 1566 anonymous Ottoman author

File:Siege of Esztergom 1543.jpg *Source*: http://en.wikipedia.org/w/index.php?title=File:Siege_of_Esztergom_1543.jpg *License*: unknown *Contributors*: User:Uploadalt

Image:Sueleymanname nahcevan.jpg *Source*: http://en.wikipedia.org/w/index.php?title=File:Sueleymanname_nahcevan.jpg *License*: unknown *Contributors*: Fethullah Çelebi Arifi (historian, poet and painter) and/or Matrakçı Nasuh (painter of landscape) and/or other painters at the court of Sultan Suleiman the magnificent, 16th century

File:Hans Eworth Osmanischer Wurdentrager zu Pferd.jpg *Source*: http://en.wikipedia.org/w/index.php?title=File:Hans_Eworth_Osmanischer_Wurdentrager_zu_Pferd.jpg *License*: unknown *Contributors*: AndreasPraefcke, Bukk, DIREKTOR, Emijrp, Kersti Nebelsiek, Masur, Mattes, Pitke, Skipjack, Wolfmann, Wst, 1 anonymous edits

File:Ottoman fleet Indian Ocean 16th century.jpg *Source*: http://en.wikipedia.org/w/index.php?title=File:Ottoman_fleet_Indian_Ocean_16th_century.jpg *License*: unknown *Contributors*: Anonymous Ottoman painter 16th century

Image:Battle of Preveza (1538).jpg *Source*: http://en.wikipedia.org/w/index.php?title=File:Battle_of_Preveza_(1538).jpg *License*: unknown *Contributors*: Ohannes Umed Behzad

File:Francois I Suleiman.jpg *Source*: http://en.wikipedia.org/w/index.php?title=File:Francois_I_Suleiman.jpg *License*: unknown *Contributors*: AlperenGezer, Auntof6, Penubag, Pieter Kuiper, Sailko, Shakko, Thib Phil, World Imaging, Wst, 1 anonymous edits

Image:Siege of malta 1.jpg *Source*: http://en.wikipedia.org/w/index.php?title=File:Siege_of_malta_1.jpg *License*: unknown *Contributors*: Bohème, Gaius Cornelius, Kirill Lokshin, Longbow4u, Pe-Jo, Saibo, Sherool, Sioux, Wutsje

Image:Suleiman bas-relief in the U.S. House of Representatives chamber.jpg *Source*: http://en.wikipedia.org/w/index.php?title=File:Suleiman_bas-relief_in_the_U.S._House_of_Representatives_chamber.jpg *License*: unknown *Contributors*: Sculpture by Joseph Kiselewski; photo by the Architect of the Capitol

File:Istanbul - Süleymaniye camii dal Corno d'oro - Foto G. Dall'Orto 28-5-2006.jpg *Source*: http://en.wikipedia.org/w/index.php?title=File:Istanbul_-_Süleymaniye_camii_dal_Corno_d'oro_-_Foto_G._Dall'Orto_28-5-2006.jpg *License*: unknown *Contributors*: user:G.dallorto

File:Szulejmán a sátrában Buda alatt (1529).JPG *Source*: http://en.wikipedia.org/w/index.php?title=File:Szulejmán_a_sátrában_Buda_alatt_(1529).JPG *License*: unknown *Contributors*: Doncsecz

File:İstanbul 5622.jpg *Source*: http://en.wikipedia.org/w/index.php?title=File:İstanbul_5622.jpg *License*: unknown *Contributors*: User:Darwinek

File:Sultan Murad III.jpg *Source*: http://en.wikipedia.org/w/index.php?title=File:Sultan_Murad_III.jpg *License*: unknown *Contributors*: Belli değil

File:Tughra of Murad III.JPG *Source*: http://en.wikipedia.org/w/index.php?title=File:Tughra_of_Murad_III.JPG *License*: unknown *Contributors*: Unknown court calligrapher.

File:I Mustafa.jpg *Source*: http://en.wikipedia.org/w/index.php?title=File:I_Mustafa.jpg *License*: unknown *Contributors*: G.dallorto, Gryffindor

File:Tughra of Mustafa I.JPG *Source*: http://en.wikipedia.org/w/index.php?title=File:Tughra_of_Mustafa_I.JPG *License*: unknown *Contributors*: Unknown court calligrapher.

File:Murad IV.jpg *Source*: http://en.wikipedia.org/w/index.php?title=File:Murad_IV.jpg *License*: unknown *Contributors*: Aguzer, Gryffindor, Shaolin128

File:Tughra of Murad IV.svg *Source*: http://en.wikipedia.org/w/index.php?title=File:Tughra_of_Murad_IV.svg *License*: unknown *Contributors*: User:Mnmazur

File:IV. Murat.jpg *Source*: http://en.wikipedia.org/w/index.php?title=File:IV._Murat.jpg *License*: unknown *Contributors*: Belli değil

File:Vera Effigies Turcorvm Imperatoris Ibrahim Filii, Et Sultanæ, Eiusdem Matris (1707).jpg *Source*: http://en.wikipedia.org/w/index.php?title=File:Vera_Effigies_Turcorvm_Imperatoris_Ibrahim_Filii,_Et_Sultanæ,_Eiusdem_Matris_(1707).jpg *License*: unknown *Contributors*: Cooalope, Cplakidas, Jacklee, Rapsar, Takabeg

File:Edwin Lord Weeks - The Taj Mahal - Walters 37316.jpg *Source*: http://en.wikipedia.org/w/index.php?title=File:Edwin_Lord_Weeks_-_The_Taj_Mahal_-_Walters_37316.jpg *License*: unknown *Contributors*: Jarekt, Yaniv256, ZxxZxxZ

File:Mahmud1.jpg *Source*: http://en.wikipedia.org/w/index.php?title=File:Mahmud1.jpg *License*: unknown *Contributors*: Frederico, Gryffindor, Osmanlı98

File:Tughra of Mahmud I.JPG *Source*: http://en.wikipedia.org/w/index.php?title=File:Tughra_of_Mahmud_I.JPG *License*: unknown *Contributors*: Unknown court calligrapher.

File:Mahmud II.jpg *Source*: http://en.wikipedia.org/w/index.php?title=File:Mahmud_II.jpg *License*: unknown *Contributors*: Original uploader was Flavius Belisarius at en.wikipedia

File:Tughra of Mahmud II.JPG *Source*: http://en.wikipedia.org/w/index.php?title=File:Tughra_of_Mahmud_II.JPG *License*: unknown *Contributors*: Unknown court calligrapher.

File:Tugra Mahmuds II.png *Source*: http://en.wikipedia.org/w/index.php?title=File:Tugra_Mahmuds_II.png *License*: unknown *Contributors*: Baba66, BomBom, Gryffindor, MPF, Micha L. Rieser, Saperaud, 2 anonymous edits

File:Sultan tombs Divan Yolu March 2008.JPG *Source*: http://en.wikipedia.org/w/index.php?title=File:Sultan_tombs_Divan_Yolu_March_2008.JPG *License*: unknown *Contributors*: User:Gryffindor

File:Mahmud II by John Young.jpg *Source*: http://en.wikipedia.org/w/index.php?title=File:Mahmud_II_by_John_Young.jpg *License*: unknown *Contributors*: John Young (1755-1825)

Image:Mahmud II.jpg *Source*: http://en.wikipedia.org/w/index.php?title=File:Mahmud_II.jpg *License*: unknown *Contributors*: Original uploader was Flavius Belisarius at en.wikipedia

Image:Mahmudiye (1829).jpg *Source*: http://en.wikipedia.org/w/index.php?title=File:Mahmudiye_(1829).jpg *License*: unknown *Contributors*: Historic image from the archives of the Turkish Navy.

File:Tahmasp, Humayun Meeting.jpg *Source*: http://en.wikipedia.org/w/index.php?title=File:Tahmasp,_Humayun_Meeting.jpg *License*: unknown *Contributors*: Muhammad Mahdi Karim

File: .jpg *Source*: http://en.wikipedia.org/w/index.php?title=File:شاه_عباس_ی_م.jpg *License*: unknown *Contributors*: مولعمان

File:Shah Abbas I engraving by Dominicus Custos - Antwerp artist printer and engraver.jpg *Source*: http://en.wikipedia.org/w/index.php?title=File:Shah_Abbas_I_engraving_by_Dominicus_Custos_-_Antwerp_artist_printer_and_engraver.jpg *License*: unknown *Contributors*: AndreasPraefcke, Ashrf1979, Haiduc, Martin H., Qizilbash, Takabeg, 2 anonymous edits

File:Shah Abbas Horse.jpg *Source*: http://en.wikipedia.org/w/index.php?title=File:Shah_Abbas_Horse.jpg *License*: unknown *Contributors*: Zereshk

File:Jahangir's dream-Abu'l Hasan-c.1618-22.jpg *Source*: http://en.wikipedia.org/w/index.php?title=File:Jahangir's_dream-Abu'l_Hasan-c.1618-22.jpg *License*: unknown *Contributors*: Abu'l Hasan

File:Muhammad Ali Beg.jpg *Source*: http://en.wikipedia.org/w/index.php?title=File:Muhammad_Ali_Beg.jpg *License*: unknown *Contributors*: Foroa, Roland zh, Sridhar1000

File:Fortress of Hormuz.jpg *Source*: http://en.wikipedia.org/w/index.php?title=File:Fortress_of_Hormuz.jpg *License*: unknown *Contributors*: Anonymous 17th century author

File:Baptistère kelisa-e-vank esfahan.jpg *Source*: http://en.wikipedia.org/w/index.php?title=File:Baptistère_kelisa-e-vank_esfahan.jpg *License*: unknown *Contributors*: User:Fabienkhan

File:Embassy to Europe.jpg *Source*: http://en.wikipedia.org/w/index.php?title=File:Embassy_to_Europe.jpg *License*: unknown *Contributors*: Carlo Caliari and Gabriele Calliari

File:Shah Abbas and Wine Boy.jpg *Source*: http://en.wikipedia.org/w/index.php?title=File:Shah_Abbas_and_Wine_Boy.jpg *License*: unknown *Contributors*: Bibi Saint-Pol, Calame, DenghiùComm, Fabienkhan, Fg68at, G.dallorto, Johnbod, Lysis, Mani1, Nevit, Robert Prummel, Surena, 1 anonymous edits

File:Shah Safi.png *Source*: http://en.wikipedia.org/w/index.php?title=File:Shah_Safi.png *License*: unknown *Contributors*: Alborz Fallah

File:Nader Shah Afshar.jpg *Source*: http://en.wikipedia.org/w/index.php?title=File:Nader_Shah_Afshar.jpg *License*: unknown *Contributors*: Andres rojas22, FSII, Hohum, Look2See1, Sridhar1000, Yone Fernandes, ی‌ن‌وگ‌ن‌اب‌ن‌وگ‌ن

File:Nadershahtomb.jpg *Source*: http://en.wikipedia.org/w/index.php?title=File:Nadershahtomb.jpg *License*: unknown *Contributors*: Bontenbal, Kasir, Shauni, Wikiacc, ی‌ن‌وگ‌ن‌اب‌ن‌وگ‌ن, 4 anonymous edits

File: .jpg *Source*: http://en.wikipedia.org/w/index.php?title=File:نادر_شاه.jpg *License*: unknown *Contributors*: مولعمان

File: .JPG *Source*: http://en.wikipedia.org/w/index.php?title=File:نادر_و_فرزندان_ش.JPG *License*: unknown *Contributors*: مولعمان

File:A Nawab of Awadh, Lucknow, India. 19th century.jpg *Source*: http://en.wikipedia.org/w/index.php?title=File:A_Nawab_of_Awadh,_Lucknow,_India._19th_century.jpg *License*: unknown *Contributors*: Abhishekjoshi, Ashrf1979, Innotata, Johnbod, Roland zh

File:Nader Shah afshar.jpg *Source*: http://en.wikipedia.org/w/index.php?title=File:Nader_Shah_afshar.jpg *License*: unknown *Contributors*: User:Adel-adili

File:Nadir Shah.jpg *Source*: http://en.wikipedia.org/w/index.php?title=File:Nadir_Shah.jpg *License*: unknown *Contributors*: Jonas Hanway: Zuverläßige Geschichte der Englischen Handlung durch Rußland, über die Caspische See, nach Persien, der Tartarey und Türkey, Armenien und China. Samt einer Beschreibung der Landesbeschaffenheit dieser Länder.

File:Nader Shah Jewels 3 - edited.png *Source*: http://en.wikipedia.org/w/index.php?title=File:Nader_Shah_Jewels_3_-_edited.png *License*: unknown *Contributors*: Siroos777

File:Naderafshartomb.jpg *Source*: http://en.wikipedia.org/w/index.php?title=File:Naderafshartomb.jpg *License*: unknown *Contributors*: Original uploader was Zereshk at en.wikipedia

File:OttomanEmpireIn1683.png *Source*: http://en.wikipedia.org/w/index.php?title=File:OttomanEmpireIn1683.png *License*: unknown *Contributors*: Atilim Gunes Baydin

File:Shir & Khorshid1.svg *Source*: http://en.wikipedia.org/w/index.php?title=File:Shir_&_Khorshid1.svg *License*: unknown *Contributors*: User:Stannered

File:An Early Painting of Fath Ali Shah.jpg *Source*: http://en.wikipedia.org/w/index.php?title=File:An_Early_Painting_of_Fath_Ali_Shah.jpg *License*: unknown *Contributors*: Alexanderps, FSII, Shakko, Wayiran, ی‌ن‌وگ‌ن‌اب‌ن‌وگ‌ن

File:Fath Ali Shah(Saad Abad).jpg *Source*: http://en.wikipedia.org/w/index.php?title=File:Fath_Ali_Shah(Saad_Abad).jpg *License*: unknown *Contributors*: Mihr Ali

File:17.5-30-2003-Portraet-af-Fath-Ali-Shah-Qajar.jpg *Source*: http://en.wikipedia.org/w/index.php?title=File:17.5-30-2003-Portraet-af-Fath-Ali-Shah-Qajar.jpg *License*: unknown *Contributors*: Alborzagros, Kleuske, Peter23, Thierry Caro, 4 anonymous edits

File:Gardane Jaubert Jouanin.jpg *Source*: http://en.wikipedia.org/w/index.php?title=File:Gardane_Jaubert_Jouanin.jpg *License*: unknown *Contributors*: Anne97432, Takabeg, Wayiran, World Imaging, ی‌ن‌وگ‌ن‌اب‌ن‌وگ‌ن

File:Persia1808.JPG *Source*: http://en.wikipedia.org/w/index.php?title=File:Persia1808.JPG *License*: unknown *Contributors*: Chipmunkdavis, Mrostami, Phatom87, Wareh, Wayiran, Zaccarias, Zykasaa

File:Fath Ali Shah Qajar Firman in Shikasta Nastaliq script January 1831.jpg *Source*: http://en.wikipedia.org/w/index.php?title=File:Fath_Ali_Shah_Qajar_Firman_in_Shikasta_Nastaliq_script_January_1831.jpg *License*: unknown *Contributors*: User:PHGCOM

File:Golestan-qajar.jpg *Source*: http://en.wikipedia.org/w/index.php?title=File:Golestan-qajar.jpg *License*: unknown *Contributors*: User:Leo71538

MIX
Papier aus verantwortungsvollen Quellen
Paper from responsible sources
FSC® C105338

Printed by Books on Demand GmbH, Norderstedt / Germany